Biomass for Energy Country Specific Show Case Studies

Special Issue Editor
Tariq Al-Shemmeri

MDPI • Basel • Beijing • Wuhan • Barcelona • Belgrade

MDPI

Special Issue Editor
Tariq Al-Shemmeri
Staffordshire University
UK

Editorial Office
MDPI
St. Alban-Anlage 66
Basel, Switzerland

This edition is a reprint of the Special Issues published online in the open access journal *Energies* (ISSN 1996-1073) from 2015–2017 (available at: http://www.mdpi.com/journal/energies/special_issues/advances-in-biomass; http://www.mdpi.com/journal/energies/special_issues/biomass-for-country).

For citation purposes, cite each article independently as indicated on the article page online and as indicated below:

Lastname, F.M.; Lastname, F.M. Article title. *Journal Name* **Year**, *Article number*, page range.

First Edition 2018

ISBN 978-3-03842-912-8 (Pbk)
ISBN 978-3-03842-911-1 (PDF)

Table of Contents

About the Special Issue Editor

Tarik Al-Shemmeri, Prof. Dr., is the Director of the European Centre of Excellence "Biomass for Energy" at Staffordshire University. He leads a big European consortium of partners across the different disciplines all involved in the acceleration of the use of biomass to provide a clean and reliable source of energy. This project is funded by INTERREG, part of the European Community programme. Professor Al-Shemmeri has extensive research and teaching experience on the subject and has supervised many doctorate and master's students on the subject.

Preface to "Biomass for Energy Country Specific Show Case Studies"

Global warming describes the current rise in the average temperature of the Earth's air and oceans. Global warming has been recognised as the biggest threat to humans, animals and plants on earth.

A long series of scientific research and international studies has shown, with more than 90% certainty, that there has been an increase in overall temperatures due to the greenhouse gases produced by humans. The use of fossil fuels is the main source of these emissions.

There has been an outcry by scientists and politicians to combat this serious threat, and a number of international agreements have eventually been reached to act on measures to reduce its impact. In Europe, there has been almost a competition between governments to set ambitious levels of reductions to gain popularity and to demonstrate how serious these countries are about finding a solution to this huge problem.

"A reduction of greenhouse gas emissions in the EU by at least 40% by 2030 (compared to 1990 levels) is one of the targets agreed by the European Council as part of the 2030 climate and energy framework. Since the EU ETS will be the main instrument for achieving this target, its reform is necessary to ensure a well-functioning system."

On 15 July 2015, the Commission presented a second proposal, which represents a broader review of the EU ETS. The aim of the proposal is to take the European Council's guidance on the role the EU ETS should play in achieving the EU's 2030 greenhouse gas emissions reduction target and make it law. The proposed changes also aim to foster innovation and the use of low-carbon technologies, helping to create new opportunities for jobs and growth, while maintaining the necessary safeguards to protect industrial competitiveness in Europe."

The technology used and sources of energy adopted depend on the availability of sources, the technology to utilise these sources and the demand/supply match. Renewables such as solar, wind, hydro and biomass are available to a varying extent in different parts of the world. Solar without storage is only available during daylight, wind availability is also not guaranteed all the time, and the availability of hydraulic energy is limited to locations where a river or lake can be utilised to provide such energy. Biomass is the only source which can provide a steady and uninterrupted supply of energy.

A secure and cost-effective supply of low carbon energy is one of the main goals of a sustainable carbon strategy. Biomass can make an important contribution to decarbonisation. However, there are risks such as food security and deforestation if biomass is used excessively. Biomass is the only renewable source that can be used across all three energy sectors (transport, heat and electricity).

The technology to provide these three forms of energy from biomass is experiencing a revival and there are numerous examples to demonstrate this by applications. This book is a collection of case studies that have been developed worldwide to provide the reader with very useful and real life examples of how biomass is becoming more efficient and economically and environmentally superior to some of the common forms of energy conversion technology, such as solar and wind; furthermore, biomass is not prone to weather variations or intermittent sunshine or wind.

It has been predicted that bioenergy will allow the UK to meet its energy and climate change objectives, including the 2020 renewables targets and 2050 carbon reductions targets.

The use of biomass to produce electricity, for example, has been extensively demonstrated using steam turbines, gas turbines or external gas engines, also known as Stirling engines. The challenges facing biomass include determining which types of organic materials should be used and which processes would be necessary to harness its energy into an inexpensive and efficient method.

In this text, a number of research papers are presented to demonstrate some case studies on the use of biomass in its different forms.

Tariq Al-Shemmeri
Special Issue Editor

energies

MDPI

Review

Biohydrogen Production from Lignocellulosic Biomass: Technology and Sustainability

Anoop Singh [1,*], Surajbhan Sevda [2], Ibrahim M. Abu Reesh [2], Karolien Vanbroekhoven [3], Dheeraj Rathore [4] and Deepak Pant [3,*]

1   Government of India, Ministry of Science and Technology, Department of Scientific and Industrial Research (DSIR), Technology Bhawan, New Mehrauli Road, New Delhi 110016, India

2   Department of Chemical Engineering, Qatar University, Doha 2713, Qatar; sevdasuraj@gmail.com (S.S.); abureesh@qu.edu.qa (I.M.A.R.)

3   Separation and Conversion Technologies, Flemish Institute for Technological Research (VITO), Mol 2400, Belgium; karolien.vanbroekhoven@vito.be

4   School of Environment and Sustainable Development, Central University of Gujarat, Gandhinagar 382030, Gujarat, India; dheeraj.rathore@cug.ac.in

*   Correspondence: apsinghenv@gmail.com (A.S.); deepak.pant@vito.be (D.P.); Tel.: +91-828-579-7386 (A.S.); +32-1433-6969 (D.P.); Fax: +32-1432-6586 (D.P.)

Academic Editor: Tariq Al-Shemmeri
Received: 16 June 2015 ; Accepted: 10 November 2015 ; Published: 17 November 2015

Abstract: Among the various renewable energy sources, biohydrogen is gaining a lot of traction as it has very high efficiency of conversion to usable power with less pollutant generation. The various technologies available for the production of biohydrogen from lignocellulosic biomass such as direct biophotolysis, indirect biophotolysis, photo, and dark fermentations have some drawbacks (e.g., low yield and slower production rate, *etc.*), which limits their practical application. Among these, metabolic engineering is presently the most promising for the production of biohydrogen as it overcomes most of the limitations in other technologies. Microbial electrolysis is another recent technology that is progressing very rapidly. However, it is the dark fermentation approach, followed by photo fermentation, which seem closer to commercialization. Biohydrogen production from lignocellulosic biomass is particularly suitable for relatively small and decentralized systems and it can be considered as an important sustainable and renewable energy source. The comprehensive life cycle assessment (LCA) of biohydrogen production from lignocellulosic biomass and its comparison with other biofuels can be a tool for policy decisions. In this paper, we discuss the various possible approaches for producing biohydrogen from lignocellulosic biomass which is an globally available abundant resource. The main technological challenges are discussed in detail, followed by potential solutions.

Keywords: biohydrogen; biofuels; lignocellulosic biomass; technology; sustainability; life cycle assessment

1. Introduction

Recent years have seen a rapid surge in research activities focusing intensely on alternative fuels in order to reduce the dependency on fossil fuels, mainly by providing local energetic resources. This is mainly due to two reasons, the first being that new fuels are needed to supplement and ultimately replace depleting oil reserves and secondly, fuels capable of low or nil CO_2 emissions are urgently required to reduce the impact of global warming [1–7]. Hydrogen (H_2), which can be used in fuel cells mainly to operate machines, is a fascinating alternative, particularly because its combustion provides high amounts of energy and water is the only reaction product. Among all biofuels, H_2 has the highest gravimetric energy density at 141 MJ/kg. Despite this, its volumetric energy density, at only 12 MJ/m^3 (at normal temperature and pressure) is low. This is an important aspect particularly in reference to transportation fuel. It is considered to be one of the cleanest energy carriers if produced using energy generated from renewable sources. In summary, H_2 is interesting due to its potentially high efficiency of conversion to usable power, low generation of pollutants and high energy density [8]. Global H_2 production today amounts to around 700 billion Nm^3 and is based almost exclusively on fossil fuels [9]. However, for H_2 to be accepted as a sustainable substitute for fossil fuels, it has to be produced from renewable feedstock other than fossil fuels [10].

Hydrogen has been suggested as the ideal fuel of the future. It is considered as one of the cleanest energy carriers to be generated from renewable sources [11]. It has a high energy yield (122 kJ/g) which is 2.75 times greater than hydrocarbon fuels. It can be easily used in fuel cells for generation of electricity. Though not a primary energy source, it serves as a medium through which primary energy sources (such as H_2 produced from nuclear power and/or solar energy) can be stored, transported and utilized to fulfill our energy needs. The major problem facing H_2 as a fuel is its unavailability in Nature. H_2 can be produced safely, is environmentally friendly when combusted, and versatile *i.e.*, has many potential energy uses, including powering non-polluting vehicles, heating homes and offices, and fueling aircraft. Current H_2 production technologies such as steam reforming of natural gas, thermal cracking or coal gasification are not environmentally friendly. Biological H_2 production is a promising alternative. There are two methods to produce H_2 from microorganisms. The first method uses photosynthetic microorganisms such as bacteria or algae (photofermentative processes) and the second method uses fermentative organisms (dark fermentation processes). Fermentative H_2 production has the advantage of producing H_2 under mild conditions with the additional benefit of allowing residual biomass valorization. The dark fermentation process is more attractive as it has the potential to use wastewater and organic wastes and has higher production rates compared to photofermentative processes. So far, few studies have used real wastewater for the production of H_2 due to inhibition by both substrate and/or product in the fermentation process [12]. Studies on bioH$_2$ production have been focused on photodecomposition of organic compounds by photosynthetic bacteria, dark fermentation from organic compounds with anaerobes and biophotolysis of water using algae and cyanobacteria [13–17].

Lignocellulosic biomass is the most abundant in Nature and it is present in hardwood, softwood, grasses, and agricultural residues. The global annual yields of lignocellulosic biomass residues were estimated to exceed 220 billion tons, equivalent to about 60–80 billion tons of crude oil [12]. Lignocellulosic feedstocks consist mainly of glucose and xylose and thus microbial strains that can effectively degrade glucose and xylose are important for development of renewable H_2 production processes [18]. Direct conversion of lignocellulosic biomass to H_2 needs pretreatment to hydrolyze the incorporated heterogeneous and crystalline structure [19,20]. The lignocellulosic biomass hence presents an attractive, low-cost feed stock for H_2 production.

Aim of the Paper

In recent past, several reviews have appeared which have discussed the prospects and challenges of biomass-based H_2 [21,22]. Earlier, Kraemer and Bagley gave a thorough description of the yield improvement approaches in fermentative H_2 production [23]. Wang and Wan summarized the main factors influencing fermentative H_2 production [24]. A special issue of the journal *International Journal of Hydrogen Energy*, recently dealt with "Biohydrogen: From Basic Concepts to Technology" [25]. Biohydrogen can be generated by adopting different technologies and different technologies can perform differently. Thus, the aim of this paper is to discuss specifically the technological aspect of biohydrogen production from lignocellulosic biomass and its sustainability on the basis of a life cycle assessment (LCA).

2. Feed Stock for Biohydrogen Production

Glucose is the ideal substrate, but it is too costly at present. Many agricultural residues and food wastes are rich in carbohydrates that could serve as feedstock. Lignocellulosic biomass is another sustainable feedstock for H_2 production [26]. The criteria for an ideal feedstock for sustainable H_2 production, which include high carbohydrate content, minimum pre-treatment requirement, sustainable resources, low cost and sufficient concentration of carbohydrate for fermentative conversion, have been suggested by Bartacek [27]. The substrates usable for fermentative H_2 production were further divided into four main groups, namely, pure substrates such as glucose; energy crops such as *Miscanthus*, solid wastes like food waste and industrial wastewaters such as wastewater from the pulp and paper industry.

A variety of substrates has been used as feedstock for H_2 production. For example, the fermentation of household wastes under different temperature conditions has been well studied [28–30]. An increase in H_2 yields as temperature increased to thermophilic regimes was reported.

Wastewaters and residual biomass with high carbohydrate content have also been demonstrated to be a suitable candidate for dark fermentation. This includes molasses [31,32] and cheese whey [33,34], which have been evaluated under continuous stirred tank reactor (CSTR) and immobilized system configurations. Besides, H_2 production from soluble and particulate starch and cellulose [35,36], xylose [37], sugar beet [38], wastewater from a sugar beet refinery [39] and the bottom layer from a beer manufacturing plant [40] has also been demonstrated.

3. Technology

3.1. Biohydrogen Production Systems

The conventional methods for producing H_2 gas include steam reforming of methane and hydrocarbons, non-catalytic partial oxidation of fossil fuels and autothermal reforming. However, most of these methods are energy intensive processes requiring high temperatures (>850 °C). A general scheme of H_2 production from renewable sources is shown in Figure 1. Biological methods of H_2 production are preferable to chemical methods because of the possibility to use sunlight, CO_2 and organic wastes as substrates for environmentally benign conversions, under moderate conditions.

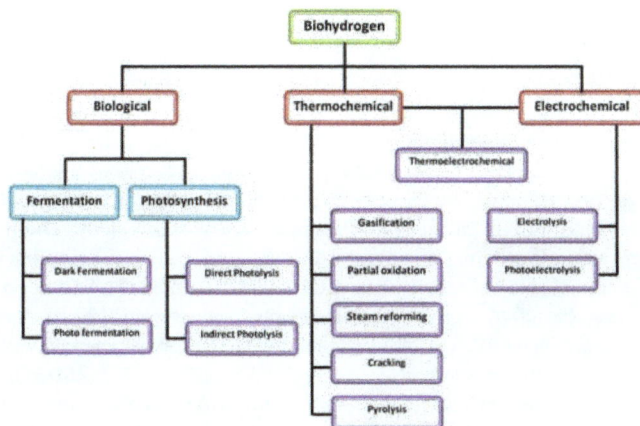

Figure 1. The main alternative methods of H_2 production from energy sources.

The biological production of H_2 involves light-dependent methods: direct and indirect biophotolysis, and photo fermentation. The other routes are light-independent methods, including the dark fermentation process and water-gas shift reaction of photoheterotrophic bacteria. This biologically produced H_2, generally referred to as "biohydrogen", is characterized by low H_2 yields which present a challenge for commercial applications.

3.1.1. Dark/Anaerobic Fermentation

Dark fermentation is one of the most common processes for bio-H_2 production. Although only 15%–20% of the theoretical H_2 potential of carbohydrates can be harvested, dark fermentation is considered as a promising process in a two phase anaerobic treatment system [23]. Also, since the CO_2 produced in dark fermentation has already been fixed by the waste treated originally from the atmosphere, the emissions associated with global climate change are virtually zero [41]. Microbial species analysis of hydrogen-producing cultures (using anaerobic sludge as inoculum) shows the presence of *Clostridium cellulosi*, *Clostridium acetobutylicum*, *Clostridium tyrobutyricum*, *Enterobacteriaceae* and *Streptococcus bovis* [42] Fermentative H_2 production usually proceeds from the anaerobic glycolytic breakdown of sugars. The theoretical complete oxidation of 1 mole of hexose to CO_2 can produce 12 moles of H_2. Nevertheless, the theoretical yield of H_2 via acetic acid fermentation cannot be higher than 4 moles. Actual H_2 yields are quite lower, typically ranging from 1.0 to 2.5 moles per mole of hexose consumed. Recently, Varanasi *et al.* reported production of 2.95 mol H_2/mol hexose equivalent by thermophillic dark fermentation using cellubiose as substrate [43]. If butyric acid is produced as the major fermentation product instead of acetic acid, only 2 moles of H_2 can be produced [44]. H_2 yield is even lower when more reduced organic compounds such as lactic acid, propionic acid and ethanol are produced, because these metabolites represent end products of metabolic pathways that bypass the major H_2-producing reaction [45]. Recently, it was concluded that to maximize net energy gain via dark fermentation, appropriate cultures capable of high-H_2 yield have to be employed and the process has to be operated at near-ambient temperatures with the lowest feedstock concentration as possible [10]. In an experiment with *Thermotoga neapolitana* sparged with N_2 and supplemented with 40 mM sodium bicarbonate a 2.8 and 2.7 mol/mol glucose yield of hydrogen with a lactic acid/acetic acid ratio of 0.26 was obtained, challenging the currently accepted dark fermentation model that predicts reduction of this gas when glucose is converted into organic products different from acetate [46]. Pradhan *et al.* reviewed the hydrogen production efficiency of a similar bacterium (*Thermotoga neapolitana*) with different feedstocks and found 1.9–3.5 mol H_2/mol hexose yields achievable with a range of feedstocks and variable substrate loads [47]. Byproducts of the reactions are acetic acid, lactic acid and ethanol.

3.1.2. Photo Fermentation

Photo fermentation is carried out by purple non-sulfur (PNS) photosynthetic bacteria which can grow as photoheterotrophs, photoautotrophs or chemoheterotrophs [48]. These bacteria produce H_2 under photoheterotrophic conditions (light, anaerobiosis, organic electron donor) [49]. The advantages of this process over photolysis of water using green algae and cyanobacteria, are that oxygen does not inhibit the process and that these bacteria can be used in a wide variety of conditions (*i.e.*, batch processes, continuous cultures, and immobilized systems) [50]. The hydrogenase and nitrogenase enzymes produced in photosynthesis by green algae and photosynthetic bacteria, respectively, play a crucial role in biohydrogen production. The main PNS bacteria that participate in H_2 production are *Rhodospirillum rubrum*, *Rhodopseudomonas palustris*, *Rhodobacter sphaeroides* O.U 001, *Rhodobacter sphaeroides* RV, *Rhodobacter sulfidophilus* and *Rhodobacter capsulatus*. Kapdan *et al.* used three different pure strains of *Rhodobacter sphaeroides* (RV, NRLL and DSZM) in batch experiments to select the most suitable strain [51]. *R. sphaeroides* RV resulted in the highest cumulative hydrogen gas formation (178 mL), hydrogen yield (1.23 mol·H_2·mol^{-1} glucose) and specific hydrogen production rate (46 mL·H_2·g^{-1} biomass·h^{-1}) at 5 g·L^{-1} initial total sugar concentration among the other pure cultures. Using *Rhodobacter capsulatus* JP91, Keskin and Hallenbeck compare the photofermentative biohydrogen yield of different feedstocks in a batch culture experiment [52]. Overall yield of biohydrogen was 10.5, 8 and 14.9 mol H_2/mol sucrose using beet molasses, black strap molasses and

4

sucrose respectively. Optimization of process parameters such as availability of solar light, bioreactor configuration and proper C/N ratio in substrate (synthetic and derived from waste products) still needs to be studied at higher scale.

3.1.3. Combined Biotechnologies

Combination of two or more of the abovementioned techniques have also been studied for improved H_2 yields. Theoretically 12 moles of H_2 per mole of glucose can be generated by combining dark fermentation with photo fermentation (using PNS bacteria) [48]. For instance, Nath *et al.* studied combined dark and photo fermentation using glucose as substrate [53]. The effluent from the dark process (containing unconverted metabolites, mainly acetic acid) underwent photo fermentation by *Rhodobacter sphaeroides* in a column photo-bioreactor demonstrating the feasibility of this combination to achieve higher yields of H_2 by complete utilization of the chemical energy stored in the substrate. A sequential process using glucose as substrate and an immobilized system for the photo fermentation step evaluating key factors such as diluted ratio of dark fermentation effluent, ratio of dark and photo fermentation bacteria, light intensity, and light/dark cycle has also been studied [54]. During the combined process, maximum total H_2 yield was 5.374 moles of H_2/moles of glucose. However, the sterilization step applied to the dark fermentation effluent may pose a constraint to a scale-up the process. The combined system can also be run in continuous mode and achieve more combined H_2 yield [49]. These further combinations will reduce the overall cost of H_2 production but more field studies are required to obtain an economical H_2 production process.

3.1.4. Bioelectrochemical Production

Bioelectrochemical production of H_2 is the latest technology using systems called microbial electrolysis cells (MEC). This is an emerging field where the oxidation of organic material is carried out by the bacteria present at the anode and results in formation of protons, CO_2 and electrons (Figure 2). Protons migrate through a proton exchange membrane (PEM) to the cathode and the electrons are transported through the external circuit to the cathode [55]. By applying an external voltage of approximately 0.5–0.9 V, these electrons combine at the cathode with protons producing H_2 gas (Table 1). The advantage here is the low energy consumption (0.3–0.9 V) necessary for microbial electrohydrogenesis to produce H_2 in comparison to the theoretical minimum voltage of 1.23 V required for water electrolysis [56]. An overall scheme of H_2 production from lignocellulosic biomass is shown in Figure 3. Figure 3 summarized the pathways for biohydrogen production by using lignocellulosic biomass, depending on the nature of feed (solid or liquid), pre-treatment methods were used and then followed by the dark fermentation.

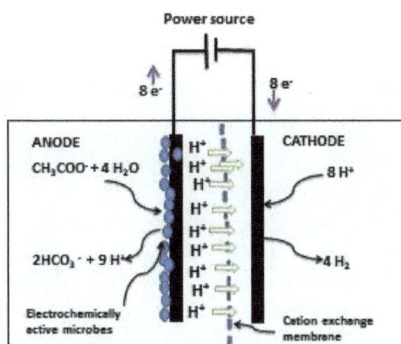

Figure 2. Schematic of hydrogen production in MEC (Adapted from [56]).

5

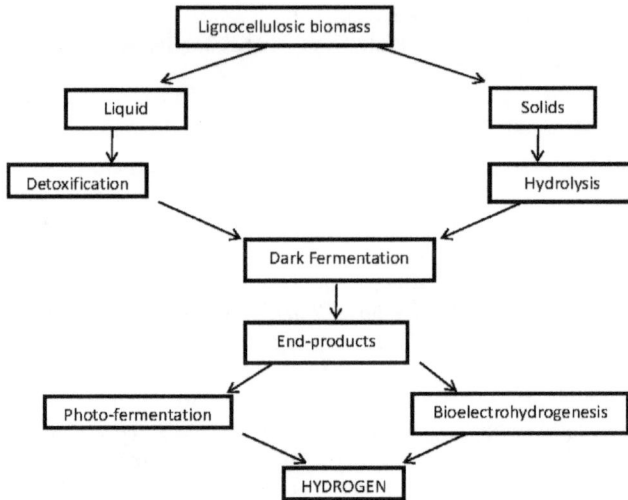

Figure 3. General scheme for biohydrogen production from lignocellulosic biomass (adapted from [12]).

Table 1. A comparison between the three major routes for biological hydrogen production (adapted from [57]).

Production Routes	Main Reaction	H_2 Production Rates (mmol/h· L)	Remark
Direct photolysis	$2H_2O$ + "light energy" $\rightarrow 2H_2 + O_2$	0.07	Similar to the processes found in plants and algal photosynthesis.
Photo fermentation	$C_6H_{12}O_6 + 6H_2O$ + "light energy" $\rightarrow 12H_2 + 6CO_2$, $\Delta G_0 = +3.2$ kJ	145–160	Bacteria evolve molecular H_2 catalyzed by nitrogenase under N-deficient conditions using light energy and reduced compounds (organic acids).
Dark fermentation	Pyruvate + CoA $\rightarrow$ acetyl-CoA + formate OR Pyruvate + CoA + $2Fd(ox) \rightarrow$ Acetyl-CoA + CO_2 + 2Fd (red)	77	H_2 is produced by anaerobic bacteria, grown in the dark on carbohydrate rich substrate.

Table 2 shows the various substrates used for H_2 production in different MEC volumes. The rate of H_2 production differs with substrate due to their degradation pathways. To date MECs have shown H_2 production from initial volumes ranging from 5 mL to 1000 L (pilot plant) reactors, which shows that MECs can be a better solution for producing H_2 in the cathodic chamber while treating wastewater in the anodic chamber [58]. Still, there are many issues to be addressed for the long term real time application such as electrode and membrane stability for longer duration and reactor configuration design for higher volumes. Before scale up, mathematical models are required, which need to be validated first for the present lab scale studies and then on the basis of data obtained, higher volume MECs may be designed and validated.

Table 2. Various substrates used in MEC for hydrogen production (adapted from [59]).

Substrate	Concentration (g/L)	Applied Voltage (V)	MEC Volume (mL)	Hydrogen Production Rate (m^3 H$_2$/m^3/day)	Reference
A de-oiled refinery wastewater	04–1	0.7	5	79% (Hydrogen production based on COD removal)	[60]
Sodium Acetate	1	0.6	18	2.0	[61]
Glucose	2	0.6	26	0.25 ± 0.03	[62]
Glucose	2	0.8	26	0.37 ± 0.04	[62]
Fermentation effluent	1	0.6	26	1.41	[63]
Sodium Acetate	1	0.6	28	1.99 ± 0.02	[64]
Sodium Acetate	1	0.8	28	3.12 ± 0.002	[64]
Sodium Acetate	1	0.5	28	1.7	[65]
Glucose	1	0.5	28	0.83 ± 0.18	[66]
Glucose	1	0.9	28	1.87 ± 0.30	[66]
Potato wastewater	1.9–2.5 (COD)	0.9	28	0.74	[67]
Swine wastewater	2 (COD)	0.5	28	0.9–1.0	[68]
Sodium Acetate	1	0.6	48	0.76	[65]
Sodium Acetate	1	0.7	76	-	[69]
Sodium Acetate	1	0.8	240	0.0231 ± 0.003	[70]
Sodium Acetate	1	1	400	1.58	[71]
Sodium Acetate	2	0.6	500	0.53	[72]
Winery wastewater	8	0.9	1000 Lt	0.19 ± 0.04	[58]
Sodium Acetate	1	0.5	6600	0.02	[56]

3.2. Microbiology of Biohydrogen Production

Perera *et al.* evaluated three main routes for biological H$_2$ production [10]. These are (1) direct photolysis, in which cyanobacteria decomposes water to generate hydrogen and oxygen in presence of light; (2) photo fermentation, where anoxygenic photoheterotrophic bacteria utilizes organic feedstock to produce H$_2$ in presence of light and (3) dark fermentation, in which anaerobic heterotrophic bacteria utilizes organic feedstock without any light to produce H$_2$. A comparison of these three main routes is shown in Table 1.

The microbiology and biochemistry of dark fermentative H$_2$ production was discussed in detail by Hawkes *et al.* [42]. H$_2$ production in *Clostridia* is due to the presence of hydrogenase enzymes. These transfer electrons from reduced ferredoxin or NADH to protons to regenerate the oxidized forms (Fd$_{ox}$ and NAD+) required so that glycolysis and oxidative decarboxylation of pyruvate can proceed to generate ATP.

Pure microbial cultures have mainly been used in lab-scale reactors for studying the effect of environmental and operational parameters on fermentation profiles and carbon metabolism. One of the successful tests using pure culture in a pilot-scale bioreactor using a non-sterilized feedstock employed *Caldicellulosiruptor saccharolyticus* [73]. However, most studies on H$_2$ production on biowaste have been performed using mixed cultures under mesophilic conditions [74,75]. Only a few studies have focused on mixed thermophilic consortia [76,77]. It has been demonstrated that the extreme thermophile *C. saccharolyticus* can produce H$_2$ from mono- and disaccharides [78]. Hexose is the predominant component in the cellulose hydrolysates. A highest H$_2$ yield of approximately 83% of the theoretical value (4 mol· mol^{-1} hexose) has been reported using thermophilic anaerobic bacteria [78].

3.3. Limiting Factors in Biohydrogen Production Systems

The most challenging barrier of fermentative H_2 production is its low H_2 molar yield [26]. Thauer *et al.* predicted that 4 moles of H_2 per mole of glucose is the biological maximum in *Clostridial* microbes if acetate is the only waste by-product [79]. In practice, even that figure is rarely achieved. A number of factors adversely affect and inhibit H_2 fermentation [44]. H_2 itself, when it reaches high concentrations not only makes its production thermodynamically unfavourable but also acts as an inhibitory agent as do other metabolic products, such as acetic acid and propionic acid [17,80]. Partial pressure of H_2 is one of the most critical parameters in fermentative production of H_2 as high H_2 partial pressures make H_2 production thermodynamically unfavourable. Removal of produced H_2 from the liquid phase lowers the H_2 partial pressure which in turn increases H_2 yield [81]. Moreover, the H_2 remaining in the system might be consumed by some bacteria [82]. Removal of dissolved H_2 and reduction of H_2 partial pressure can be achieved by nitrogen flushing, adsorption of H_2 by metals and H_2 stripping by boiling or by introduction of steam [83–85]. Low H_2 partial pressure also needs to be maintained because hydrogenases (such as NiFe-hydrogenase) may re-oxidize the produced hydrogen into protons and electrons [86]. Gas sparging has proved to be an efficient method to maintain maximum hydrogen production even though it leads to biogas dilution and higher cost for hydrogen recovery [87]. Depending on the nature of the flushing gas, the flow rate and the reactor configuration, volumetric production of biogas up to 120% has been achieved [85,88]. Non-sparging techniques such as headspace modification under vacuum, high pressure or gas adsorption (reviewed in [87]), hydrogen-separating membranes [83] and using mechanical stirring [89,90], have also showed significant improvements in hydrogen yield. Argon has been often used to flush both oxygen and nitrogen and to keep a low H_2 partial pressure in the reactors, but it increases production costs and hinders H_2 purification [91]. Some researchers have reported reduced pressure and CO_2 for flushing the headspace and maintaining low H_2 partial pressure in dark fermentation [92,93], but the information on photofermentation is deficite. Montiel-Corona *et al.* suggested that flushing with Ar could be replaced with reduced pressure, which can be less expensive and practical for hydrogen recuperation [91]. Coupling the dark and photo fermentation showed an increased total hydrogen yield. One of the major drawbacks in coupling the dark and photo fermentation processes is the need of keeping apart the H_2-producing microflora and the presence of NH_4^+, which may be naturally present in wastewater and may also be generated in the dark fermentation process when hydraulic retention time (HRT) is high enough to achieve protein degradation, especially when particulate substrates (as in the case of food wastes) are being considered, since HRT may be as high as 5 days [94].

In case of bioelectrochemical production of H_2 in MEC, the main challenge is avoiding methane formation via methanogenesis [24], though researchers are now shifting more towards methane formation rather than H_2 in these systems [95,96]. Another issue limiting the large-scale application of this technology is the use of precious metal catalyst such as platinum which is usually used on the cathode [97]. Though there have been efforts to use low cost materials such as stainless steel [98] and Ni-based electrodes [61], the results are much lesser from the targets.

3.4. Role of Metabolic Engineering

The application of genomic and molecular tools has made it possible to steer the metabolic pathways towards maximal H_2 production and avoid waste and by-product accumulation. This is especially true when genetic engineering is conducted on cellulolytic microbes [26]. The main principles of genetic engineering include: (1) overexpression of cellulases, hemicellulases and lignases to maximize substrate availability, (2) elimination of H_2-consuming hydrogenases and (3) overexpression of H_2-producing hydrogenases [53]. Metabolic engineering modifications have been used to increase H_2 production in fermentative systems [99]. These include over-expression of H_2-evolving enzymes [100], the knockout of metabolic pathways that compete for reducing equivalents [81] and the introduction/over-expression of genes (cellulases, hemicellulases and lignases) to enhance carbohydrate availability to the cell [101]. Inactivation of the gene lactate dehydrogenase (ldhA) in *E. coli* by introducing mutations could lead to a modest increase (20%–45%) in net hydrogen production (reviewed in [102]).

Ryu *et al.* combined several known approaches to construct a superior hydrogen-producing strain of the purple nonsulfur photosynthetic bacterium *Rhodobacter sphaeroides* HPCA* (mutant expressing NifA L62Q) [103]. In this strain maximum hydrogen levels are reached almost twice as fast as in wild type cells and final hydrogen levels are ~39% higher than in the wild type as well. As increased number of genomes for H_2 producing microorganisms are sequenced and compared and as more specific enzymes are functionally characterized, the distinctive metabolic strategies used and enzymological contexts through which H_2 evolution is controlled in different organisms will become clearer. This will allow researchers to construct more effective strategies to modulate competing pathways, and help in the designing molecular engineering strategies leading to enhanced H_2 evolution.

4. Kinetic Models for Hydrogen Production by Fermentation

Different factors such as substrate and inhibitor concentrations, temperature, pH and reactor type affect H_2 production by fermentation. Modeling of the H_2 production is very important to improve, analyze and predict H_2 production during fermentation. Mathematical models include the kinetic of cell growth and product(s) formation, substrate utilization and inhibition. In addition some models are developed to describe the effect of pH, temperature and dilution rate on H_2 production. The obtained model kinetic constants can be used in the design, operation and optimization of the fermentative H_2 production process. Different kinetic models have been proposed to describe growth of H_2 producing bacteria, substrate degradation and H_2 production. H_2 production is reported as growth associated product.

Monod (or Michaelis–Menten equation) (Equation (1)) which is an unstructured, non-segregated model of microbial growth, fits a wide range of data. The kinetic constants of this equation, K_S and μ_{max}, can be obtained by linear regression. Wang and Wan reported on previous studies using a Monod model to describe H_2 production with time in bio-H_2 fermentation [104]:

$$\mu = \frac{1}{X}\frac{dX}{dt} = \mu_m \frac{S}{K_s + S} \tag{1}$$

where μ is the specific growth rate, X is the biomass concentration, S is the substrate concentration, K_s is the saturation constant, μ_m is the maximum specific growth rate.

Recently, the logistic model (Equation (2)) became the most popular in describing cell growth. This equation has a sigmoidal shape that includes the lag phase, exponential and stationary phase of the batch growth:

$$\mu = \frac{1}{X}\frac{dX}{dt} = \mu_m \left(1 - \frac{X}{X_m}\right) \tag{2}$$

where X_m is the maximum biomass concentration.

At high substrate concentration, the cell growth is inhibited and production of H_2 is reduced. Different substrate inhibition models have been proposed. The Haldane-Andrew model (Equation (3)) is widely used to describe the substrate dependence of the specific growth rate of H_2 fermentations. Wang and Wan have reported that previous studies used an Andrews model to describe H_2 production with time [104]. Other substrate inhibition models are used in the literature such as modified Han-Levenspiel model (Equation (4)):

$$\mu = \frac{1}{X}\frac{dX}{dt} = \mu_m \frac{S}{K_s + S + \frac{S^2}{K_i}} \tag{3}$$

where K_i is the inhibition constant.

The presence of other inhibitors such as salts and the product cause reduction of H_2 production. Some models have been proposed to describe the effect of inhibitors such as the modified Han-Levenspiel model (Equation (4)):

9

$$\mu = \frac{1}{X}\frac{dX}{dt} = \mu_m\left(1 - \frac{C}{C_m}\right) \tag{4}$$

where C is the inhibitor concentration, C_m is the maximum inhibitor concentration or the concentration of inhibitor above which there is no biomass growth

The modified Gompertz model (Equation (5)) is widely used to describe the progress of cumulative H_2 production in batch fermentations [104]:

$$H_t = H_{max}\exp\left\{-\exp\left[\frac{R_{max} \times e}{H_{max}}(\lambda - t) + 1\right]\right\} \tag{5}$$

where H_t is the cumulative volume of H_2 produced at any time (mL), H_{max} is the gas production potential (mL), R_{max} is the maximum gas production rate (mL/h), λ is the lag time (h). t is the incubation time (h)

The Luedeking-Piret model (Equation (6)) has been widely used to describe the relation between cell growth rate and H_2 production:

$$\frac{dP}{dt} = Y_{P/X}\frac{dX}{dt} + \beta X \tag{6}$$

where P is the product, $Y_{P/X}$ is the growth associated yield coefficient; β is the non-growth associated product yield coefficient.

Wang and Wan reported that previous studies used the Luedeking–Piret model to relate cell growth rate and H_2 production rate [104]. The effect of temperature on the fermentative H_2 production has been widely described using the Arrhenius model, while the effect of pH on the substrate consumption rate is described by an Andrew model using the concentration of H^+ as the limiting substrate concentration. According to this model, the rate of substrate consumption passes through maximum with increasing H^+ concentration.

5. Sustainability and Life Cycle Assessment

The concept of sustainable development is an attempt to combine growing concerns about a range of environmental issues with socio-economic issues and implies smooth transition to more effective technologies from a point view of an environmental impact and energy efficiency [105,106]. H_2 can be considered one of the pillars of a future sustainable energy system [107]. H_2 production could be a possible avenue for the large-scale sustainable generation of H_2 needed to fuel a future H_2 economy [106]. Despite its many obvious advantages, there remains a problem with storage and transportation. Pressurized H_2 gas occupies a great deal of volume compared with other fuels. For example, gasoline that with equal energy content, needs about 30 times less volume at 100 bar gas pressure. Due to its high explosivity there are also obvious safety concerns with the use of pressurized or liquefied H_2 in vehicles as well as additional energy use for pressurizing or liquefaction. Furthermore, the overall energy balance of using H_2 as vehicle fuel does indeed seem to be less beneficial than gasoline, but being the only non-carbon fuel it may still make sense to produce H_2 from waste streams if some of the obstacles can be solved and it can be used effectively for energy production to feed into grid or to use in stationary requirements, e.g., industries, *etc.*

Though this paper is focused on bio-H_2 production from lignocellulosic biomass, it is important to compare it to other production methods using various substrates. Such a comparison has been made in Table 3 by presenting the various H_2 production systems, which show different H_2 yields from different feedstocks by adopting different production systems. Therefore, life cycle assessment (LCA) could be a tool to scrutinize the best H_2 production system for a particular feedstock in terms of environmental impact and indirect natural resource costs towards different services and commodities [108]. LCA allows the possibility of comparing different H_2 production approaches and identifying the environmental "hot spot" of the whole process, which helps in development of a sustainable H_2 production process [106,109]. Investigations of the environmental benefits and impacts

10

from a life cycle perspective are scarce. Only a few LCA-studies have been performed specifically on H_2 production. The feedstocks investigated so far are steamed potato peel, wheat straw and sweet sorghum stalks [110–112].

Table 3. Comparison of different biohydrogen production systems.

Reactor	Feed Stock	Maximum H_2 Yield	Reference
	Fermentation		
	Dark fermentation		
CSTR	Starch	0.52 L/h/L and 13.2 mmol H_2/g total sugar	[113]
Batch	Glycerol	0.41 mol H_2/mol glycerol	[114]
FBR	Sucrose	4.26 mol H_2/mol sucrose	[115]
Batch	Food waste	593 mL H_2/g carbohydrate	[116]
Fed-batch	Swine manure	18.7×10^{-3} g H_2 per g TVS	[117]
Batch	Sucrose	4.3 mol H_2/mol sucrose	[118]
Batch	Fructose, sorbitol, glucose	1.27, 1.46 and 1.51 mol H_2/substrate	[119]
Fed-batch	Starch, glucose	465 mL H_2/g starch, 3.1 mol H_2/mol glucose	[120]
Batch	Food waste	39.14 mL H_2/g food waste (219.91 mL H_2/VS$_{added}$)	[121]
Batch	Crude Glycerol	64.24 mmol H_2/L and 5.74 mmol H_2/g COD consumed	[122]
Batch	Distillery wastewaters	1 L H_2/L medium	[123]
Batch	Cheese whey	94.2 L H_2/kgvs	[124]
Batch	Water hyacinth (leaves and stems)	76.7 mL H_2/g TVS was obtained at 20 g/L of water hyacinth	[125]
Batch	waste ground wheat solution	SHPR = 25.7 mL H_2/g cells/h	[126]
	Photo fermentation		
CSTR	Sucrose	5.81 mol H_2/mol hexose	[127]
Fed-batch operation	Wheat starch	201 mL H_2 g/L starch	[128]
Batch	Molasses	0.50 mmol H_2/L$_c$ h	[129]
Batch	Beet molasses	10.5 mol H_2/mol sucrose	[52]
Batch	Black strap	8 mol H_2/mol sucrose	[52]
Batch	Sucrose	14 mol H_2/mol sucrose	[52]
Batch	Ground wheat starch	46 mL H_2/g biomass/h, 1.23 mol H_2/mol glucose	[51]
Batch	lignocellulose-derived organic acids	7 mL H_2/mL of the fermentation effluent	[130]
	Photosynthesis		
	Direct Photolysis		
Batch	Lactate	0.07 mmol H_2 (l × h) or 54 mL/h· g dry weight	[131]
	Indirect Photolysis		
Batch	arabinose and xylose	14.55 mmol/g (arabinose); 13.73 mmol/g (xylose)	[132]
	Thermochemical		
	Gasification		
Continuous supercritical water gasification	glucose	10.5–11.2 mol/mol glucose	[133]
	Partial Oxidation		
Batch	municipal sludge	Not reported the amount	[134]
	Steam reforming		
molten carbonate fuel cell (MCFC) system	ethanol	5 mol H_2/mol fed ethanol	[135]
	Cracking		
fixed-bed quartz micro reactor	Methane	500 μmoles/min	[136]
	Pyrolysis		
stainless steel tank reactor	Biomass (redwood sawdust; cole stalk and rice husk) feed	65.39 g/Kg biomass for redwood sawdust; 40.0 g/Kg biomass for cole stalk and rice husk	[137]
	Thermoelectrochemical		
membrane electrode assembly	sulfur dioxide	0.4 A/cm^2 at 0.835 V (H_2 production rate did not reported)	[138]
membrane electrode assembly	anhydrous hydrogen bromide	2.0 A/cm^2 at 1.91 V (H_2 production rate did not reported)	[138]

Table 3. *Cont.*

Reactor	Feed Stock	Maximum H_2 Yield	Reference
	Electrochemical		
	Electrolysis		
The BiOx–TiO$_2$ electrode and stainless steel (SS, Hastelloy C-22) were used as an anode and a cathode in the electrochemical system, respectively	arsenite (As(III))	9.4 μmoles/min	[139]
	Photoelectrolysis		
The TiO$_2$(ns) was prepared in the form of a sol-gel	photoelectrode system TiO$_2$(ns)–VO$_2$	$6\ L \cdot h^{-1} \cdot m^{-2}$ for the TiO$_2$(ns); $13.0\ L \cdot h^{-1} \cdot m^{-2}$ for the TiO$_2$(ns)–VO$_2$ photoelectrode	[140]

In connection with a European research study, HYVOLUTION, the life cycle environmental impacts of pilot production of H_2 through thermophilic fermentation, and photo fermentation of potato peel was compared to production of H_2 from natural gas through steam methane reforming (SMR) [112]. It was demonstrated that the bio-H_2 production had approximately 5.7 times higher environmental impacts (negative impacts on the environment) than a centralized SMR. The processes involved in steam (pretreatment), phosphate buffer (used in photo fermentation) and potassium hydroxide (used in thermophilic fermentation), were the main causes of the environmental impact (98.3%). Recirculation of the sewage reduces the environmental impacts considerably to having only approximately two times more environmental impact than SMR. If instead biomethane were produced for use in the SMR the environmental impact would be reduced to less than 1/3 of the traditional SMR [112]. On the other hand alternative use of the peel would be as animal feed and Djomo *et al.* showed that the production of bio-H_2 is more beneficial than the use as animal feed by a factor of 2–3 [110]. In a more recent study Djomo and Blumberga investigated potential differences in environmental performance between the three different feedstocks [111]. They performed a "well-to-tank" study *i.e.*, the system boundary is at supplying H_2 to road vehicles meaning that the combustion and transportation of H_2 in the vehicles, was not included. Further, the production of feedstock was excluded as they are considered wastes. Their conclusion is in contrast to the earlier study they find that H_2 produced from any of the feedstock reduced GHG-emissions by approximately 55% compared to SMR and a few percent less for gasoline. When the subsequent use of the remains from the H_2 production were considered as animal feed, an environmental benefit could be observed. The energy ratio calculated was 1.08–1.17, *i.e.*, the energy gain is between 8 and 17%. Though steamed potato peel was slightly better, no significant environmental differences were observed between the feedstocks [111]. The results compare well with those of Manish and Banerjee who investigated the energy balance of H_2 and found an energy ratio of 3.1 (excluding the gas treatment and the compressing) [57].

The conclusion from these studies from an environmental view point is that the production of H_2 for renewable energy production from potato peel could be preferred to using SMR or as direct animal feed due to the lesser environmental impacts. The LCA studies can further be used for identification of the main environmental improvements in the technology development (e.g., recirculation of the sewage and reuse of the remains for animal feed). The LCA of H_2 is very important before taking them into consideration for commercial scale production and policy decisions on H_2 promotion.

6. Future Directions and Perspectives

One option proposed to lower feedstock costs is to identify microbes that can directly utilize hemicellulose and cellulose [26]. This would eliminate the need for cellulase enzymes and simplify biomass pretreatment. As cellulose is the most abundant biopolymer in the world [141], its bioconversion provides a viable approach to produce renewable H_2 from organic matter. The combined dark fermentation coupling with photo fermentation, or dark fermentation coupling with bioelectrohydrogenesis is a promising H_2 production process from lignocellulosic biomass if the technological barriers can be overcome [12]. Overall, to develop a mature H_2 production technology,

Energies **2015**, *8*, 13062–13080

bioconversion performance from lignocellulosic biomass need to be further improved in terms of production rates, cost-effectiveness, and system scale-up. Based on the limited number of LCA studies done on H_2 production, it can be assumed that the bioconversion of lignocelluloses-to-H_2 on industrial scale is a feasible option to produce H_2 via biotechnology. However, more in-depth studies need to be carried out to confirm this.

7. Conclusions

Although considerable progress has been made on H_2 production from lignocellulosic biomass, several challenges remain for its commercial application. Among the various techniques available for H_2 production from lignocellulosic biomass, dark fermentation seems to have an edge over the others and is the closest to commercialization. Photo fermentation is the next best option, though it has to overcome the problems associated with reactor design and operation. Bioelectrochemical H_2 production is still in its infancy and needs much more research and development. The kinetic models for H_2 production provide insights on substrate utilization and factors limiting higher yields. The models will help in scale up studies for validating the proposed data and later on with the experimental data. The few environmental assessment studies performed from a LCA perspective show that H_2 production from lignocellulosic biomass also may be preferable to other renewable energy production pathways. Such studies can furthermore help identifying technological improvement options. The results of LCA studies could also help policy makers in taking decision on policies related to promotion of renewable energy.

Author Contributions: Anoop Singh and Deepak Pant originated the manuscript idea, created the outline, wrote parts of the manuscript and did the final editing. Surajbhan Sevda and Ibrahim M. Abu Reesh wrote about the modelling aspect and prepared the final revised draft. Karolien Vanbroekhoven and Dheeraj Rathore contributed to different sections of the manuscript mainly microbial electrolysis and sustainability.

Conflicts of Interest: The authors declare no conflict of interest.

References

1. Prasad, S.; Singh, A.; Joshi, H.C. Ethanol as an alternative fuel from agricultural, industrial and urban residues. *Resour. Conserv. Recycl.* **2007**, *50*, 1–39. [CrossRef]
2. Prasad, S.; Singh, A.; Joshi, H.C. Ethanol production from sweet sorghum syrup for utilization as automotive fuel in India. *Energy Fuels* **2007**, *21*, 2415–2420. [CrossRef]
3. Pant, D.; Van Bogaert, G.; Diels, L.; Vanbroekhoven, K. A review of the substrates used in microbial fuel cells (MFCs) for sustainable energy production. *Bioresour. Technol.* **2010**, *101*, 1533–1543. [CrossRef] [PubMed]
4. Pant, D.; Singh, A.; van Bogaert, G.; Gallego, Y.A.; Diels, L.; Vanbroekhoven, K. An introduction to the life cycle assessment (LCA) of bioelectrochemical systems (BES) for sustainable energy and product generation: Relevance and key aspects. *Renew. Sustain. Energy Rev.* **2011**, *15*, 1305–1313. [CrossRef]
5. Singh, A.; Smyth, B.M.; Murphy, J.D. A biofuel strategy for Ireland with an emphasis on production of biomethane and minimization of land-take. *Renew. Sustain. Energy Rev.* **2010**, *14*, 277–288. [CrossRef]
6. Singh, A.; Pant, D.; Korres, N.E.; Nizami, A.; Prasad, S.; Murphy, J.D. Key issues in life cycle assessment of ethanol production from lignocellulosic biomass: Challenges and perspectives. *Bioresour. Technol.* **2010**, *101*, 5003–5012. [CrossRef] [PubMed]
7. Singh, A.; Nigam, P.S.; Murphy, J.D. Renewable fuels from algae: An answer to debatable land based fuels. *Bioresour. Technol.* **2011**, *102*, 10–16. [CrossRef] [PubMed]
8. Hallenbeck, P.C.; Ghosh, D. Advances in fermentative biohydrogen production: The way forward? *Trends Biotechnol.* **2009**, *27*, 287–297. [CrossRef] [PubMed]
9. Ball, M.; Wietschel, M. The future of hydrogen—Opportunities and challenges. *Int. J. Hydrog. Energy* **2009**, *34*, 615–627. [CrossRef]
10. Perera, K.R.J.; Ketheesan, B.; Gadhamshetty, V.; Nirmalakhandan, N. Fermentative biohydrogen production: Evaluation of net energy gain. *Int. J. Hydrog. Energy* **2010**, *35*, 12224–12233. [CrossRef]
11. Kovacs, K.; Maroti, G.; Rakhely, G. A novel approach for biohydrogen production. *Int. J. Hydrog. Energy* **2006**, *31*, 1460–1468. [CrossRef]

12. Ren, N.; Wang, A.; Cao, G.; Xu, J.; Gao, L. Bioconversion of lignocellulosic biomass to hydrogen: Potential and challenges. *Biotechnol. Adv.* **2009**, *27*, 1051–1060. [CrossRef] [PubMed]

13. Kumar, N.; Das, D. Enhancement of hydrogen production by Enterobacter cloacae IIT-BT 08. *Process Biochem.* **2000**, *35*, 589–593. [CrossRef]

14. Chen, C.C.; Lin, C.Y.; Chang, J.S. Kinetics of hydrogen production with continuous anaerobic cultures utilizing sucrose as the limiting substrate. *Appl. Microbiol. Biotechnol.* **2001**, *57*, 56–64. [PubMed]

15. Ginkel, S.V.; Sung, S.; Lay, J.J. Biohydrogen production as a function of pH and substrate concentration. *Environ. Sci. Technol.* **2001**, *35*, 4726–4730. [CrossRef] [PubMed]

16. Xing, D.; Ren, N.; Li, Q.; Lin, M.; Wang, A.; Zhao, L. Ethanoligenens harbinense gen. nov., sp. nov., isolated from molasses wastewater. *Int. J. Syst. Evolut. Microbiol.* **2006**, *56*, 755–760. [CrossRef] [PubMed]

17. Wang, L.; Zhou, Q.; Li, F. Avoiding propionic acid accumulation in the anaerobic process for biohydrogen production. *Biomass Bioenergy* **2006**, *30*, 177–182. [CrossRef]

18. Ren, N.; Cao, G.; Guo, W.; Wang, A.; Zhu, Y.; Liu, B.; Xu, J. Biological hydrogen production from corn stover by moderately thermophile Thermoanaerobacterium thermosaccharolyticum W16. *Int. J. Hydrog. Energy* **2010**, *35*, 2708–2712. [CrossRef]

19. De Vrije, T.; de Haas, G.; Tan, G.B.; Keijsers, E.R.; Claassen, P.A.M. Pretreatment of Miscanthus for hydrogen production by Thermotoga elfii. *Int. J. Hydrog. Energy* **2002**, *27*, 1381–1390. [CrossRef]

20. Ntaikou, I.; Gavala, H.N.; Kornaros, M.; Lyberatos, G. Hydrogen production from sugars and sweet sorghum biomass using Ruminococcus albus. *Int. J. Hydrog. Energy* **2008**, *33*, 1153–1163. [CrossRef]

21. Kapdan, I.K.; Kargi, F. Bio-hydrogen production from waste materials. *Enzym. Microb. Technol.* **2006**, *38*, 569–582. [CrossRef]

22. Kalinci, Y.; Hepbasli, A.; Dincer, I. Biomass-based hydrogen production: A review and analysis. *Int. J. Hydrog. Energy* **2009**, *34*, 8799–8817. [CrossRef]

23. Kraemer, J.T.; Bagley, D.M. Improving the yield from fermentative hydrogen production. *Biotechnol. Lett.* **2007**, *29*, 685–695. [CrossRef] [PubMed]

24. Wang, J.; Wan, W. Factors influencing fermentative hydrogen production: A review. *Int. J. Hydrog. Energy* **2009**, *34*, 799–811. [CrossRef]

25. Cournac, L.; Sarma, P.M.; Fontecave, M. Biohydrogen: From Basic Concepts to Technology. *Int. J. Hydrog. Energy* **2010**, *35*, 10638. [CrossRef]

26. Turner, J.; Sverdrup, G.; Mann, M.K.; Maness, P.C.; Kroposki, B.; Ghirardi, M.; Evans, R.J.; Blake, D. Renewable hydrogen production. *Int. J. Energy Res.* **2008**, *32*, 379–407. [CrossRef]

27. Bartacek, J.; Zabranska, J.; Lens, P.N. Developments and constraints in fermentative hydrogen production. *Biofuels Bioprod. Biorefining* **2007**, *1*, 201–214. [CrossRef]

28. Nielsen, A.T.; Amandusson, H.; Bjorklund, R.; Dannetun, H.; Ejlertsson, J.; Ekedahl, L.G.; Lundström, I.; Svensson, B.H. Hydrogen production from organic waste. *Int. J. Hydrog. Energy* **2001**, *26*, 547–550. [CrossRef]

29. Kim, S.H.; Shin, H.S. Effects of base-pretreatment on continuous enriched culture for hydrogen production from food waste. *Int. J. Hydrog. Energy* **2008**, *33*, 5266–5274. [CrossRef]

30. Kim, D.H.; Kim, S.H.; Shin, H.S. Hydrogen fermentation of food waste without inoculum addition. *Enzyme Microb. Technol.* **2009**, *45*, 181–187. [CrossRef]

31. Ren, N.; Li, J.; Li, B.; Wang, Y.; Liu, S. Biohydrogen production from molasses by anaerobic fermentation with a pilot-scale bioreactor system. *Int. J. Hydrog. Energy* **2006**, *31*, 2147–2157. [CrossRef]

32. Li, J.; Li, B.; Zhu, G.; Ren, N.; Bo, L.; He, J. Hydrogen production from diluted molasses by anaerobic hydrogen producing bacteria in an anaerobic baffled reactor (ABR). *Int. J. Hydrog. Energy* **2007**, *32*, 3274–3283. [CrossRef]

33. Yang, P.; Zhang, R.; Mcgarvey, J.; Benemann, J. Biohydrogen production from cheese processing wastewater by anaerobic fermentation using mixed microbial communities. *Int. J. Hydrog. Energy* **2007**, *32*, 4761–4771. [CrossRef]

34. Azbar, N.; Çetinkaya Dokgöz, F.T.; Keskin, T.; Korkmaz, K.S.; Syed, H.M. Continuous fermentative hydrogen production from cheese whey wastewater under thermophilic anaerobic conditions. *Int. J. Hydrog. Energy* **2009**, *34*, 7441–7447. [CrossRef]

35. Lay, J.J. Modeling and optimization of anaerobic digested sludge converting starch to hydrogen. *Biotechnol. Bioeng.* **2000**, *63*, 269–278. [CrossRef]

36. Ueno, Y.; Haruta, S.; Ishii, M.; Igarashi, Y. Microbial community in anaerobic hydrogen-producing microflora enriched from sludge compost. *Appl. Microbiol. Biotechnol.* **2001**, *57*, 555–562. [PubMed]
37. Lin, C.Y.; Cheng, C.H. Fermentative hydrogen production from xylose using anaerobic mixed microflora. *Int. J. Hydrog. Energy* **2006**, *31*, 832–840. [CrossRef]
38. Hussy, I.; Hawkes, F.R.; Dinsdale, R.; Hawkes, D.L. Continuous fermentative hydrogen production from sucrose and sugarbeet. *Int. J. Hydrog. Energy* **2005**, *30*, 471–483. [CrossRef]
39. Wang, C.H.; Lin, P.J.; Chang, J.S. Fermentative conversion of sucrose and pineapple waste into hydrogen gas in phosphate-buffered culture seeded with municipal sewage sludge. *Process Biochem.* **2006**, *41*, 1353–1358. [CrossRef]
40. Fan, K.S.; Kan, N.R.; Lay, J.J. Effect of hydraulic retention time on anaerobic hydrogenesis in CSTR. *Bioresour. Technol.* **2006**, *97*, 84–89. [CrossRef] [PubMed]
41. Hawkes, F.R.; Dinsdale, R.; Hawkes, D.L.; Hussy, I. Sustainable fermentative hydrogen production: Challenges for process optimisation. *Int. J. Hydrog. Energy* **2002**, *27*, 1339–1347. [CrossRef]
42. Hawkes, F.R.; Hussy, I.; Kyazze, G.; Dinsdale, R.; Hawkes, D.L. Continuous dark fermentative hydrogen production by mesophilic microflora: Principles and progress. *Int. J. Hydrog. Energy* **2007**, *32*, 172–184. [CrossRef]
43. Varanasi, J.L.; Roy, S.; Pandit, S.; Das, D. Improvement of energy recovery from cellobiose by thermophillic dark fermentative hydrogen production followed by microbial fuel cell. *Int. J. Hydrog. Energy* **2015**, *40*, 8311–8321. [CrossRef]
44. Angenent, L.T.; Karim, K.; Al-Dahhan, M.H.; Wrenn, B.A.; Domíguez-Espinosa, R. Production of bioenergy and biochemicals from industrial and agricultural wastewater. *Trends Biotechnol.* **2004**, *22*, 477–485. [CrossRef] [PubMed]
45. Levin, D. Biohydrogen production: Prospects and limitations to practical application. *Int. J. Hydrog. Energy* **2004**, *29*, 173–185. [CrossRef]
46. Dipasquale, L.; D'Ippolito, G.; Fontana, A. Capnophilic lactic fermentation and hydrogen synthesis by Thermotoga neapolitana: An unexpected deviation from the dark fermentation model. *Int. J. Hydrog. Energy* **2014**, *39*, 4857–4862. [CrossRef]
47. Pradhan, N.; Dipasquale, L.; D'Ippolito, G.; Panico, A.; Lens, P.N.L.; Esposito, G.; Fontana, A. Hydrogen production by the thermophilic bacterium thermotoga neapolitana. *Int. J. Mol. Sci.* **2015**, *16*, 12578–12600. [CrossRef] [PubMed]
48. Basak, N.; Das, D. The prospect of purple non-sulfur (PNS) photosynthetic bacteria for hydrogen production: The present state of the art. *World J. Microbiol. Biotechnol.* **2007**, *23*, 31–42. [CrossRef]
49. Redwood, M.D.; Paterson-Beedle, M.; MacAskie, L.E. Integrating dark and light bio-hydrogen production strategies: Towards the hydrogen economy. *Rev. Environ. Sci. Biotechnol.* **2009**, *8*, 149–185. [CrossRef]
50. Holladay, J.D.; Hu, J.; King, D.L.; Wang, Y. An overview of hydrogen production technologies. *Catal. Today* **2009**, *139*, 244–260. [CrossRef]
51. Kapdan, I.K.; Kargi, F.; Oztekin, R.; Argun, H. Bio-hydrogen production from acid hydrolyzed wheat starch by photo-fermentation using different Rhodobacter sp. *Int. J. Hydrog. Energy* **2009**, *34*, 2201–2207. [CrossRef]
52. Keskin, T.; Hallenbeck, P.C. Hydrogen production from sugar industry wastes using single-stage photofermentation. *Bioresour. Technol.* **2012**, *112*, 131–136. [CrossRef] [PubMed]
53. Nath, K.; Das, D. Improvement of fermentative hydrogen production: Various approaches. *Appl. Microbiol. Biotechnol.* **2004**, *65*, 520–529. [CrossRef] [PubMed]
54. Liu, B.F.; Ren, N.Q.; Xie, G.J.; Ding, J.; Guo, W.Q.; Xing, D.F. Enhanced bio-hydrogen production by the combination of dark- and photo-fermentation in batch culture. *Bioresour. Technol.* **2010**, *101*, 5325–5329. [CrossRef] [PubMed]
55. Logan, B.E. *Microbial Fuel Cell*; Wiley & Sons, Inc.: Hoboken, NJ, USA, 2008.
56. Rozendal, R.; Hamelers, H.; Euverink, G.; Metz, S.; Buisman, C. Principle and perspectives of hydrogen production through biocatalyzed electrolysis. *Int. J. Hydrog. Energy* **2006**, *31*, 1632–1640. [CrossRef]
57. Manish, S.; Banerjee, R. Comparison of biohydrogen production processes. *Int. J. Hydrog. Energy* **2008**, *33*, 279–286. [CrossRef]
58. Cusick, R.D.; Bryan, B.; Parker, D.S.; Merrill, M.D.; Mehanna, M.; Kiely, P.D.; Liu, G.; Logan, B.E. Performance of a pilot-scale continuous flow microbial electrolysis cell fed winery wastewater. *Appl. Microbiol. Biotechnol.* **2011**, *89*, 2053–2063. [CrossRef] [PubMed]

59. Kadier, A.; Simayi, Y.; Kalil, M.S.; Abdeshahian, P.; Hamid, A.A. A review of the substrates used in microbial electrolysis cells (MECs) for producing sustainable and clean hydrogen gas. *Renew. Energy* **2014**, *71*, 466–472. [CrossRef]

60. Ren, L.; Siegert, M.; Ivanov, I.; Pisciotta, J.M.; Logan, B.E. Treatability studies on different refinery wastewater samples using high-throughput microbial electrolysis cells (MECs). *Bioresour. Technol.* **2013**, *136*, 322–328. [CrossRef] [PubMed]

61. Hu, H.; Fan, Y.; Liu, H. Hydrogen production in single-chamber tubular microbial electrolysis cells using non-precious-metal catalysts. *Int. J. Hydrog. Energy* **2009**, *34*, 8535–8542. [CrossRef]

62. Lu, L.; Xing, D.; Ren, N.; Logan, B.E. Syntrophic interactions drive the hydrogen production from glucose at low temperature in microbial electrolysis cells. *Bioresour. Technol.* **2012**, *124*, 68–76. [CrossRef] [PubMed]

63. Lu, L.; Ren, N.; Xing, D.; Logan, B.E. Hydrogen production with effluent from an ethanol-H2-coproducing fermentation reactor using a single-chamber microbial electrolysis cell. *Biosens. Bioelectron.* **2009**, *24*, 3055–3060. [CrossRef] [PubMed]

64. Call, D.; Logan, B.E. Hydrogen production in a single chamber microbial electrolysis cell lacking a membrane. *Environ. Sci. Technol.* **2008**, *42*, 3401–3406. [CrossRef] [PubMed]

65. Call, D.F.; Merrill, M.D.; Logan, B.E. High surface area stainless steel brushes as cathodes in microbial electrolysis cells. *Environ. Sci. Technol.* **2009**, *43*, 2179–2183. [CrossRef] [PubMed]

66. Selembo, P.A.; Perez, J.M.; Lloyd, W.A.; Logan, B.E. High hydrogen production from glycerol or glucose by electrohydrogenesis using microbial electrolysis cells. *Int. J. Hydrog. Energy* **2009**, *34*, 5373–5381. [CrossRef]

67. Wang, X.; Cheng, S.; Feng, Y.; Merrill, M.D.; Saito, T.; Logan, B.E. Use of carbon mesh anodes and the effect of different pretreatment methods on power production in microbial fuel cells. *Environ. Sci. Technol.* **2009**, *43*, 6870–6874. [CrossRef] [PubMed]

68. Wagner, R.C.; Regan, J.M.; Oh, S.-E.; Zuo, Y.; Logan, B.E. Hydrogen and methane production from swine wastewater using microbial electrolysis cells. *Water Res.* **2009**, *43*, 1480–1488. [CrossRef] [PubMed]

69. Pisciotta, J.M.; Zaybak, Z.; Call, D.F.; Nam, J.Y.; Logan, B.E. Enrichment of microbial electrolysis cell biocathodes from sediment microbial fuel cell bioanodes. *Appl. Environ. Microbiol.* **2012**, *78*, 5212–5219. [CrossRef] [PubMed]

70. Xiao, L.; Wen, Z.; Ci, S.; Chen, J.; He, Z. Carbon/iron-based nanorod catalysts for hydrogen production in microbial electrolysis cells. *Nano Energy* **2012**, *1*, 751–756. [CrossRef]

71. Guo, K.; Tang, X.; Du, Z.; Li, H. Hydrogen production from acetate in a cathode-on-top single-chamber microbial electrolysis cell with a mipor cathode. *Biochem. Eng. J.* **2010**, *51*, 48–52. [CrossRef]

72. Hu, H.; Fan, Y.; Liu, H. Hydrogen production using single-chamber membrane-free microbial electrolysis cells. *Water Res.* **2008**, *42*, 4172–4178. [CrossRef] [PubMed]

73. Van Groenestijn, J.W.; Geelhoed, J.S.; Goorissen, H.P.; Meesters, K.P.; Stams, A.J.; Claassen, P.A. Performance and population analysis of a non-sterile trickle bed reactor inoculated with caldicellulosiruptor saccharolyticus, a thermophilic hydrogen producer. *Biotechnol. Bioeng.* **2009**, *102*, 1361–1367. [CrossRef] [PubMed]

74. Kleerebezem, R.; van Loosdrecht, M.C. Mixed culture biotechnology for bioenergy production. *Curr. Opin. Biotechnol.* **2007**, *18*, 207–212. [CrossRef] [PubMed]

75. Kengen, S.W.M.; Goorissen, H.P.; Verhaart, M.; Stams, A.J.M.; van Niel, E.W.J.; Claassen, P.A.M. Biological Hydrogen Production by Anaerobic Microorganisms. In *Biofuels*; John Wiley & Sons, Ltd.: Chichester, UK, 2009; pp. 197–221.

76. Kotsopoulos, T.A.; Zeng, R.J.; Angelidaki, I. Biohydrogen production in granular up-flow anaerobic sludge blanket (UASB) reactors with mixed cultures under hyper-thermophilic temperature (70 C). *Biotechnol. Bioeng.* **2006**, *94*, 296–302. [CrossRef] [PubMed]

77. De Vrije, T.; Bakker, R.R.; Budde, M.A.; Lai, M.H.; Mars, A.E.; Claassen, P.A. Efficient hydrogen production from the lignocellulosic energy crop Miscanthus by the extreme thermophilic bacteria Caldicellulosiruptor saccharolyticus and Thermotoga neapolitana. *Biotechnol. Biofuels* **2009**, *2*, 12. [CrossRef] [PubMed]

78. Van Niel, E.W.; Budde, M.A.; De Haas, G.; Van der Wal, F.J.; Claassen, P.A.; Stams, A.J. Distinctive properties of high hydrogen producing extreme thermophiles, Caldicellulosiruptor saccharolyticus and Thermotoga elfii. *Int. J. Hydrog. Energy* **2002**, *27*, 1391–1398. [CrossRef]

79. Thauer, R.K.; Jungermann, K.; Decker, K. Energy conservation in chemotrophic anaerobic bacteria. *Bacteriol. Rev.* **1977**, *41*, 100–180. [PubMed]

80. Vanginkel, S.; Oh, S.; Logan, B. Biohydrogen gas production from food processing and domestic wastewaters. *Int. J. Hydrog. Energy* **2005**, *30*, 1535–1542. [CrossRef]
81. Liu, D.; Liu, D.; Zeng, R.J.; Angelidaki, I. Hydrogen and methane production from household solid waste in the two-stage fermentation process. *Water Res.* **2006**, *40*, 2230–2236. [CrossRef] [PubMed]
82. Oh, S.-E.; van Ginkel, S.; Logan, B.E. The relative effectiveness of pH control and heat treatment for enhancing biohydrogen gas production. *Environ. Sci. Technol.* **2003**, *37*, 5186–5190. [CrossRef] [PubMed]
83. Liang, T.M.; Cheng, S.S.; Wu, K.L. Behavioral study on hydrogen fermentation reactor installed with silicone rubber membrane. *Int. J. Hydrog. Energy* **2002**, *27*, 1157–1165. [CrossRef]
84. Van Groenestijn, J.W.; Hazewinkel, J.H.; Nienoord, M.; Bussmann, P.J. Energy aspects of biological hydrogen production in high rate bioreactors operated in the thermophilic temperature range. *Int. J. Hydrog. Energy* **2002**, *27*, 1141–1147. [CrossRef]
85. Kim, D.H.; Han, S.K.; Kim, S.H.; Shin, H.S. Effect of gas sparging on continuous fermentative hydrogen production. *Int. J. Hydrog. Energy* **2006**, *31*, 2158–2169. [CrossRef]
86. Koku, H. Aspects of the metabolism of hydrogen production by Rhodobacter sphaeroides. *Int. J. Hydrog. Energy* **2002**, *27*, 1315–1329. [CrossRef]
87. Beckers, L.; Masset, J.; Hamilton, C.; Delvigne, F.; Toye, D.; Crine, M.; Thonart, P.; Hiligsmann, S. Investigation of the links between mass transfer conditions, dissolved hydrogen concentration and biohydrogen production by the pure strain Clostridium butyricum CWBI1009. *Biochem. Eng. J.* **2015**, *98*, 18–28. [CrossRef]
88. Mizuno, O.; Dinsdale, R.; Hawkes, F.R.; Hawkes, D.L.; Noike, T. Enhancement of hydrogen production from glucose by nitrogen gas sparging. *Bioresour. Technol.* **2000**, *73*, 59–65. [CrossRef]
89. Chou, C.; Wang, C.; Huang, C.; Lay, J. Pilot study of the influence of stirring and pH on anaerobes converting high-solid organic wastes to hydrogen. *Int. J. Hydrog. Energy* **2008**, *33*, 1550–1558. [CrossRef]
90. Fontes Lima, D.M.; Zaiat, M. The influence of the degree of back-mixing on hydrogen production in an anaerobic fixed-bed reactor. *Int. J. Hydrog. Energy* **2012**, *37*, 9630–9635. [CrossRef]
91. Montiel-Corona, V.; Revah, S.; Morales, M. Hydrogen production by an enriched photoheterotrophic culture using dark fermentation effluent as substrate: Effect of flushing method, bicarbonate addition, and outdoor–indoor conditions. *Int. J. Hydrog. Energy* **2015**, *40*, 9096–9105. [CrossRef]
92. Clark, I.C.; Zhang, R.H.; Upadhyaya, S.K. The effect of low pressure and mixing on biological hydrogen production via anaerobic fermentation. *Int. J. Hydrog. Energy* **2012**, *37*, 11504–11513. [CrossRef]
93. Kim, D.-H.; Shin, H.-S.; Kim, S.-H. Enhanced H_2 fermentation of organic waste by CO_2 sparging. *Int. J. Hydrog. Energy* **2012**, *37*, 15563–15568. [CrossRef]
94. Gómez, X.; Morán, A.; Cuetos, M.J.; Sánchez, M.E. The production of hydrogen by dark fermentation of municipal solid wastes and slaughterhouse waste: A two-phase process. *J. Power Sources* **2006**, *157*, 727–732. [CrossRef]
95. Clauwaert, P.; Verstraete, W. Methanogenesis in membraneless microbial electrolysis cells. *Appl. Microbiol. Biotechnol.* **2009**, *82*, 829–836. [CrossRef] [PubMed]
96. Cheng, S.; Xing, D.; Call, D.F.; Logan, B.E. Direct biological conversion of electrical current into methane by electromethanogenesis. *Environ. Sci. Technol.* **2009**, *43*, 3953–3958. [CrossRef] [PubMed]
97. Pant, D.; van Bogaert, G.; de Smet, M.; Diels, L.; Vanbroekhoven, K. Use of novel permeable membrane and air cathodes in acetate microbial fuel cells. *Electrochimica Acta* **2010**, *55*, 7710–7716. [CrossRef]
98. Selembo, P.A.; Merrill, M.D.; Logan, B.E. The use of stainless steel and nickel alloys as low-cost cathodes in microbial electrolysis cells. *J. Power Sour.* **2009**, *190*, 271–278. [CrossRef]
99. Maeda, T.; Sanchez-Torres, V.; Wood, T.K. Metabolic engineering to enhance bacterial hydrogen production. *Microb. Biotechnol.* **2008**, *1*, 30–39. [CrossRef] [PubMed]
100. Morimoto, K.; Kimura, T.; Sakka, K.; Ohmiya, K. Overexpression of a hydrogenase gene in Clostridium paraputrificum to enhance hydrogen gas production. *FEMS Microbiol. Lett.* **2005**, *246*, 229–234. [CrossRef] [PubMed]
101. Vardar-Schara, G.; Maeda, T.; Wood, T.K. Metabolically engineered bacteria for producing hydrogen via fermentation. *Microb. Biotechnol.* **2008**, *1*, 107–125. [CrossRef] [PubMed]
102. Hallenbeck, P.C.; Ghosh, D. Improvements in fermentative biological hydrogen production through metabolic engineering. *J. Environ. Manag.* **2012**, *95*, S360–S364. [CrossRef] [PubMed]
103. Ryu, M.-H.; Hull, N.C.; Gomelsky, M. Metabolic engineering of Rhodobacter sphaeroides for improved hydrogen production. *Int. J. Hydrog. Energy* **2014**, *39*, 6384–6390. [CrossRef]

104. Wang, J.; Wan, W. Kinetic models for fermentative hydrogen production: A review. *Int. J. Hydrog. Energy* **2009**, *34*, 3313–3323. [CrossRef]
105. Dincer, I. Hydrogen and Fuel Cell Technologies for Sustainable Future. *Jordan J. Mech. Ind. Eng.* **2008**, *2*, 1–14.
106. Rathore, D.; Singh, A. Biohydrogen production from microalgae. In *Biofuels Technologies Recent Developments*; Gupta, V., Tuohy, M., Eds.; Springer: Berlin/Heidelberg, Germany, 2013; pp. 317–333.
107. Kwak, H.Y.; Lee, H.S.; Jung, J.Y.; Jeon, J.S.; Park, D.R. Exergetic and thermoeconomic analysis of a 200-kW phosphoric acid fuel cell plant. *Fuel* **2004**, *83*, 2087–2094. [CrossRef]
108. Rubio Rodríguez, M.A.; de Ruyck, J.; Díaz, P.R.; Verma, V.K.; Bram, S. An LCA based indicator for evaluation of alternative energy routes. *Appl. Energy* **2011**, *88*, 630–635. [CrossRef]
109. Romagnoli, F.; Blumberga, D.; Pilicka, I. Life cycle assessment of biohydrogen production in photosynthetic processes. *Int. J. Hydrog. Energy* **2011**, *36*, 7866–7871. [CrossRef]
110. Djomo, S.N.; Humbert, S. Dagnija Blumberga Life cycle assessment of hydrogen produced from potato steam peels. *Int. J. Hydrog. Energy* **2008**, *33*, 3067–3072. [CrossRef]
111. Djomo, S.N.; Blumberga, D. Comparative life cycle assessment of three biohydrogen pathways. *Bioresour. Technol.* **2011**, *102*, 2684–2694. [CrossRef] [PubMed]
112. Ochs, D.; Wukovits, W.; Ahrer, W. Life cycle inventory analysis of biological hydrogen production by thermophilic and photo fermentation of potato steam peels (PSP). *J. Clean. Prod.* **2010**, *18*, S88–S94. [CrossRef]
113. Chen, S.D.; Lo, Y.C.; Lee, K.S.; Huang, T.I.; Chang, J.S. Sequencing batch reactor enhances bacterial hydrolysis of starch promoting continuous bio-hydrogen production from starch feedstock. *Int. J. Hydrog. Energy* **2009**, *34*, 8549–8557. [CrossRef]
114. Seifert, K.; Waligorska, M.; Wojtowski, M.; Laniecki, M. Hydrogen generation from glycerol in batch fermentation process. *Int. J. Hydrog. Energy* **2009**, *34*, 3671–3678. [CrossRef]
115. Lin, C.N.; Wu, S.Y.; Chang, J.S.; Chang, J.S. Biohydrogen production in a three-phase fluidized bed bioreactor using sewage sludge immobilized by ethylene-vinyl acetate copolymer. *Bioresour. Technol.* **2009**, *100*, 3298–3301. [CrossRef] [PubMed]
116. Yasin Nazlina, H.M.; Aini, R.; Ismail, F.; Zulkhairi, M.; Hassan, M.A. Effect of different temperature, initial pH and substrate composition on biohydrogen production from food waste in batch fermentation. *Asian J. Biotechnol.* **2009**, *1*, 42–50.
117. Zhu, J.; Li, Y.; Wu, X.; Miller, C.; Chen, P.; Ruan, R. Swine manure fermentation for hydrogen production. *Bioresour. Technol.* **2009**, *100*, 5472–5477. [CrossRef] [PubMed]
118. Gadhamshetty, V.; Johnson, D.C.; Nirmalakhandan, N.; Smith, G.B.; Deng, S. Feasibility of biohydrogen production at low temperatures in unbuffered reactors. *Int. J. Hydrog. Energy* **2009**, *34*, 1233–1243. [CrossRef]
119. Ghosh, D.; Hallenbeck, P.C. Fermentative hydrogen yields from different sugars by batch cultures of metabolically engineered *Escherichia coli* DJT135. *Int. J. Hydrog. Energy* **2009**, *34*, 7979–7982. [CrossRef]
120. Kargi, F.; Pamukoglu, M.Y. Dark fermentation of ground wheat starch for bio-hydrogen production by fed-batch operation. *Int. J. Hydrog. Energy* **2009**, *34*, 2940–2946. [CrossRef]
121. Han, W.; Ye, M.; Zhu, A.J.; Zhao, H.T.; Li, Y.F. Batch dark fermentation from enzymatic hydrolyzed food waste for hydrogen production. *Bioresour. Technol.* **2015**, *191*, 24–29. [CrossRef] [PubMed]
122. Chookaew, T.; O-Thong, S.; Prasertsan, P. Biohydrogen production from crude glycerol by two stage of dark and photo fermentation. *Int. J. Hydrog. Energy* **2015**, *40*, 7433–7438. [CrossRef]
123. Wicher, E.; Seifert, K.; Zagrodnik, R.; Pietrzyk, B.; Laniecki, M. Hydrogen gas production from distillery wastewater by dark fermentation. *Int. J. Hydrog. Energy* **2013**, *38*, 7767–7773. [CrossRef]
124. Moreno, R.; Escapa, A.; Cara, J.; Carracedo, B.; Gómez, X. A two-stage process for hydrogen production from cheese whey: Integration of dark fermentation and biocatalyzed electrolysis. *Int. J. Hydrog. Energy* **2015**, *40*, 168–175. [CrossRef]
125. Su, H.; Cheng, J.; Zhou, J.; Song, W.; Cen, K. Hydrogen production from water hyacinth through dark- and photo- fermentation. *Int. J. Hydrog. Energy* **2010**, *35*, 8929–8937. [CrossRef]
126. Argun, H.; Kargi, F. Effects of sludge pre-treatment method on bio-hydrogen production by dark fermentation of waste ground wheat. *Int. J. Hydrog. Energy* **2009**, *34*, 8543–8548. [CrossRef]
127. Laurinavichene, T.V.; Belokopytov, B.F.; Laurinavichius, K.S.; Tekucheva, D.N.; Seibert, M.; Tsygankov, A.A. Towards the integration of dark- and photo-fermentative waste treatment. 3. Potato as substrate for sequential dark fermentation and light-driven H_2 production. *Int. J. Hydrog. Energy* **2010**, *35*, 8536–8543. [CrossRef]

128. Ozmihci, S.; Kargi, F. Effects of starch loading rate on performance of combined fed-batch fermentation of ground wheat for bio-hydrogen production. *Int. J. Hydrog. Energy* **2010**, *35*, 1106–1111. [CrossRef]
129. Avcioglu, S.G.; Ozgur, E.; Eroglu, I.; Yucel, M.; Gunduz, U. Biohydrogen production in an outdoor panel photobioreactor on dark fermentation effluent of molasses. *Int. J. Hydrog. Energy* **2011**, *36*, 11360–11368. [CrossRef]
130. Zhu, Z.; Shi, J.; Zhou, Z.; Hu, F.; Bao, J. Photo-fermentation of Rhodobacter sphaeroides for hydrogen production using lignocellulose-derived organic acids. *Process Biochem.* **2010**, *45*, 1894–1898. [CrossRef]
131. Francou, N.; Vignais, P.M. Hydrogen production by Rhodopseudomonas capsulata cells entrapped in carrageenan beads. *Biotechnol. Lett.* **1984**, *6*, 639–644. [CrossRef]
132. Taguchi, F.; Mizukami, N.; Hasegawa, K.; Saito-Taki, T. Microbial conversion of arabinose and xylose to hydrogen by a newly isolated *clostridium* sp. No. 2. *Can. J. Microbiol.* **1994**, *40*, 228–233. [CrossRef]
133. Susanti, R.F.; Dianningrum, L.W.; Yum, T.; Kim, Y.; Lee, B.G.; Kim, J. High-yield hydrogen production from glucose by supercritical water gasification without added catalyst. *Int. J. Hydrog. Energy* **2012**, *37*, 11677–11690. [CrossRef]
134. Qinming, Z.; Shuzhong, W.; Liang, W.; Donghai, X. Catalytic Hydrogen Production from Municipal Sludge in Supercritical Water with Partial Oxidation. *Chall. Power Eng. Environ.* **2007**, *1*, 1252–1255.
135. Frusteri, F.; Freni, S.; Chiodo, V.; Spadaro, L.; Bonura, G.; Cavallaro, S. Potassium improved stability of Ni/MgO in the steam reforming of ethanol for the production of hydrogen for MCFC. *J. Power Sour.* **2004**, *132*, 139–144. [CrossRef]
136. Aiello, R.; Fiscus, J.E.; Loye, H.; Amiridis, M.D. Hydrogen production via the direct cracking of methane over Ni/SiO$_2$: Catalyst deactivation and regeneration. *Appl. Catal. A Gen.* **2000**, *192*, 227–234. [CrossRef]
137. Deng, W.; Jiang, H.; Wu, Y.; Fan, H.; Ji, J. Hydrogen production from biomass pyrolysis in molten alkali. *AASRI Procedia* **2012**, *3*, 217–223.
138. Sivasubramanian, P.; Ramasamy, R.P.; Freire, F.J.; Holland, C.E.; Weidner, J.W. Electrochemical hydrogen production from thermochemical cycles using a proton exchange membrane electrolyzer. *Int. J. Hydrog. Energy* **2007**, *32*, 463–468. [CrossRef]
139. Kim, J.; Kwon, D.; Kim, K.; Hoffmann, M.R. Electrochemical Production of Hydrogen Coupled with the Oxidation of Arsenite. *Environ. Sci. Technol.* **2014**, *48*, 2059–2066. [CrossRef] [PubMed]
140. Karn, R.K.; Srivastava, O.N. On the synthesis and photochemical studies of nanostructured TiO 1 and TiO 1 admixed VO 1 photoelectrodes in regard to hydrogen production through photoelectrolysis. *Int. J. Hydrog. Energy* **1999**, *24*, 965–971. [CrossRef]
141. Perlack, R.D.; Wright, L.L.; Turhollow, A.F.; Graham, R.L.; Stokes, B.J.; Erbach, D.C. *Biomass as Feedstock For a Bioenergy and Bioproducts Industry: The Technical Feasibility of a Billion-Ton Annual Supply*; U.S. Department of Energy: Oak Ridge, TN, USA, 2005.

energies

MDPI

Article

Bioenergy and Food Supply: A Spatial-Agent Dynamic Model of Agricultural Land Use for Jiangsu Province in China

Kesheng Shu [1,2,3,*], Uwe A. Schneider [4] and Jürgen Scheffran [3]

[1] Institute of Geographic Sciences and Natural Resources Research, Chinese Academy of Sciences, Beijing 100101, China
[2] University of Chinese Academy of Sciences, Beijing 100049, China
[3] Research Group Climate Change and Security, Institute of Geography, Center for Earth System Research and Sustainability, University of Hamburg, Hamburg 20144, Germany; juergen.scheffran@uni-hamburg.de
[4] Research Unit Sustainability and Global Change, Center for Earth System Research and Sustainability, University of Hamburg, Hamburg 20144, Germany; uwe.schneider@uni-hamburg.de
* Correspondence: kesheng.shu@uni-hamburg.de; Tel.: +49-40-42838-9193; Fax: +49-40-42838-9211

Academic Editor: Tariq Al-Shemmeri
Received: 31 July 2015; Accepted: 10 November 2015; Published: 24 November 2015

Abstract: In this paper we develop an agent-based model to explore a feasible way of simultaneously providing sufficient food and bioenergy feedstocks in China. Concerns over the competition for agricultural land resources between food and bioenergy supply hinder the further development of bioenergy, especially in China, the country that needs to feed the world's largest population. Prior research has suggested the introduction of energy crops and reviewed the resulting agricultural land use change in China. However, there is a lack of quantitative studies which estimate the value, contribution, and impact of bioenergy for specific conditions at the county level and provide adequate information to guide local practices. To fill this gap, we choose the Jiangsu Province in China as a case study, build up a spatial-agent dynamic model of agricultural land use, and perform a sensitivity analysis for important parameters. The simulation results show that straw from conventional crops generally dominates Jiangsu's biomass supply with a contribution above 85%. The sensitivity analyses reveal severe consequences of bioenergy targets for local land use. For Jiangsu Province, reclaimed mudflats, an alternative to arable lands for energy crop plantation, help to secure the local biomass supply and to alleviate the land use conflict between food and biomass production.

Keywords: bioenergy supply; agent-based model; land use; general algebraic modeling system (GAMS); China

1. Introduction

Modern bioenergy has experienced a worldwide boost in the last decades in response to concerns over energy security and climate change [1–3]. For some developing countries, bioenergy may also help to revive their agricultural sectors and rural areas [4,5]. However, rising food prices have sparked a debate as to whether and to what extent bioenergy development would put food supply at risk [6–9]. To simultaneously satisfy the demand for bioenergy feedstock and food, especially in the context of a fast growing bioenergy industry, previous studies have suggested a variety of options. These include improved agricultural practices, plantation of dedicated energy crops, use of cellulosic biomass conversion technology, exploitation of marginal or degraded lands, and joint production of energy and animal feed [10–15]. For China, a country accommodating the world's largest population, finding a feasible compromise between food and bioenergy production is of great importance. Earlier qualitative studies addressed the introduction of energy crops to China's cropping system [16–18] and the reclamation of mudflats for energy crop plantations [19–21]. Insights from these studies were integrated by the administrative bodies in a series of official development plans [22–24]. On the global and country level, a cluster of pioneering quantitative work has been done to explore and mitigate the resource competition between food and bioenergy. Yamamoto *et al.* [25] used a global-land-use-and-energy (GLUE-11) model to evaluate the global bioenergy potential. Schneider and McCarl [26] developed the Agricultural Sector and Mitigation of Greenhouse Gas (ASMGHG) model to examine the potential of biofuel for reducing greenhouse gas (GHG) emissions in the U.S., considering both food crops and energy crops. Johansson and Azar [27] created the Land Use Change Energy and Agriculture (LUCEA) model to analyze the competition between agricultural and energy systems in the same country. Havlík *et al.* [28] and Kraxner *et al.* [29] applied the Global Biosphere Management (GLOBIOM) model to estimate land use impacts of bioenergy targets at a global scale.

While previous research has presented ideas of how to mitigate the conflict between bioenergy feedstock supply and food supply at aggregated scales, there is limited analysis of the value, contribution, and impact of bioenergy strategies for specific local conditions, *i.e.*, at the county level. Results from global or national models, however, may not be adequate to guide local practices. This is especially true for China, which is rarely touched on in existing quantitative research. To fill this gap, this study develops a spatial-agent dynamic model of agricultural land use to simulate the annual land use patterns after the introduction of energy crops. By choosing Jiangsu Province in China as a case study, this model provides specific results to guide the regional development of bioenergy. Furthermore, by using sensitivity analysis, we evaluate the impact of alternative bioenergy targets on the local agricultural land use and measure the role of reclaimed mudflats in alleviating the land use conflict between biomass and food supply.

The paper is structured as follows: first, we provide background information on the study area relevant to the construction of our model. Next, we present the analytical framework and briefly explain the individual components. More detailed information can be found in the Appendix A. Subsequently, we present the simulation results including optimal patterns of land use and bioenergy feedstock supply. Finally, we provide a concluding discussion.

2. The Bioenergy Development in Jiangsu Province

2.1. The Study Area

Jiangsu Province is located on the eastern coast of China. Together with Shanghai City and Zhejiang Province, it constitutes one of the most economically advanced areas of China, often referred to as the Yangtze River Delta Economic Circle (YRDEC). Jiangsu Province contains 13 prefecture-level cities, and each city administrates several county-level units with a total number of 102 (Figure 1). Due to differences in topography and socio-economic development, Jiangsu Province is often divided into three sub-regions: Southern Jiangsu, Central Jiangsu, and Northern Jiangsu.

Rapid economic growth in Jiangsu Province over the last 30 years has come at the expense of high GHG emissions. In 2007, the area's estimated carbon emission totaled 144 million tonnes, with an annual growth rate of 14% during the period of 1996–2007 [30]. To curb CO_2 emissions, the provincial government issued an action plan, which includes the increasing use of renewable energy [31]. In 1990, up to 92% of Jiangsu's carbon emissions came from the consumption of energy, with coal contributing the most [32]. Given that more than 92% of coal, 93% of crude oil and 99% of natural gas are imported from outside of Jiangsu [33], the use of domestic renewable energy sources, in particular, bioenergy can improve the region's energy security. Besides reducing the use of fossil fuels, bioenergy can also stimulate rural development through bringing in investments in infrastructure and additional income opportunities for farmers. These benefits can help to reduce the migration flows from rural to urban areas and curtail the regional disparity within Jiangsu.

While facing a high demand for bioenergy, Jiangsu Province also enjoys favorable local conditions for biomass production. The high output of grains implies a large amount of crop residues that could be used as a biomass feedstock. Furthermore, its particular location on the eastern coast enables access to additional land resources, *i.e.*, reclaimed mudflats, which can be used for large-scale biomass production (Figure 2).

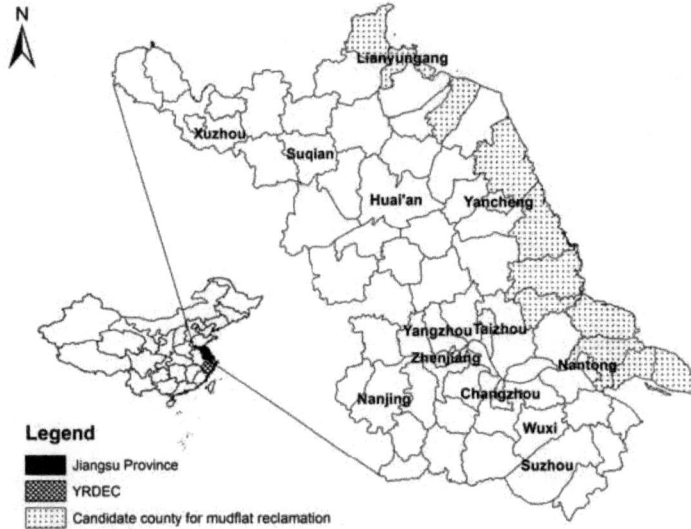

Figure 1. The location of Jiangsu Province. The list of potential counties for mudflat reclamation is identified in [23]. YRDEC is the acronym for the Yangtze River Delta Economic Circle.

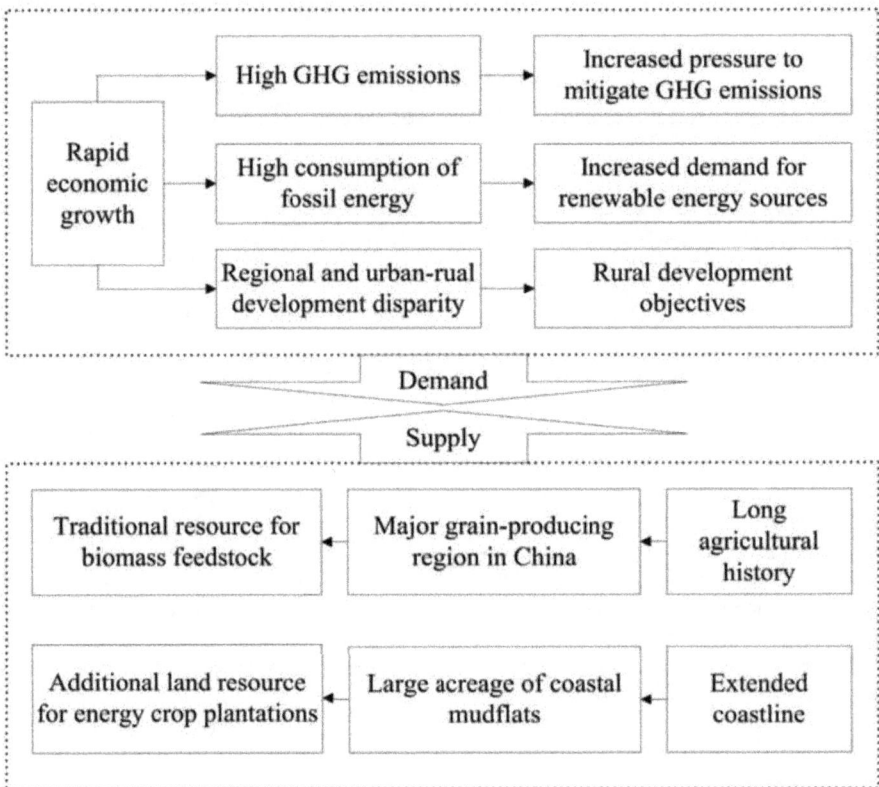

Figure 2. Supply and demand conditions for bioenergy development in Jiangsu Province.

2.2. Conventional Crops

In order to distinguish dedicated energy crops, this study uses the term "conventional crops" for annual crops planted for the purpose of food, fodder, and industrial material production. In addition to grains or seeds, these crops offer straw as a source of bioenergy feedstock. However, we restrict the utilization ratio of this source to be less than 70% [34], due to the concerns over soil degradation and erosion caused by the excess removal of straw.

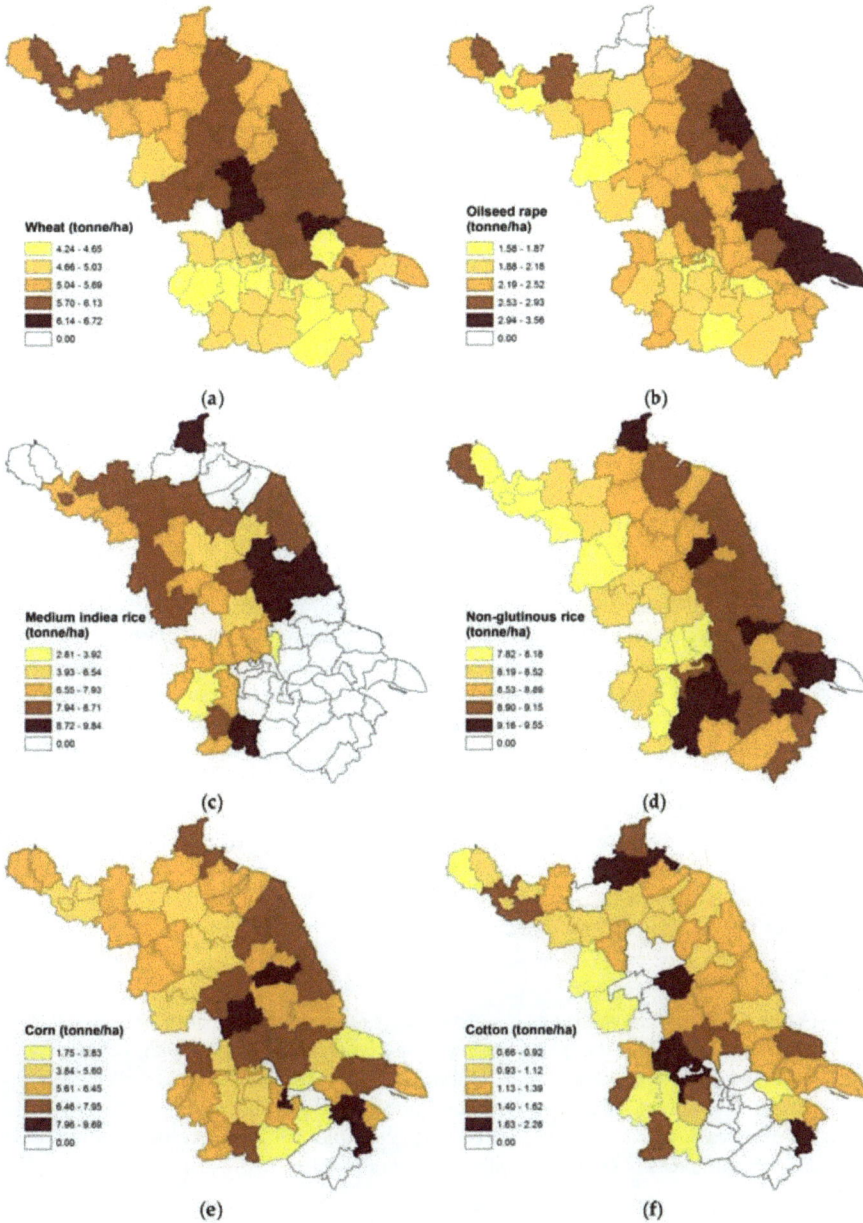

Figure 3. *Cont.*

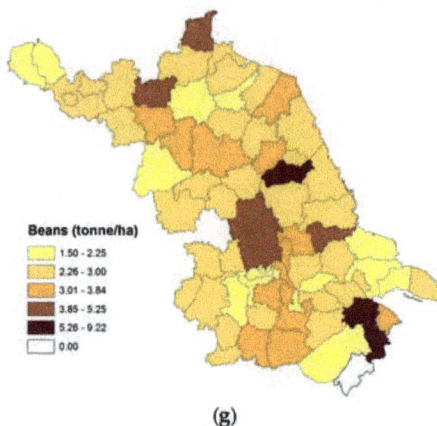

(g)

Figure 3. Crop yields in Jiangsu Province in 2010. (1) Data source: [35]; (2) The areas with yield data of 0.00 mean no plantation for that crop in practice. Data are presented for (**a**) wheat; (**b**) oil seed rape; (**c**) medium indiea rice; (**d**) non-glutinous rice; (**e**) corn; (**f**) cotton and (**g**) beans.

According to the official statistics [34], wheat, oilseed rape, medium indiea rice, non-glutinous rice, corn, cotton, and beans are the top cash crops in Jiangsu Province. Their respective planting areas account for more than 70% of Jiangsu's total arable land. The straw from those crops contributes up to 88.5% of the provincial straw supply. Our model takes these seven plants as conventional crops and further categorizes them into summer crops and autumn crops, with respect to the local rotation cropping system. Figure 3 demonstrates the crop yields per hectare in 2010 on the county-level, with a five-scale color code representing five different levels of the yield. Cultivation cost data are obtained from the annual survey "Cost-Benefit Investigation of Jiangsu Agricultural Products" conducted by the Cost Investigation Supervision Branch of the Jiangsu Commodity Price Bureau. The available data span five years from 2006 to 2010 and were retrieved from answers to a questionnaire randomly distributed to the farmers across 57 representative counties belonging to 13 prefecture-level cities. Figure 4 illustrates the average costs for six main crops cultivated in Jiangsu Province in 2010. In the model, the per-hectare cost data which are reported in monetary values are converted to physical input requirements for specific production factors (including land, labour, fertilizer, pesticide and others), in accordance with the individual market price (or equivalent price).

Since the local cost-benefit survey does not include beans, historical county-level data are unavailable. To fill this gap, we employ the provincial-level data retrieved from the "Collection of Cost-Benefit Data of China's Agricultural Products". Similarly, for counties where certain crops are planted but cost-benefit data are not collected, we interpolate the data from neighboring areas by adopting the OrdinaryKriging Method offered in ArcGIS.

2.3. Energy Crops

In recent decades, a variety of lignocellulosic energy crops, including both perennial herbaceous crops (*i.e.*, switchgrass, *Miscanthus*) and woody crops (such as willow, poplar, eucalyptus), have been introduced in many countries [36]. In China, however, large-scale commercialized plantation of energy crops is still absent. While we don't have observed cultivation data for commercial energy crops, several research institutes have engaged in species selection and conducted small-scale field experiments in certain parts of China, including Jiangsu Province [16,37–41].

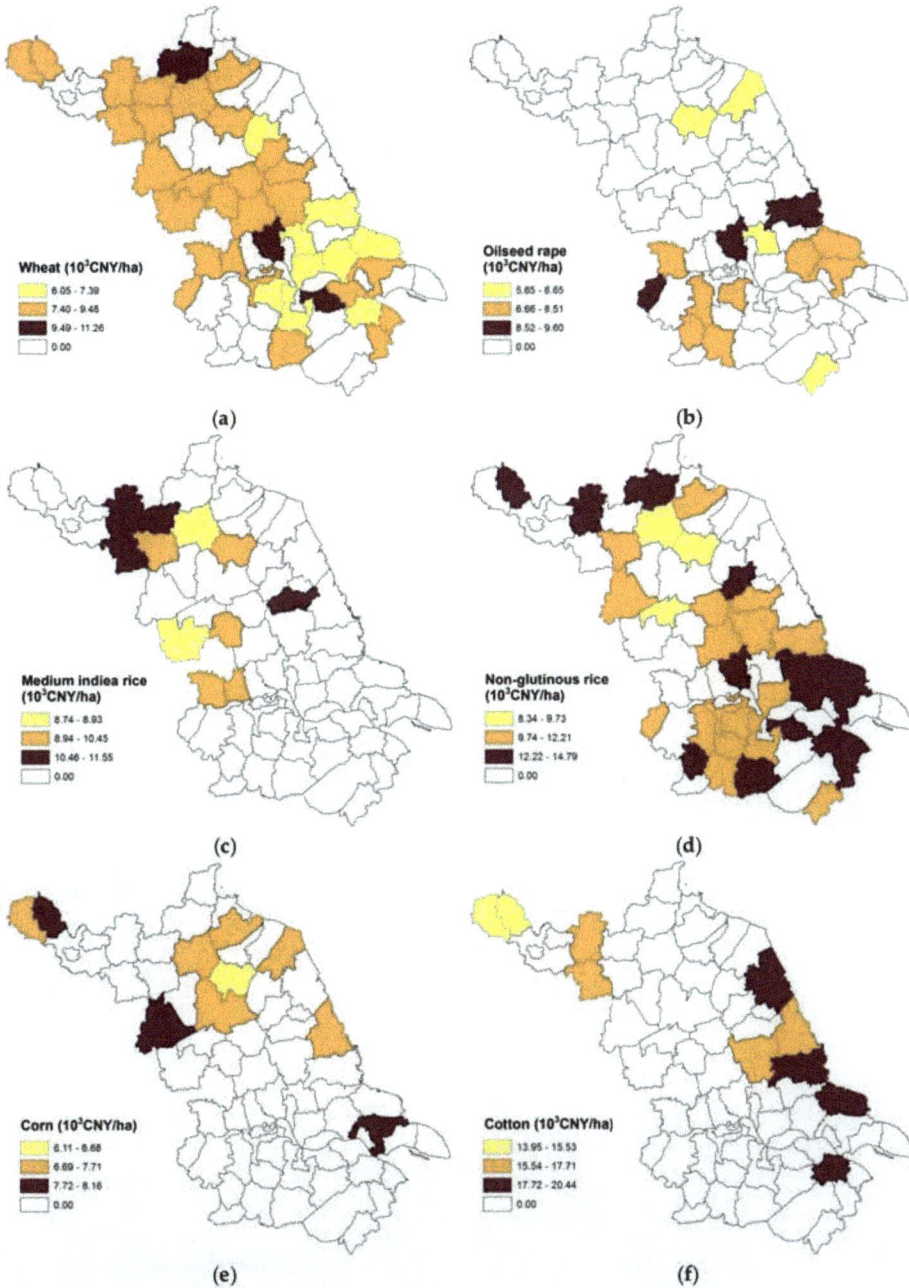

Figure 4. Crop cultivation cost in Jiangsu Province in 2010. (1) Data source: Data Collection of Cost-Benefit Investigation of Jiangsu Agricultural Products (2011); (2) Since the crop of beans is not covered by this Data Collection, county-level cost data are not available; (3) For areas showing a value of 0.00, data were not collected. Data are presented for (**a**) wheat; (**b**) oil seed rape; (**c**) medium indiea rice; (**d**) non-glutinous rice; (**e**) corn; (**f**) cotton.

In this study, we rely on the results of these experiments and discuss the selection from four candidate crops: switchgrass, silver reed, giant reed and miscanthus. Because these data are obtained from small-scale experiments under controlled conditions, they may not reflect all uncertainties of real agricultural operations and may not fully consider the potential environmental side effects.

Previous research reveals that the yield of energy crops reaches a plateau in the third year [42]. We adopt this finding and also define the total life span for each crop. In particular, switchgrass, silver reed and giant reed are set to live for 10 years, miscanthus for 20 years. After specifying the yield of each energy crop, we need to calculate its cultivation cost. First, we translate the experimental cost data of switchgrass, silver reed and giant reed measured from monetary values to physical input requirements for each production factor using individual factor prices. Second, we apply a reseeding rate of 25% to switchgrass and 0% to the other three crops [39,41]. Third, we extrapolate the cost data for switchgrass as a proxy for the unavailable cost data for miscanthus plantation. Particularly, we assume that the ratio of the cost of switchgrass to the cost of miscanthus is the same both in China and in the U.S. Then we apply this ratio obtained from the study in Illinois, USA [41] to the experimental data of switchgrass in China [40]. The resulting adaptive yields and cultivation costs of energy crops are shown in Table 1.

Table 1. Yield and cost data for potential energy crops in Jiangsu Province [16,37–41].

Items							Crops					
	Switchgrass			Silver Reed			Giant Reed			Miscanthus		
	Ages											
	1	2	3–10	1	2	3–10	1	2	3–10	1	2	3–20
Yield (tonne/ha)	6.8	15.4	28.1	7	17.7	29	16.2	30.5	34.4	9.55	19.4	30.0
Cost (per ha) Seedling (10^3 CNY)	135	33.8	0	90	0	0	90	0	0	423.7	0	0
Planting (d)	7.5	1.8	0	7.5	0	0	7.5	0	0	30.5	0	0
Maintenance (d)	7.5	7.5	7.5	7.5	7.5	7.5	7.5	7.5	7.5	10.4	10.4	10.4
Harvest (d)	15	15	15	15	15	15	15	15	15	44.7	44.7	44.7
Irrigation (CNY)	600	0	0	600	0	0	600	0	0	600	0	0
Water (CNY)	180	0	0	180	0	0	180	0	0	180	0	0
Electricity (kWh)	846.8	0	0	846.8	0	0	846.8	0	0	846.8	0	0
N-Fertilizer (kg)	150	150	150	150	150	150	150	150	150	165	69	103.5
Herbicide (10^3 kg)	3.8	0	0	3.8	0	0	3.8	0	0	4.1	0	0

2.4. Mudflats

The use of mudflats in Jiangsu Province as a potential land resource for energy crops has been thoroughly explored in many studies [43–45]. This option is considered by the local government in its "Outline of Reclamation and Utilization Plan for the Jiangsu Coastal Mudflats Resource (2010–2020)". This guideline specified the concrete location, acreage, and utilization of reclaimed mudflats throughout three execution stages as shown in Table 2. Our model conforms to this guideline and assumes that reclaimed mudflats are exclusively used for energy crops. Thus, conventional crops can only be planted on arable lands, while energy crops can be planted either on arable lands or on reclaimed mudflats.

2.5. Biomass and Food Demand

In this study, we translate the objective of reconciling bioenergy feedstock provision and food supply into a set of mathematical restrictions. Particularly, the production of biomass and food is required to meet the respective demand levels. Therefore, the simulation results on land use patterns reflect the level of the two demands. The values of biomass demand in the years 2012 and 2015 are taken from [34], which considers four utilization options of biomass for energy purposes including biofuels, electric power, biochemicals, and solidification.

Table 2. The distribution of mudflats in Jiangsu Province and the extent of energy crop plantations in different periods (10^3 ha) [22].

No.	Bank Section (Shoal)	County	Area Suitable for Reclamation			
			Total	Period 1	Period 2	Period 3
A01	Xiuzhen estuary–Youwang estuary	Ganyu	1.00	0.00	0.47	0.00
A02	Xingzhuang estuary–Linhongkou	Ganyu	1.67	0.00	0.00	0.00
A03	Linhongkou–Xishu	Lianyungang	2.33	0.00	0.00	0.00
A04	Xuwei port	Lianyungang	4.67	0.00	0.00	0.00
A05	Xiaodong port–Xintan port	Xiangshui	1.33	0.60	0.00	0.67
A06	Shuangyang port–Yunliang estuary	Sheyang	1.00	0.00	0.00	0.93
A07	Yunliang estuary–Sheyang estuary	Sheyang	1.67	0.73	0.00	0.00
A08	Simaoyou estuary– Wanggang estuary	Dafeng	6.00	1.00	0.00	1.60
A09	Wanggang estuary–Chuandong port	Dafeng	5.00	2.53	0.00	2.20
A10-1	Chuandong port–Dongtai estuary	Dafeng	1.17	0.00	1.10	0.00
A10-2	Chuandong port–Dongtai estuary	Dongtai	1.17	0.00	1.10	0.00
A11	Tiaozini	Dongtai	26.67	8.00	9.33	0.00
A12-1	Fangtang estuary–Xinbeiling estuary	Dongtai	3.33	1.28	1.87	0.00
A12-2	Fangtang estuary–Xinbeiling estuary	Hai'an	2.00	1.92	0.00	0.00
A13	Xinbeiling estuary–Xiaoyangkou	Rudong	4.00	0.00	3.67	0.00
A14	Xiaoyangkou–Juejukou	Rudong	12.00	1.27	0.93	1.60
A15	Juejukou–Dongling port	Rudong	21.33	2.60	2.60	8.67
A16	Yaosha–Lengjiasa	Tongzhou	29.33	0.00	3.47	15.60
A17-1	Yaowang port–Haozhi port	Tongzhou	1.92	0.45	0.40	0.00
A17-2	Yaowang port–Haozhi port	Haimen	1.92	0.45	0.40	0.00
A17-3	Yaowang port–Haozhi port	Qidong	3.83	0.90	0.80	0.00
A18	Haozhi port–Tanglu port	Qidong	3.33	0.00	1.80	0.00
A19	Xiexing port–Yuantuojiao	Qidong	3.33	0.00	1.07	0.00
A20	Dongsha	Dongtai	21.33	0.00	0.00	13.87
A21	Gaoni	Dongtai	18.67	0.00	0.00	12.13
	Total		180.00	21.73	29.01	57.27

Period 1: 2010–2012, Period 2: 2013–2015, Period 3: 2016–2020.

To curb the potential uncertainties of demand projections, we also collect other years' data from the literature [46–48]. For intermediate years, the demand values are estimated through regression analysis. Food demand levels from 2011 to 2020 are based on previous study on Jiangsu's food consumption [49]. The final demand data used in our study are listed in Table 3. To address the data uncertainty, we perform a sensitivity analysis.

Table 3. Demand targets for biomass and food demand between 2011 and 2030 (10^3 t) [46–49].

Year	2011	2012	2013	2014	2015	2016	2017	2018	2019	2020
Biomass	6302	7467	8632	9797	10,962	12,127	13,292	14,457	15,622	16,787
Food	29,720	29,885	30,050	30,214	30,379	30,544	30,708	30,873	31,038	31,203

Year	2021	2022	2023	2024	2025	2026	2027	2028	2029	2030
Biomass	17,952	19,117	20,282	21,447	22,612	23,777	24,942	26,107	27,272	28,437
Food	31,367	31,532	31,697	31,861	32,026	32,191	32,355	32,520	32,685	32,850

3. Model Design and Structure

3.1. Model Framework

We develop a modelling framework to mimic the annually recurring decision-making process of farmers, *i.e.*, the type and intensity of crops, which meets both food demand and biomass demand (Figure 5). The model framework integrates a spatial-agent system dynamic model and a partial equilibrium model of the agricultural sector. The former model was applied by Scheffran and BenDor in their simulation of the energy crop production in Illinois in the U.S. [1]. The advantage of this model is the maintained geographical characteristics of the individual farmers through using a spatial array of uniform grid cells to index their positions in the landscape. The model portrays the spatial relationships between farmers and depicts their particular geographical characteristics such as altitude, climate, shading, slope, and soil conditions. We aggregate Jiangsu's 102 county-level administrative units into 70 counties by considering the administrative relations between neighboring units and the data accessibility. The conglomeration of all farmers in a county is considered as one agent in our regional agricultural sector model. This model component in our study was derived from the ASMGHG model developed by Schneider and McCarl [50], which links agricultural commodity markets to regionalized cropping systems. Similarly, our model couples the aggregation of each agent's cropping system with the provincial bioenergy feedstock market.

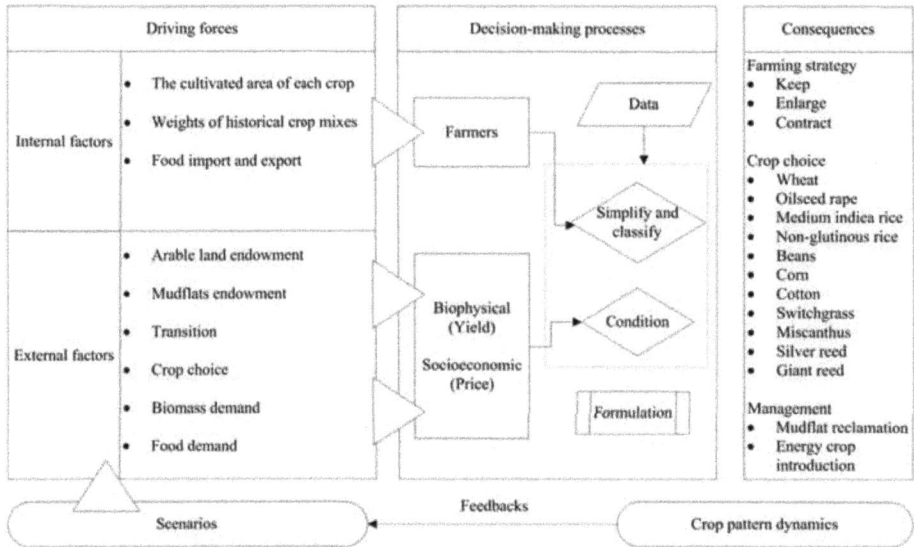

Figure 5. The framework of the spatial-agent dynamic model of optimized agricultural land use.

The spatial-agent dynamic model of agricultural land use developed for this study combines the strength of both models: a) the depiction of heterogeneous geographical features of each agent leading to non-uniform opportunity costs for energy crops and b) Jiangsu's bioenergy feedstock market, which determines the price of biomass through the intersection of aggregate supply from all farmers and governmental demand targets. Thus, a single farmer's decision process is affected by the decisions of other farmers in the same market.

3.2. Model Structure

Our model determines the cost-efficient land use pattern, which simultaneously meets the demand for food and the demand bioenergy feedstock. This model is programmed in GAMS and consists of an objective function, a group of decision variables and a set of constraining equations. The objective function maximizes total agricultural economic surplus over a 20-year horizon with an annual time step. The values of the decision variables (internal factors in Figure 5) are endogenously determined through the optimization process. The constraining equations integrate environmental limits and market demands (external factors in Figure 5), which influences the farmers' decision-making process. Mathematically, these equations define the convex feasible region for all decision variables. Each model element is briefly described in Table 4.

28

Table 4. The significance of model equations and variables.

Model Equation	Mathematical Structure	Description
Objective function	$WELFARE = REVENUE - COST$	The sum of producer revenue in all commodity markets, minus specific and unspecific production cost and the cost of mudflat reclamation.
Resource limits	$LAND^{concrop} + LAND^{enecrop} \leq endowment^{arableland}$	The cultivated land in each region and time period cannot exceed given endowments.
	$LAND^{mudflat} \leq endowment^{mudflat}$	According to Jiangsu's official directive, a limited area of reclaimed mudflats mainly scattered in the coastal counties can be devoted to energy crop plantation.
	$LAND^{enecrop}_{a,t} \leq LAND^{enecrop}_{a-1,t-1}$ $LAND^{mudflat}_{a,t} \leq LAND^{mudflat}_{a-1,t-1}$	The area of energy crop plantation in higher age classes cannot exceed the area of the corresponding previous age class in the previous period.
	$\sum_{his} \left(landuse^{concrop}_{his} \times CMIX_{his} \right) = LAND^{concrop}$	Cropping activities are restricted to a linear combination of historically observed choices. Onal and McCarl [51] find that historical crop mix restrictions implicitly embody numerous farming constraints, which are difficult to observe. These include crop rotation considerations, perceived risk reactions, and a variety of natural conditions.
	$demand^{biomass} \leq yield^{biomass} \times LAND^{concrop}$ $+yield^{biomass} \times (LAND^{enecrop} + LAND^{mudflat})$	Biomass production needs to satisfy minimum biomass demand.
	$demand^{food} \leq yield^{food} \times LAND^{concrop}$ $+FOODTRADE$	Food production needs to satisfy minimum food demand.
Decision variables	$LAND^{concrop}, LAND^{enecrop}, LAND^{mudflat}$	Cultivated area includes arable lands and mudflats; Crops in the model are divided into conventional crops and energy crops.
	$CMIX$	The weights of historical land use patterns for decisions on land use in future years.
	$FOODTRADE$	Inter-provincial food trade.

Solving the model requires finding an optimal level for all decision variables subject to compliance with all constraining equations and, in the meantime, maximizing the objective function. As discussed by McCarl and Spreen [52], maximization of consumer and producer surplus yields the competitive market equilibrium. Thus, the optimal variable levels can be interpreted as likely equilibrium levels for agricultural activities under given economic, political and technological conditions. Simultaneously, the shadow prices, identical to the marginal values of the biomass and food demand constraint equations, determine market-clearing prices of food and bioenergy feedstock. A detailed description of our model is presented in the Appendix A.

4. Simulation Results

4.1. Land Use Patterns before and after the Introduction of Energy Crops

Starting from historically observed land use patterns in Jiangsu Province, the model results provide a cost-effective supply path of bioenergy feedstock for a steadily growing biomass demand. While the land use decisions are simulated for every year, we only show the cumulative change over the model's entire time horizon.

Figure 6 shows the change in cultivated area between the initial and final year. Given that there is no commercial energy crop plantation in the base year, the land use changes for conventional crops and energy crops are presented in different ways: For conventional crops, relative land use changes are computed (Figure 6a–g), which are obtained by dividing the projected cultivated area for each crop in 2030 by the observed area in 2010. For energy crops, the absolute land use changes are shown, which are equal to the land areas in 2030 (Figure 6h,i). The land use change of conventional crops reveals two things: (1) The expansion of summer crops, *i.e.*, wheat and oilseed rape, mainly occurs in Southern Jiangsu where oilseed rape expands extraordinarily fast (for example, its cultivated area has expanded 20 times in Wuxi City, 13 times in Kunshan County), thanks to its higher biomass output than wheat. (2) The land use changes for autumn crops differ across regions. Non-glutinous rice is mainly increased in Southern Jiangsu. By contrast, Northern Jiangsu favors more cotton, which is planted along the coastal line. Medium indiea rice is planted primarily in the eastern part of Jiangsu Province. For beans, on the other hand, they are found across the entire Jiangsu province.

Besides simulating the land use patterns for seven conventional crops, our model selects giant reed from four potential energy crops and suggests it to be planted on the arable lands in four out of 70 counties (namely in Jiangyin, Donghai, Huai'an and Jurong County) and on the reclaimed mudflats in four out of 10 counties (they are Hai'an, Ganyu, Dongtai and Dafeng County). In all counties where giant reed increases, conventional crops decrease and confirm the land use conflict between conventional and energy crops.

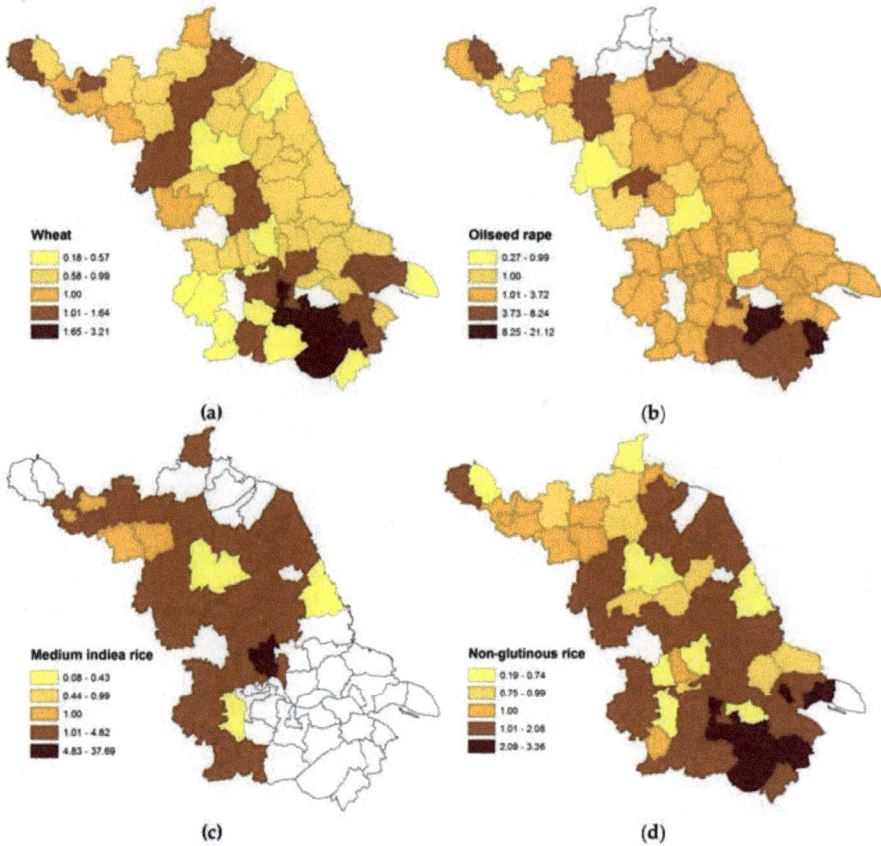

(a)

(b)

(c)

(d)

Figure 6. *Cont.*

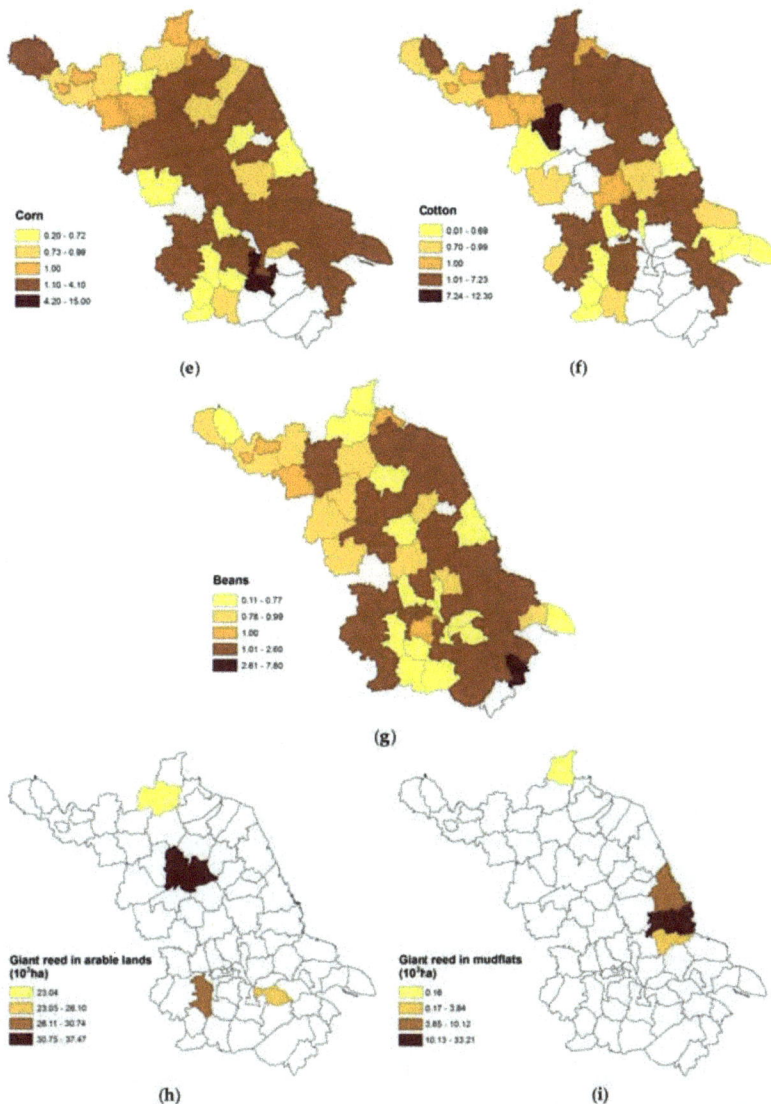

Figure 6. Projected land use change in Jiangsu Province in 2030 after the introduction of energy crops. Data are shown for (**a**) wheat; (**b**) oil seed rape; (**c**) medium indiea rice; (**d**) non-glutinous rice; (**e**) corn; (**f**) cotton and (**g**) beans; (**h**) giant reed planted in arable lands and (**i**) giant reed planted in mudflats. (1) Land use data for 2010 are retrieved from [35]; (2) White areas indicate zero plantations for particular crops.

4.2. Bioenergy Feedstock Supply

The simulation results disclose the sources of bioenergy feedstock supply (Figure 7). Straw from conventional crops constitutes the main source of bioenergy feedstock, contributing more than 85% of the total supply. The highest contribution comes from straw residues of two large-scale planted crops (wheat in summer and non-glutinous rice in autumn). Their relative shares remain similar throughout the twenty years of projection, reflecting a stable reproduction of the crop mix described in the item of crop choice in Table 4.

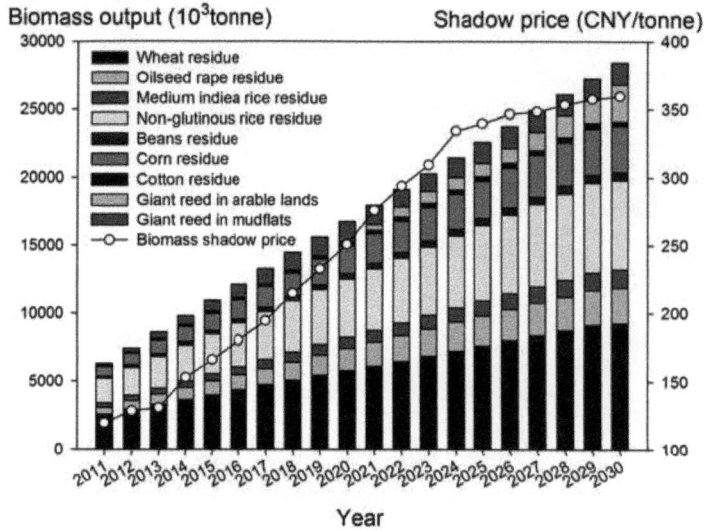

Figure 7. The projected bioenergy feedstock supply and its shadow price in Jiangsu Province for the period 2011–2030.

Up to 15% of the bioenergy supply comes from energy crops. Our results suggest giant reed first to be planted on reclaimed mudflats starting in 2011 and then extended to arable lands in 2020. Although giant reed starts on arable land at later phases, the contribution to biomass output is expected to surpass the contribution from mudflats in the year 2029. The area of giant reed on mudflats reaches a plateau at around 47 thousand ha with a production of 1.5 million tonnes of biomass in 2024 and afterwards.

Our model also determines the shadow price of bioenergy feedstock, *i.e.*, the price society must pay to farmers in order to induce a sufficient amount of supply. To smooth out price fluctuations caused by the model's dynamic structure, we use a simple 10-year moving average of each year's shadow price. Figure 7 shows a steady increase of the biomass shadow price, which reflects the increase in the opportunity cost of biomass production induced by the rising bioenergy demand over time. As more biomass is demanded, less productive resources have to be gradually put into use and, therefore, lead to higher production cost. This upwards trend is sustained across the whole time horizon; however, in 2024 a turning point can be observed. A possible reason for this phenomenon is the use of the mudflats, whose reclamation cost outweighs other factors and thus dominates the total opportunity cost of biomass production. After 2024, there is no additional mudflat reclamation for the purpose of energy crop production. Thus, the expenditure on mudflat reclamation diminishes and reduces the rate of increase in opportunity cost.

4.3. Sensitivity Analysis 1: Bioenergy Feedstock Demand

To find out how sensitive the optimal land use pattern is to the exogenously specified biomass demand targets, we analyze four alternative scenarios. These include two amplified demand scenarios (1.2 and 1.5 times as much as the biomass demand in the basic scenario) and two reduced demand scenarios (0.5 and 0.8 times as much as the biomass demand in the basic scenario).

4.3.1. The Introduction of Energy Crops

Our model computes optimal pathways for the introduction of energy crops in Jiangsu Province for a given biomass demand level. As shown in Figure 8, the curve representing the basic scenario suggests energy crops to be introduced early but with a small contribution. As biomass demand increases, the importance of energy crops jumps from 2% in 2011 to 15% in 2030, playing an auxiliary role in local biomass supply. For Jiangsu Province, this simulation result is a cost-effective way to simultaneously supply sufficient bioenergy feedstock and food by exploiting the dual-use of conventional crops (both food and energy use) and the single-use of energy crops (only energy use). A gradual and small-scale introduction of energy crops at the initial stage suggested by the model helps to realize a smooth transition process in the local agricultural sector towards the coordinated food and energy production, and to avert the potential resistance from local farmers who are used to traditional crop management practices.

Share ratio (%)

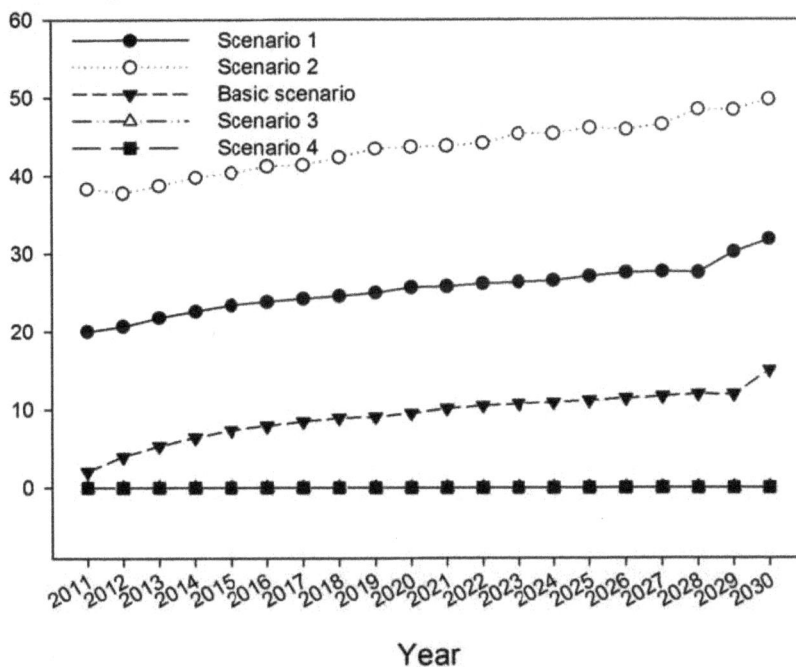

Figure 8. The projected share of biomass from energy crops in bioenergy feedstock supply in Jiangsu Province between 2011 and 2030. Note: Under scenario 1 and 2 (the amplified demand scenarios), the biomass demands are set to 1.2 and 1.5 times the demand level in the basic scenario, respectively. In scenario 3 and 4 (the reduced demand scenarios), the respective demand multipliers are 0.5 and 0.8. Please note that the symbols representing scenario 3 and 4 always overlap each other.

However, we need to note that the above analysis is highly dependent on the given biomass demand level in the basic scenario. It is possible that different biomass demand levels may lead to differentiated land use patterns. As demonstrated in Figure 8, when the demand rises by 20% (scenario 1), the share of energy crops in biomass supply will ascend by more than 15%. If the demand climbs up for another 30% (scenario 2), the energy crops react by expanding their plantation and offering another 20% share. In this case, energy crops meet half of the total biomass demand in the final years. Instead, once the demand curtails by 20% or more (scenarios 3 and 4), energy crops will be completely ruled out from local biomass supply. These results conclude that in terms of the time and scale, the introduction of energy crops to Jiangsu Province is extremely sensitive to the level of biomass demand. The setting of bioenergy development targets would have far-reaching consequences, for example the timing of energy crop introduction, the plantation scale of energy crops and correspondingly, the distribution of other conventional crops.

4.3.2. Differentiated Land Use Patterns in Jiangsu's Three Sub-Regions

To echo the disparities in Jiangsu's regional development, in this section we examine the introduction of energy crops on the level of the province's three sub-regions, other than taking the whole province. Particularly, aiming at exploring the relationships between food and biomass supply, we only focus on the use of arable lands.

The allocations of arable lands between conventional crops and energy crops in three sub-regions perform quite different from each other: (1) Central Jiangsu always concentrates on the cultivation of conventional crops, no matter which level of biomass demand is applied. (2) The higher the biomass demand, the earlier and larger amounts of energy crops are to be planted on arable lands. As shown in Figure 9, energy crops are firstly introduced to the arable lands in the year of 2020 with the share of 2% in Southern Jiangsu under the basic scenario. By contrast, under the scenario 1, the crops firstly appear in 2011 with the share of 10% in Northern Jiangsu and 4% in Southern Jiangsu.

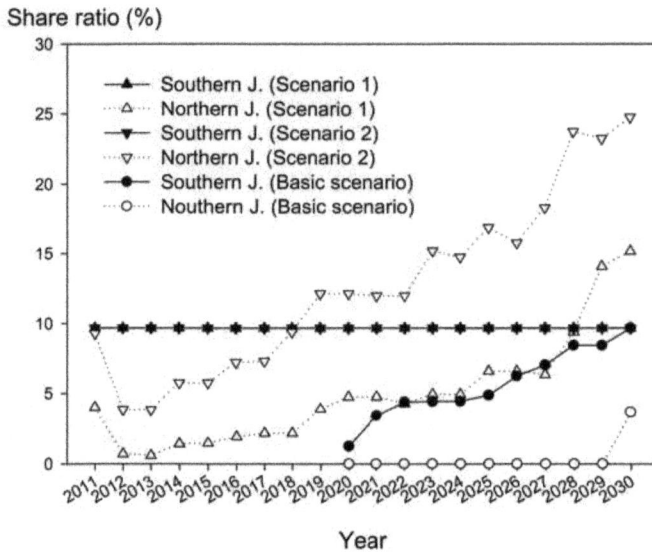

Figure 9. The projected share of energy crops on arable land in autumn in Jiangsu Province between 2011 and 2030. (1) Under scenarios 1 and 2 (the amplified demand scenarios), the biomass demands are set to 1.2 and 1.5 times as much as the demand level in the basic scenario, respectively; (2) As pointed out before, there are no energy crops in the reduced demand scenarios (scenarios 3 and 4), *i.e.*, the share ratio of energy crops stays at 0. Thus, these scenarios are not shown in this figure; (3) Since there is no energy crop planted on arable lands in Central Jiangsu under all scenarios, this area is not included in the figure.

(3) The share of energy crops on the arable lands in Southern Jiangsu faces a distinct hurdle at around 10%. This finding implies that the arable land in Southern Jiangsu is the pioneer for energy crop plantation in the short term, though compared with Northern Jiangsu this area does not have enough potential to accommodate a large-scale plantation in the long term. This evidence corroborates the Jiangsu's official orientation towards Northern Jiangsu as a bioenergy production basis, which was mentioned in the "Development Plan for the Coastal area of Jiangsu Province" in 2009.

4.4. Sensitivity Analysis 2: The Role of Mudflats

This section evaluates the role of mudflats in meeting the soaring biomass demand. We compare the basic scenario, where a certain fraction of mudflats is available for energy crops, with a situation in which all mudflats are excluded. Figure 10 uses three indicators to measure their role.

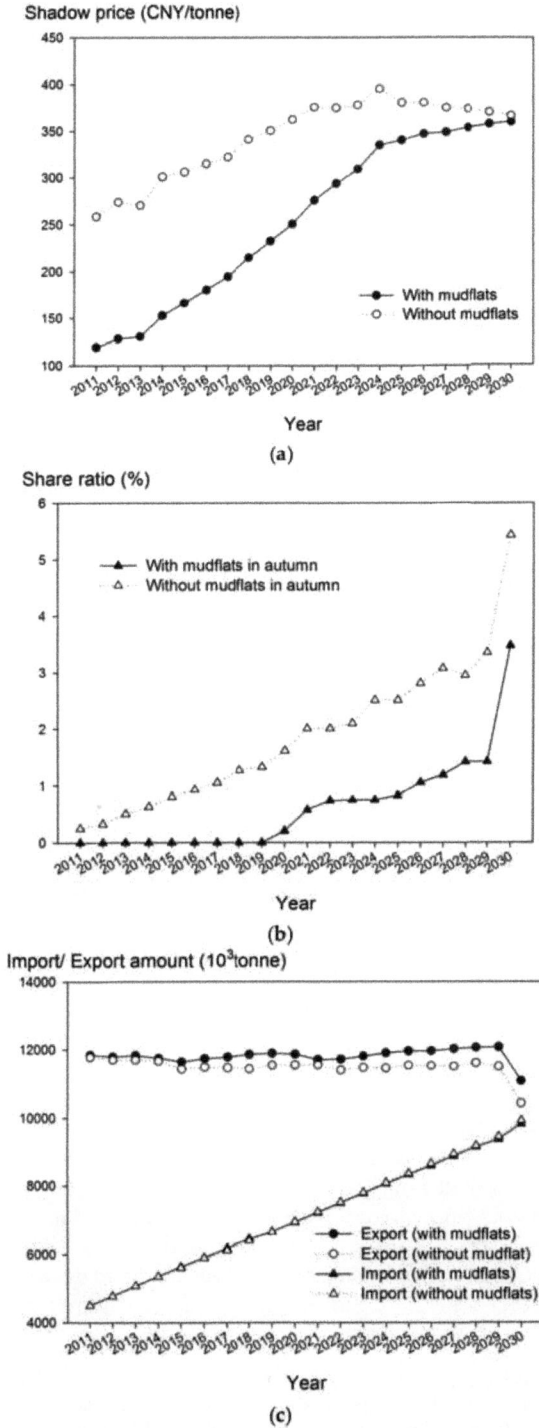

Figure 10. The role of mudflats measured in (**a**) biomass shadow price; (**b**) the share of energy crops on arable land; (**c**) food trade in Jiangsu Province between 2011 and 2030.

First, the presence of mudflats substantially lowers the shadow prices of bioenergy feedstock. For example, in the first ten years, the introduction of mudflats brings down the biomass price by more than 100 CNY/t (Figure 10a), which significantly decreases the purchasing cost of bioenergy feedstock and facilitates the development of bioenergy industry in Jiangsu Province, especially in the initial years. However, this effect diminishes along with the increasing biomass demand, because the share of energy crops on reclaimed mudflats in the biomass supply becomes small in the later years, especially after the year of 2024, the turning point of the use of reclaimed mudflats.

Second, the participation of mudflats alleviates the land use conflict between conventional crops and energy crops. In 2030, up to 2% of the arable land resources in the province could be saved and used for food production (Figure 10b).

Third, the mudflats help to produce an additional amount of about 0.75 million tonnes of food for cross-boundary food trade in 2030 (0.65 million tonnes more wheat and medium indiea rice for export, 0.10 million tonnes less beans and corn for import, as shown in Figure 10c), accounting for 3.59% of total amount of food trade in the same year.

Based on the results from the above analysis, it is safe to say that the accessibility to reclaimed mudflats can enhance the robustness of local biomass supply and relieve land use conflict between food and biomass production in Jiangsu Province, at the cost of mudflat losses.

5. Discussion and Conclusions

We develop a model to explore a cost-effective way to simultaneously reach food and bioenergy feedstock targets for Jiangsu Province. We apply this model to evaluate the impact of a bioenergy development plan on local agricultural land use and to measure the role of reclaimed mudflats in alleviating the land use conflict between bioenergy and food supply. The simulation results reveal the following insights: (1) The introduction of energy crops leads to seasonally differentiated land use patterns for conventional crops: the cultivation area of summer crops mainly expands in Northern Jiangsu. For autumn crops, such a pattern does not exist. (2) Energy crops are introduced first on arable lands in Jiangyin, Donghai, Huai'an and Jurong Counties and on reclaimed mudflats in the counties of Hai'an, Ganyu, Dongtai, and Dafeng. The economically preferred energy crop is giant reed. (3) The straws from conventional crops contribute more than 85% of the total biomass supply. Among these crops, wheat and non-glutinous rice top the bioenergy feedstock supply, followed by corn, oilseed rape, medium indiea rice, beans, and cotton. (4) On the provincial level, the introduction of energy crops is highly sensitive to the level of biomass demand. Therefore, alternative bioenergy development plans will result in different land use patterns. (5) On the sub-regional level, energy crops are employed first in Southern Jiangsu. However, over time, Northern Jiangsu cultivates energy crops on a larger area, supporting its status as a renewable energy production basis outlined in Jiangsu's official development plan. (6) Reclaimed mudflats, as an alternative resource for energy crops, help secure the local biomass supply and alleviate the land use conflict between food and biomass production.

While the model offers useful insights into the bioenergy development in Jiangsu Province, two limitations should be noted: First, this research does not depict detailed biophysical characteristics of individual crop management systems. Thus, impacts on local biodiversity and ecosystem services are not considered. Second, this model suggests mudflats are an economically attractive resource for the cultivation of energy crops. While we consider possible environmental impacts of this resource and exclude ecologically sensitive areas, this study does not account for social side effects. For example, the model ignores the marginalization of smallholder farmers caused by the large-scale introduction of energy crops on reclaimed mudflats. Further research is needed to address the full spectrum of environmental and social side effects of bioenergy development.

Acknowledgments: This research was sponsored by the Chinese Scholarship Council (CSC) and the Austrian Federal Ministry of Science and Research (BMWF) through the Ernst Mach Grant, and in part by the German Science Foundation (DFG) through Hamburg University's Cluster of Excellence "Integrated Climate System Analysis and Prediction". The authors thank involved governmental departments and bioenergy practitioners in Jiangsu Province for support in data collection.

Author Contributions: All authors together conceived and designed the model; Kesheng Shu performed the model, analyzed the data and wrote the paper with feedback by the co-authors.

Conflicts of Interest: The authors declare no conflict of interest.

Appendix A. The Spatial-Agent Dynamic Model Specification

In the general formulation of the county-level dynamic agent-based model, the present value of the total profit is maximized across the whole time frame of a system covering the cultivation of both conventional crops and energy crops, subject to constraints on resource endowment, energy crop transition, cultivation selection and product demand.

Appendix B. Indices

u	county-level regions (Nanjing, Pukou, Liuhe, Shushui, Gaochun, Wuxi, Jiangyin, Yixing, Xuzhou, Fengxian, Peixian, Tongshan, Suining, Xinyi, Pizhou, Changzhou, Wujin, Liyang, Jintan, Suzhou, Changshu, Zhangjiagang, Kunshan, Wujiang, Taicang, Nantong, Tongzhou, Hai-an, Rudong, Qidong, Rugao, Haimen, Lianyungang, Ganyu, Donghai, Guanyun, Guannan, Huai-an, Lianshui, Hongze, Xuyi, Jinhu, Yancheng, Yandu, Xiangshui, Binhai, Funing, Sheyang, Jianhu, Dongtai, Dafeng, Yangzhou, Baoying, Yizheng, Gaoyou, Jiangdu, Zhenjiang, Dantu, Danyang, Yangzhong, Jurong, Taizhou, Xinghua, Jingjiang, Taixing, Jiangyan, Suqian, Shuyang, Siyang, Sihong)
fc	food crops (wheat, oilseed-rape, medium indiea rice, non-glutinous rice, beans, corn, cotton)
sc	summer crops (wheat, oilseed-rape)
ac	autumn crops (medium indiea rice, non-glutinous rice, beans, corn, cotton)
pc	perennial crops (switchgrass, miscanthus, silver reed, giant reed)
pr	products (wheat, seeds, medium indiea rice, non-glutinous rice, beans, corn, cotton, straw)
pr_food	non-bioenergy products (wheat, seeds, medium indiea rice, non-glutinous rice, beans, corn, cotton)
pr_energy	products as bioenergy feedstock (straw)
t	time horizon (2006–2020)
s	the current policy scenario (s1)
n	crop season (summer, autumn)
a	crop age (1,2, ... ,20)
ht	historical year (2000–2010)
inp	the input factors during crop field management (land, family labor, hired labour, water, n-fertilizer, p-fertilizer, k-fertilizer, o-fertilizer, pesticide, agricultural-film, diesel, electricity)

Appendix C. Exogenous Data

$y_{u,fc,pr,t,n}^{foodcrop}$	yield of food crop (tonne/ha)
$y_{pc,pr\_energy,a}^{perennial\ crop}$	yield of perennial crop (tonne/ha)
$ps_{pr,t,s}$	price subsidy (CNY/tonne)
$v_{u,pr\_food,t}^{grains}$	price of non-biomass product (CNY/tonne)
$sub_{u,fc,t,s}^{food\ crop}$	land subsidy for conventional crops (CNY/ha)
$sub_{u,pc,t,s}^{perennial\ crop}$	land subsidy for perennial crops (CNY/ha)
$c^{mudflat}$	reclamation cost of mudflats (CNY/ha)
$b_{u,t}^{land}$	total arable land area (10^3 ha)
$b_{u,t}^{mudflat}$	mudflats resource potential (10^3 ha)
$h_{u,fc,ht,n}$	historical cultivation data (10^3 ha)
k_{pc}	expected lifespan of perennial crops (year)
$dema_t^{biomass}$	demand of bioenergy feedstock (10^3 tonne)
$dema_t^{grains}$	demand of grains (10^3 tonne)
r	discount rate
η_u	proportion of straw for energy-end use (electricity and biofuels) to its total amount (%)

α	ratio of straw from main food crops (wheat, oilseed-rape, medium indiea rice, non-glutinous rice corn, cotton, beans) for its potential in the province (%)
$terminalvalue_{u,pc,a}^{arableland}$	terminal value of perennial crop in arable land (CNY/ha)
$terminalvalue_{u,pc,a}^{mudflat}$	terminal value of perennial crop in mudflats (CNY/ha)
$terminalvalue_{u,fc,n}^{food\ crop}$	terminal value of food crop (CNY/ha)
$consumption_{fc,t,inp,u}^{conventional\ crop}$	consumption of input factors for cultivation of conventional crops (ha/ha, d/ha, m³/ ha, kg/ha, kWh/ha)
$consumption_{pc,a,inp}^{perennial\ crop}$	consumption of input factors for cultivation of perennial crops (ha/ha, d/ha, m³/ha, kg/ha, kWh/ha)
$v_{fc,t,inp,u}^{cc\ input}$	price of input factors for food crops (CNY/ha, CNY/d, CNY/m³, CNY/kg, CNY/kWh)
$v_{t,inp,u}^{pc\ input}$	price of input factors for perennial crops (CNY/ha, CNY/d, CNY/m³, CNY/kg, CNY/kWh)

Appendix D. Decision Variables

$LAND_{u,fc,t,n}^{food\ crop}$	cultivated area for food crops in arable land (10^3 ha)
$LAND_{u,pc,t,a}^{perennial\ crop}$	cultivated area for perennial crops in arable land (10^3 ha)
$LAND_{u,pc,t,a}^{mudflat}$	cultivated area for perennial crops in reclaimed mudflats (10^3 ha)
$CMIX_{u,t,ht,n}$	weights of historical data
$PRICE_t^{biomass}$	endogenous price of biomass (CNY/tonne)
$FOODFLOW_{fc,t}$	import or export amount of food trade (10^3 t)

Appendix E. Objective Function

Max WELF =

$$\sum_t (1+r)^{-t} \cdot$$

$$\left(\begin{array}{c} \sum_{u,fc,pr\_food,n} \left[y_{u,fc,pr\_food,t,n}^{food\ crop} \cdot LAND_{u,fc,t,n}^{food\ crop} \cdot \left(v_{pr\_food,t}^{grains} + ps_{pr\_food,t,s} \right) \right] \\ + \sum_{u,fc,pr\_energy,n} \left[y_{u,fc,pr\_energy,t,n}^{food\ crop} \cdot LAND_{u,fc,t,n}^{food\ crop} \cdot \left(PRICE_t^{biomass} + ps_{pr\_energy,t,s} \right) \right] \\ + \sum_{u,fc,n} \left[LAND_{u,fc,t,n}^{food\ crop} \cdot sub_{u,fc,t,s}^{foodcrop} \right] \\ + \sum_{u,pc,pr\_energy,a} \left[y_{pc,pr\_energy,a}^{perennial\ crop} \cdot \left(LAND_{u,pc,t,a}^{perennial\ crop} + LAND_{u,pc,t,a}^{mudflat} \right) \cdot \left(PRICE_t^{biomass} + ps_{pr\_energy,t,s} \right) \right] \\ + \sum_{u,pc,a} \left[\left(LAND_{u,pc,t,a}^{perennial\ crop} + LAND_{u,pc,t,a}^{mudflat} \right) \cdot sub_{u,pc,t,s}^{perennial\ crop} \right] \end{array} \right)$$

$$+ (1+r)^{-t} \cdot$$

$$\left\{ \begin{array}{c} \sum_{u,pc,a} \left(LAND_{u,pc,t,a}^{perennial\ crop} \cdot terminalvalue_{u,pc,a}^{arable\ land} + LAND_{u,pc,t,a}^{mudflat} \cdot terminalvalue_{u,pc,a}^{mudflat} \right) |_{t=/2030/} \\ + \sum_{u,fc,n} \left(LAND_{u,fc,t,n}^{food\ crop} \cdot terminalvalue_{u,fc,n}^{food\ crop} \right) |_{t=/2030/} \end{array} \right\} \qquad (1)$$

$$- \sum_t (1+r)^{-t} \cdot$$

$$\left(\begin{array}{c} c^{mudflat} \cdot \left(\sum_{u,pc,a} LAND_{u,pc,t,a}^{mudflat} - \sum_{u,pc,a} LAND_{u,pc,t-1,a}^{mudflat} \right) \\ + \sum_{u,fc,n} \left\{ \sum_{inp} (consumption_{fc,t,inp,u}^{conventional\ crop} \cdot v_{fc,t,inp,u}^{cc\ input}) \cdot LAND_{u,fc,t,n}^{food\ crop} \right\} \\ + \sum_{u,pc,a} \left\{ \sum_{inp} (consumption_{pc,a,inp}^{perennial\ crop} \cdot v_{t,inp,u}^{pc\ input}) \cdot \left(LAND_{u,pc,t,a}^{perennial\ crop} + LAND_{u,pc,t,a}^{mudflat} \right) \right\} \end{array} \right)$$

$$\forall s$$

The objective function Equation (1) of the model maximizes the present value of the net cash flows of the agriculture sector in Jiangsu Province across the whole time frame, as the total revenue minus costs. Specifically, the revenue of agriculture sector comprises of the sale of agricultural products, governmental agricultural subsidies and terminal values which are estimated for every crop. For energy crops, it is calculated as the Present Value of future profits of the rest of the productive life of the cultivation. This is equal to $PV = \sum_t (P_t \cdot Y_t - PC_t) \cdot (1+r)^{-t}$, where P_t is the price of the crop's product in period t, Y_t is the yield and PC_t is the production cost). The cost mainly covers land resource, labor resource, fertilizers, pesticides and other auxiliary inputs.

From line 1 to line 9, the revenue terms account for:

(1)   The sales revenue of non-biomass from conventional crops;
(2)   The sales revenue of biomass from conventional crops;
(3)   The plantation subsidy on conventional crops;
(4)   The sales revenue of biomass from energy crops;
(5)   The plantation subsidy on energy crops;
(6)   The terminal value of energy crops in the terminal year;
(7)   The terminal value of conventional crops in the terminal year;

Starting from line 10 of the objective function, the cost items are:

(8)   The reclamation cost of mudflats;
(9)   The cost of production inputs for conventional crops;
(10) The cost of production inputs for energy crops.

Appendix F. Subject to

The most fundamental physical constraint on crop cultivation arises from the use of scarce and immobile resources. Particularly, the use of agricultural land is limited by given regional endowments of arable land and mudflat resources. In the following expressions, $b_{u,t}^{land}$ denotes total arable land area in region u, year t and $b_{u,t}^{mudflat}$ is total arable land area for the costal mudflat in region u and year t:

$$\sum_{fc} LAND_{u,fc,t,n}^{food\ crop} + \sum_{pc,a} LAND_{u,pc,t,a}^{perennial\ crop} \le b_{u,t}^{land} \qquad\qquad \forall u,t,n \qquad\qquad (2)$$

$$\sum_{pc,a} LAND_{u,pc,t,a}^{mudflat} \le \sum_{t=/2006/}^{t} b_{u,t}^{mudflat} \qquad \forall u,t \qquad\qquad (3)$$

Block in Equation (2) requires the sum of the arable lands allocated to certain types of crop plantation (including both conventional crops and energy crops) in one crop season to be smaller than the amount of locally accessible arable land resources, no matter which kind of field management has been adopted. This, to some extent, reflects the fact of land use conflict between food crops and energy crops. Similarly, for block in Equation (3), it applies the same structure as block in Equation (2). The difference is that block in Equation (3) proposes the limitation on mudflat resources and reclaimed mudflats are only dedicated to pc, which refers to the energy crop. As considering Jiangsu's unique feature of having a large area of mudflats located along its coast, block in Equation (3) offers us a solution that the plantation of energy crop on mudflats may be a feasible and cost effective way to secure enough biomass provision for energy purposes while decreasing its negative influences on food security as much as possible:

$$\sum_{pc,a} LAND_{u,pc,t-1,a}^{mudflat} \le \sum_{u,pc,t,a} LAND_{u,pc,t,a}^{mudflat} \qquad \forall u,t \qquad\qquad (4)$$

Block in Equation (4) assures that the reclamation process is irreversible. That means the accumulated cultivation area for energy crops in mudflats can only be enlarged. This assumption is consistent with an increasing tendency of biomass demand:

$$-LAND_{u,pc,t-1,a-1}^{\text{perennial crop}} + LAND_{u,pc,t,a}^{\text{perennial crop}} \leq 0$$
$$-LAND_{u,pc,t-1,a-1}^{\text{mudflat}} + LAND_{u,pc,t,a}^{\text{mudflat}} \leq 0 \quad \Big|_{1<a\leq k_{pc}} \qquad \forall u, pc, t, a \qquad (5)$$

Block in Equation (5) is targeted for perennial crops' consistency. Considering its natural death or farmers' active eradication, the plantation area of certain kind of perennial crop would never be larger but only smaller than or be equal to the area of itself in the prior year.

The fifth set of constraints addresses aggregation related aspects of the farmers' decision process. These constraints force farmers' cropping activities for $LAND_{u,fc,t,n}^{\text{food crop}}$ either in summer or in autumn to fall within a convex combination of historically observed seasonal choices $h_{u,fc,ht,n}$ (Equation (6)). Based on decomposition and economic duality theory, Onal and McCarl [53] show that historical crop mixes represent rational choices embodying numerous farm resource constraints, crop rotation considerations, perceived risk reactions, and a variety of natural conditions. In Equation (6), the $h_{u,fc,ht,n}$ coefficient contains the observed crop mix levels for the latest 11 years (from 2000 to 2011). $CMIX_{u,t,ht,n}$, representing the weights of historical data in different years, are positive, endogenous variables indexed by historical year and region, whose level will be determined during the optimization process:

$$-\sum_{ht}\left(h_{u,fc,ht,n} \cdot CMIX_{u,t,ht,n}\right) + LAND_{u,fc,t,n}^{\text{food crop}} = 0\big|_{2010<t\leq2030} \quad \forall u, fc, t, n \qquad (6)$$

However, crop mix constraints are not applied to the crops, which under certain policy scenarios are expected to expand far beyond the upper bound of historical relative shares [54]. As the cultivation area of energy crops is expected to greatly expand in the future, these crops are naturally excluded from this equation block:

$$dema_t^{\text{biomass}} - \sum_{u,fc,pr\_energy,n} \eta_u \cdot \frac{y_{u,fc,pr\_energy,t,n}^{\text{food crop}} \cdot LAND_{u,fc,t,n}^{\text{food crop}}}{\alpha}$$
$$- \sum_{u,pc,pr\_energy,a} y_{u,pc,pr\_energy,t,a}^{\text{perennial crop}} \cdot \left(LAND_{u,pc,t,a}^{\text{perennial crop}} + LAND_{u,pc,t,a}^{\text{mudflat}}\right) \leq 0 \quad \forall t \qquad (7)$$

The supply and demand balance of biomass is represented in block in Equation (7). The first item denotes the biomass demand in a certain year. The second item denotes the biomass from traditional food crops, namely crop straw, and the last term represents the biomass from perennial crops grown either on arable land or reclaimed mudflats. This expression fully secures the achievement of biomass development targets in due year:

$$dema_{fc,t}^{\text{grains}} - \sum_{u,fc,pr\_food,n} \left(y_{u,fc,pr\_food,t,n}^{\text{food crop}} \cdot LAND_{u,fc,t,n}^{\text{food crop}}\right) - FOODFLOW_{fc,t} \leq 0 \quad \forall fc, t \qquad (8)$$

Paralleling, the last constraint set defines the satisfaction to the requirement of food security in the background of bioenergy introduction. The first item is the demand of certain food, the following items stand for the produced food from planted conventional crops, and the last item means the gap between food demand and supply is filled by cross-boundary food trade.

References

1. Scheffran, J.; BenDor, T. Bioenergy and land use: A spatial-agent dynamic model of energy crop production in Illinois. *Int. J. Environ. Pollut.* **2009**, 39, 4–27. [CrossRef]
2. Hall, D.O.; House, J.I. Reducing atmospheric CO_2 using biomass energy and photobiology. *Energy Convers. Manag.* **1993**, 34, 889–896. [CrossRef]
3. Jungmeier, G.; Spitzer, J. Greenhouse gas emissions of bioenergy from agriculture compared to fossil energy for heat and electricity supply. *Nutr. Cycl. Agroecosyst.* **2001**, 60, 267–273. [CrossRef]

4. Demirbas, A.H.; Demirbas, I. Importance of rural bioenergy for developing countries. *Energy Convers. Manag.* **2007**, *48*, 2386–2398. [CrossRef]
5. Silveira, S. How to realize the bioenergy prospects? In *Bioenergy—Realizing the Potential*; Elsevier: Oxford, UK, 2005; pp. 3–17.
6. Ewing, M.; Msangi, S. Biofuels production in developing countries: Assessing tradeoffs in welfare and food security. *Environ. Sci. Policy* **2009**, *12*, 520–528. [CrossRef]
7. Tirado, M.C.; Cohen, M.J.; Aberman, N.; Meerman, J.; Thompson, B. Addressing the challenges of climate change and biofuel production for food and nutrition security. *Food Res. Int.* **2010**, *43*, 1729–1744. [CrossRef]
8. Scheffran, J. Biofuel conflicts and human security: Toward a sustainable bioenergy life cycle and infrastructure. *Swords Ploughsh.* **2009**, *27*, 4–10.
9. Ajanovic, A. Biofuels versus food production: Does biofuels production increase food prices? *Energy* **2011**, *36*, 2070–2076. [CrossRef]
10. Ugarte, D.D.L.T.; He, L. Is the expansion of biofuels at odds with the food security of developing countries? *Biofuels Bioprod. Biorefining* **2007**, *1*, 92–102. [CrossRef]
11. Campbell, J.E.; Lobell, D.B.; Genova, R.C.; Field, C.B. The global potential of bioenergy on abandoned agriculture lands. *Environ. Sci. Technol.* **2008**, *42*, 5791–5794. [CrossRef] [PubMed]
12. Scheffran, J. Criteria for a sustainable bioenergy infrastructure and lifecycle. In *Plant Biotechnology for Sustainable Production of Energy and Co-Products*; Mascia, P.N., Scheffran, J., Widholm, J.M., Eds.; Springer: Heidelberg, Germany, 2010; Volume 66, pp. 409–447.
13. Thomas, V.M.; Choi, D.G.; Luo, D.; Okwo, A.; Wang, J.H. Relation of biofuel to bioelectricity and agriculture: Food security, fuel security, and reducing greenhouse emissions. *Chem. Eng. Res. Des.* **2009**, *87*, 1140–1146. [CrossRef]
14. Ceotto, E.; Candilo, M.D. Sustainable bioenergy production, land and nitrogen use. *Sustain. Agric. Rev.* **2011**, *5*, 101–122.
15. Sadhukhan, J.; Ng, K.S.; Hernandez, E.M. *Biorefineries and Chemical Processes: Design, Integration and Sustainability Analysis*; John Wiley & Sons: Hoboken, NJ, USA, 2014.
16. Zong, J.; Guo, A.; Chen, J.; Liu, J. A study on biomass potentials of perennial gramineous energy plants. *Pratac. Sci.* **2012**, *29*, 809–813.
17. Shao, H.; Chu, L. Resource evaluation of typical energy plants and possible functional zone planning in China. *Biomass Bioenergy* **2008**, *32*, 283–288. [CrossRef]
18. Zhuang, D.; Jiang, D.; Liu, L.; Huang, Y. Assessment of bioenergy potential on marginal land in China. *Renew. Sustain. Energy Rev.* **2011**, *15*, 1050–1056. [CrossRef]
19. Liu, Y.; Wu, C.; Ma, X. Studies on the development and utilization of shoal land in Jiangsu province. *J. China Agric. Resour. Reg. Plan.* **2004**, *25*, 6–9.
20. Zhang, D.; Zhang, F.-R.; An, P.-L. Potential economic supply of uncultivated arable land in China. *Resour. Sci.* **2004**, *26*, 46–52.
21. Wang, F.; Zhu, Y. Development patterns and suitability assessment of tidal flat resources in Jiangsu province. *Resour. Sci.* **2009**, *31*, 619–628.
22. *Outline of Reclamation and Utilization Plan for Jiangsu Coastal Mudflat Resources (2010–2020)*; Development and Reform Commission of Jiangsu Province: Nanjing, China, 2010.
23. *Development Plan for Coastal Area of Jiangsu Province*; China National Development and Reform Commission: Beijing, China, 2009.
24. *Development Plan for Modern Agriculture of Jiangsu Province in the 12th-five Year*; Development and Reform Commission of Jiangsu Province: Nanjing, China, 2012.
25. Yamamoto, H.; Fujino, J.; Yamaji, K. Evaluation of bioenergy potential with a multi-regional global-land-use-and-energy model. *Biomass Bioenergy* **2001**, *21*, 185–203. [CrossRef]
26. Schneider, U.A.; McCarl, B.A. Economic potential of biomass based fuels for greenhouse gas emission mitigation. *Environ. Resour. Econ.* **2003**, *24*, 291–312. [CrossRef]
27. Johansson, D.A.; Azar, C. A scenario based analysis of land competition between food and bioenergy production in the US. *Clim. Change* **2007**, *82*, 267–291. [CrossRef]
28. Havlík, P.; Schneider, U.A.; Schmid, E.; Böttcher, H.; Fritz, S.; Skalský, R.; Aoki, K.; Cara, S.D.; Kindermann, G.; Kraxner, F.; et al. Global land-use implications of first and second generation biofuel targets. *Energy Policy* **2011**, *39*, 5690–5702. [CrossRef]

29. Kraxner, F.; Nordström, E.-M.; Havlík, P.; Gusti, M.; Mosnier, A.; Frank, S.; Valin, H.; Fritz, S.; Fuss, S.; Kindermann, G.; *et al.* Global bioenergy scenarios—Future forest development, land-use implications, and trade-offs. *Biomass Bioenergy* **2013**, *57*, 86–96. [CrossRef]

30. Zhang, X.; Li, S.; Huang, X.; Li, Y. Effects of carbon emissions and their spatio-temporal patterns in Jiangsu province from 1996 to 2007. *Resour. Sci.* **2010**, *32*, 768–775.

31. *Comprehensive Activity Plan of Energy-saving and GHG Emission Reduction of Jiangsu Province in 2011–2015*; People's Government of Jiangsu Province: Nanjing, China, 2012.

32. Xu, X.; Wang, D.; Jiang, H.; Shi, H. Study on greenhouse gas emission in Jiangsu province. *Water Air Soil Poll.* **1999**, *109*, 293–301.

33. *The 12th Five-year Energy Development Plan in the Jiangsu Province*; People's Government of Jiangsu Province: Nanjing, China, 2012.

34. *Comprehensive Utilization Plan for Crop Residues in Jiangsu Province (2010–2015)*; Executive Office of People's Government of Jiangsu Province (EOPGJP): Nanjing, China, 2010.

35. Statistic Bureau of Jiangsu Province. *Statistic Yearbook of Rural Areas in Jiangsu Province (2010)*; China Statistics Press: Beijing, China, 2011.

36. Bauen, A.; Berndes, G.; Junginger, M.; Londo, M.; Vuille, F.; Ball, R.; Bole, T.; Chudziak, C.; Faaij, A.; Mozaffarian, A.H. *Bioenergy—A Sustainable and Reliable Energy Source*; International Energy Agency Bioenergy: Paris, France, 2009; pp. 1–107.

37. Fan, X.; Hou, X.; Zuo, H.; Wu, J.; Duan, L. Biomass yield and quality of three kinds of bioenergy grasses in beijing of China. *Sci. Agric. Sin.* **2010**, *43*, 3316–3322.

38. Fan, X.; Zuo, H.; Hou, X.; Wu, J. Potential of miscanthus spp. and triarrhena spp. as herbaceous energy plants. *Chin. Agric. Sci. Bull.* **2010**, *26*, 381–387.

39. Hou, X.; Fan, X.; Wu, J.; Zhang, Y.; Zuo, H. Large-scale cultivation and management technologies of cellulosic bioenergy grasses in marginal land in Beijing suburb. *Crops* **2011**, *2011*, 98–101.

40. Hou, X.; Fan, X.; Wu, J.; Zuo, H. Evaluation of economic benefits and ecological values of cellulosic bioenergy grasses in Beijing suburban areas. *Acta Pratac. Sin.* **2011**, *20*, 12–17.

41. Khanna, M.; Dhungana, B.; Clifton-Brown, J. Costs of producing miscanthus and switchgrass for bioenergy in illinois. *Biomass Bioenergy* **2008**, *32*, 482–493. [CrossRef]

42. Zuo, H.; Yang, X.; Chen, Q. The Research Proceeding of Cellulosic Herbaceous Energy Crops. In Proceedings of the China Bioenergy Technology Route Standard System Construction Forum, Shanghai, China, 30 August 2008.

43. Wang, W.; Wang, C.; Pan, Z.; Pan, Q. Using Jiangsu coastal mudflats to plant salt tolerance bioenergy crop. *Jiangsu Agric. Sci.* **2010**, *37*, 484–485. (In Chinese)

44. Ling, S. Development and utilization of biomass energy in Jiangsu coastal area. *Resour. Ind.* **2010**, *12*, 117–121.

45. Ling, S. Developing new energy industries of coastal area and cultivating new economic growth pole—A case study of Jiangsu coastal area. *J. Yancheng Teach. Univ.* **2009**, *29*, 20–23. (In Chinese)

46. Xiong, P. *The Feasibility Report on Large-scale Production of Lignocellulosic Ethanol*; Unpublished internal report: Huai'an, China, 2010. (In Chinese)

47. Wang, Y.; Zhao, Y.; Ding, M.; Ji, C.; Wang, C. The current status of Jiangsu's straw resource utilization and the discussion on multiple-layer utilization pattern. *Jiangsu Agric. Sci.* **2010**, *37*, 393–396. (In Chinese)

48. Li, J.; Hu, Y. Analysis on investment and operation of straw-fired power plants in Jiangsu province. *Electr. Power Technol. Econ.* **2009**, *21*, 18–22.

49. Jiangsu Grain Bureau. Jiangsu grain consuming characteritics and development tendency analysis. In *Grain Econ. Res.*; 2005; 74, pp. 5–15. (In Chinese)

50. Schneider, U.A.; McCarl, B.A. Greenhouse gas mitigation through energy crops in the US with implications for Asian pacific countries. In *Global Warming and the Asian Pacific*; Edward Elgar Publishing Limited: Cheltenham, UK; Northampton, MA, USA, 2003; pp. 168–184.

51. Onal, H.; McCarl, B.A. Aggregation in mathematical-programming sector models and model stability. *Am. J. Agric. Econ.* **1991**, *73*, 1545–1545.

52. McCarl, B.A.; Spreen, T.H. Price endogenous mathematical programming as a tool for sector analysis. *Am. J. Agric. Econ.* **1980**, *62*, 87–102. [CrossRef]

53. Önal, H.; McCarl, B.A. Exact aggregation in mathematical programming sector models. *Can. J. Agric. Econ.* **1991**, *39*, 319–334. [CrossRef]

54. Schneider, U.A.; McCarl, B.A.; Schmid, E. Agricultural sector analysis on greenhouse gas mitigation in US agriculture and forestry. *Agric. Syst.* **2007**, *94*, 128–140. [CrossRef]

energies

MDPI

Article

The Concept, Design and Performance of a Novel Rotary Kiln Type Air-Staged Biomass Gasifier

Huiyuan Shi [1,*,†], Wen Si [2,†] and Xi Li [1]

1   Department of Environmental Science and Engineering, Fudan University, Shanghai 200433, China; xi_li@fudan.edu.cn
2   College of Information and Computer Science, Shanghai Business School, Shanghai 200235, China; siw@sbs.edu.cn
*   Correspondence: huiyuanshi11@fudan.edu.cn; Tel.: +86-181-0181-8941
†   These authors contributed equally to this work.

Academic Editor: Tariq Al-Shemmeri
Received: 5 November 2015; Accepted: 12 January 2016; Published: 22 January 2016

Abstract: Tar formation is the main bottleneck for biomass gasification technology. A novel rotary kiln type biomass gasification process was proposed. The concept design was based on air staging and process separation. This concept was demonstrated on a pilot scale rotary kiln reactor under ambient pressure and autothermic conditions. The pilot scale gasifier was divided into three different reaction regions, which were oxidative degradation, partial oxidation and char gasification. A series of tests was conducted to investigate the effect of key parameters. The results indicate that under optimum operating conditions, a fuel gas with high heat value of about 5500 kJ/Nm^3 and gas production rate of 2.32 Nm^3/kg could be produced. Tar concentration in the fuel gas could be reduced to 108 mg/Nm^3 (at the gasifier outlet) and 38 mg/Nm^3 (after gas conditioning). The cold gas efficiency and carbon conversion rate reached 75% and 78%, respectively. The performance of this gasification system shows considerable potential for implementation in distributed electricity and heat supply projects.

Keywords: biomass gasification; air staged; rotary kiln; partial oxidation; energy balance

---

1. Introduction

Thermal conversion of biomass is a promising way to produce gaseous and liquid fuels [1]. Among all the techniques, gasification has received extensive attention because it produces valuable gaseous products, which can be used as a fuel for internal combustion engines, gas turbine, boilers, and cooking [2]. Extensive studies have been conducted on biomass gasification processes to investigate how to increase the energy transformation efficiency, reduce pollution during the gasification process and improve the device reliability [3–5]. Among all the obstacles that impede the development of gasification technology, tar formation is the most cumbersome one [6]. Tar is a complex mixture of hydrocarbons, which includes aromatic compounds, along with other oxygen-containing hydrocarbons and complex polycyclic aromatic hydrocarbons (PAHs). Tar is defined by EU/IEA/US-DOE meeting as all organic contaminants with a molecular weight larger than benzene [7]. Tar easily causes fouling in ducts, valves, and engines, and polymerizes to form more complex structures during combustion, thus decreasing the reliability of equipment and increasing pollutant emissions [8,9]. Tar limits the usage of syngas in application processes. The minimum allowable limit for tar is dependent on the end user application. As tabulated by Milne and Evans [10], the general tolerance limits for engines, turbines and other combustion devices range from 5 to 100 mg/Nm^3.

There are two types of tar control technologies, which are syngas conditioning and inside gasifier treatment. The first method is proved to be effective but it is costly and may cause secondary pollution. The second method is gaining much more attention due to its economic benefits.

As reviewed by Milne and Evans [7], the formation and conversion of tar is seriously affected by the material composition, temperature, gasification agent(s), oxidation processes and residence time. These are all important operation parameters for a biomass gasifier. Plenty of works are available on tar formation and destruction behavior under the effect of the parameters mentioned above. It is generally agreed that tar is produced during biomass pyrolysis around 300 to 500 °C. It can be divided into primary, secondary and tertiary tar, according to its reaction history. The primary tar is characterized by cellulose-derived products such as levoglucosan, hydroxyacetaldehyde and furfurals. The secondary tar species are phenolics and olefins, while tertiary tar includes methyl derivatives of aromatics, such as methyl acenaphthylene, methylnaphthalene, toluene, and indene. Tar composition varies with the temperature program used during pyrolysis. High heating rates lead to a high tar production and a high proportion of primary and secondary tar, hence, flush pyrolysis technology is applied for bio-oil production [11]. Low heating rates result in a low tar production and high tertiary tar proportion, that is the reason why slow pyrolysis is used for torrefaction and gasification. Tars produced by pyrolysis may change under certain conditions. Chen *et al.* [12] reported that biomass pyrolysis tar decomposed by more than 90% at 1000 °C. Besides decomposition, tar also experiences polymerization reactions above 800 °C. Their study also revealed that primary and secondary tar could be converted into highly polymerized and highly aromatic components, such as naphthalene and indene. Houben [13] reported that tar decomposition reactions can be accelerated under an oxidative environment. Pyrolysis tar conversion rates under a partially oxidative atmosphere are higher than under inert conditions. Moreover, oxygen-containing compounds are favored, which are mainly phenol and its homologous series.

As reviewed above, oxygen supply is a key issue for biomass gasifier design. Several gasifier concepts have been developed in the past 20 years [14]. The majority of them have a single inlet for air, no matter whether it is an updraft, downdraft or fluidized gasifier. Few of them has more than one air inlet. Pan *et al.* [15] reported a 88.7 wt % of tar reduction by injecting secondary air above the biomass feeding point in a fluidized bed configuration. Narv *et al.* [16] performed secondary air injection in the freeboard of a fluidized bed gasifier and observed a tar reduction of nearly 50 wt % with a temperature rise of about 70 °C. The Asian Institute of Technology (AIT) developed a two stage biomass gasifier [7]. Experimental results showed that the tar concentration in flue gas is about 40 times less than in a single-stage reactor under similar operating conditions. This concept involves a downdraft gasifier with two air inlets. Tar produced in first pyrolysis stage will pass through a high temperature char bed and decompose at elevated temperatures. Yoshikawa [17] combined a fixed-bed pyrolyzer with a high temperature reformer using a high temperature steam/air mixture and demonstrated that the injection of high temperature steam/air mixture into the pyrolysis gas effectively decomposes tar and soot components in the pyrolysis gas into gaseous products. Henriksen *et al.* [18] reported successful working and operating experiences with a two-stage gasifier. Although it has only one oxygen inlet, the pyrolysis and gasification process are separated by a partial oxidation zone, where the produced pyrolysis tar can be decomposed to a great extent.

From the extensive literature search, it can be observed that the temperature profile and oxygen injection are key design issues that influence tar production during gasification. A few unique gasifiers have been developed. Some of them were based on an "air staging" concept and some separated gasification into many processes through structure modification. In the current work, a novel gasifier system with a thermal capacity of 1.5 MW$_{th}$ was proposed. The conceptual configuration of the gasification reactor is illustrated in Figure 1. The gasifier is a rotary kiln with a 6 degree angle of inclination. Unlike a traditional rotary kiln, it is divided into three regions, which are oxidative degradation, partial oxidation and char gasification. Biomass is fed into the reactor from the left end and passes through the gasifier by rotating the reactor. Air used for gasification is injected into three different regions by a blower.

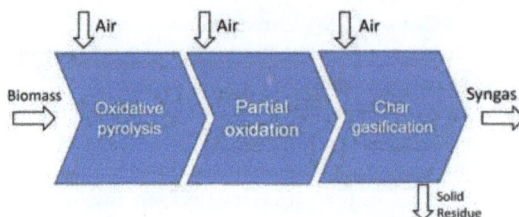

Figure 1. Conceptual schematic of the novel air staged biomass gasifier.

The pyrolysis reaction takes place in the first region under an oxidative atmosphere. Then, the pyrolytic gas with tar species is drawn into the second region by an induced fan. In this stage, homogeneous partial oxidation reactions happen. Tar components decompose into small molecule gaseous products. After the pyrolytic and partial oxidation process, the volatiles in the biomass solid fuel are released completely and this leaves a char with high carbon content. In the third stage, the char gasification reaction takes place. Besides, biomass char also acts as absorbent to adsorb tar components [19].

The air staging strategy does not just reduce tar production; it also produces an energy self-sufficient system. DTU developed a two-stage gasifier [18], where the oxidation and pyrolysis processes are separated. However, its pyrolysis stage is heated externally by exhaust gas from a gas engine. In the present gasifier, every single stage is auto-thermal. Heat for pyrolysis is provided by partial oxidation of the solid biomass fuel in an oxidative environment. Besides, radiation heat from the second region also helps maintain the temperature in the pyrolysis region. In the present study, the optimal operation parameters and main factors that influence tar production, low heat value (LHV) of the fuel gas, gas production rate, cold gas efficiency and carbon conversion rate were investigated. In addition, the heat balance was calculated and discussed.

2. Materials and Methods

2.1. Material

Mixed wood chips with an average size of $1 \times 2 \times 5$ cm were used as feedstock. The corresponding proximate and ultimate analysis results are shown in Table 1. The volatile content is as high as 60%, while the fixed carbon content is about 17%. Since the biomass used in this test is a mixture of waste wood without drying, its water and ash content is a bit higher than those of other wood materials. Air was used as gasification agent in all tests.

Table 1. Proximate and ultimate analysis of wood chips.

Proximate Analysis [1] (wt %)				Ultimate Analysis [1] (wt %)					Low Heat Value (kJ/kg)
Water	Ash	Volatile	Fixed Carbon	C	H	O	N	S	
12.40	11.28	59.36	16.96	53.24	6.36	40.14	0.12	0.14	16931

[1] As received basis.

2.2. Description of the System and Experiment Setup

2.2.1. System Description

The schematic pilot-scale configuration is illustrated in Figure 2. A pictorial view of the system is given in Figure 3. The process is performed under atmospheric pressure. Wood chips are lifted by a conveyor and stored in the hopper. A loch hopper system (two valves separated by a screw conveyer) ensures that no gas escapes and no air leaks into the gasifier. The body of gasifier is a rotary kiln type furnace, with 6 degree inclination. The reactor is made of plain carbon steel and covered with refractory material inside and an insulation layer outside. The effective inner diameter is 0.75 m, with a total length of 15 m. Two annular shape steel plates are fixed inside the gasifier to divide the inner space into three different regions. The length of each region is 5 m, 2 m and 8 m, respectively. A variable frequency motor is used to drive and control the rotation speed.

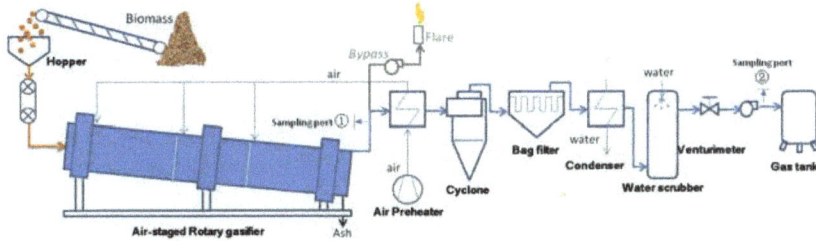

Figure 2. Flow diagram of the gasification plant.

Figure 3. A view of the1.5 MW$_{th}$ biomass gasification system.

Gasification agent is provided by an air blower. Air is preheated by the hot product gas and divided into three intakes. Three flow meters are used to control each air injection separately. The air feed system for the rotary kiln type furnace is a patented technology, which ensures no air and fuel gas leakage from the rotating furnace. First air is fed into the pyrolysis region of the reactor through tuyeres placed radically around the circumference. Secondary air is injected into the secondary part of the reactor and distributed in gaseous phase area, so that partial oxidation of the pyrolytic gas would take place and oxidation of biomass char would be prevented at the bottom of the reactor. Third air is injected into the third part of reactor through nozzles located at the bottom of the reactor. This is to ensure a fine mixture of air and hot char, and hence promote the gasification reaction.

A cyclone is used as dedusting unit for fuel gas cleaning which is heat-insulated to prevention any condensation of trace tar. An air preheater is placed after the dust remover and followed by an indirect water cooler. The cooled producer gas is then allowed to pass through a washing tower in order to remove the moisture and trace tar content in the syngas. The ash remover, gas cooler and washing tower constitute of the gas conditioning system. After gas conditioning, the syngas goes to the tank through the blower. Just before the gas tank, a water bubbler is installed in the path for safety reasons to avoid backfire. A venturimeter is installed in the path to measure the gas flow rate. In a Venturi flow meter, the mass flow rate is proportional to the square root of the pressure difference across the venture inlet and the throat.

The fuel gas exits the gasifier and enters into the following de-dusting and cooling units only if the system has been brought to a steady state. Before that, a bypass vent is used to exhaust fuel gas produced during the startup period.

2.2.2. Preparation and Procedures

Biomass fuel is prepared first. Its size and moisture content are carefully controlled according to the needs (refer to Table 1). The system is then tested for leakage by sealing the gasifier at the fuel inlet and ash tank using a water seal and subjecting it to air at a pressure of 50–70 mm water gauge with the help of an air blower. Every joint was examined carefully with soap suds. After the leakage tests, the bypass fan is started. A 200 Pa negative pressure is maintained inside the gasifier through an interlock pressure control system. Before feeding the biomass, a startup burner fueled with natural gas is used to heat the reactor. When the desired temperature distribution is achieved in the center of first region (700–750 °C), second region (650–700 °C) and third region (500–550 °C), the feeder valve is opened to feed biomass into the gasifier. The reactor rotation driving motor and air blower are turned on simultaneously. Then, the startup burner is taken out. At the same time, the assisting fuel gas (natural gas) and secondary air are injected into the second region to ignite the pyrolytic gas produced in the first region. When the required temperature (>900 °C) is reached in the second region, the assisting fuel gas is turned off and the air injection system is continuously operated. After the temperature in the second region reaches a steady state, we turn on the air injection at the third region. After that, the feedstock/air ratio is carefully adjusted to reach steady state operation conditions. In general, it takes around 8 h to make the system stable with respect to the reactor temperature. After the system reaches a steady state, product gas is shifted to the gas conditioning system. All parameters are kept constant for about 2 h for sampling and data analysis.

2.2.3. Measuring and Sampling Instrument

To position thermocouples inside the rotary gasifier is a great challenge. Calibrated S-type (platinum-rhodium) thermocouples are used for temperature measurement. The wires are electrically separated from each other by ceramic beads. All the electric wires are placed inside a 1Cr18Ni9Ti stainless steel pipe coated with insulation layer. The pipe is fixed at the center axis of the reactor with two bearings at both ends to ensure an immobile position when the reactor is rotating (Figure 4). A digital data collector is used to read out the temperature.

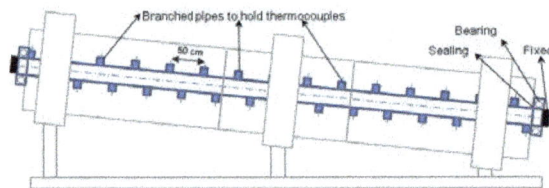

Figure 4. Position of thermocouples inside the gasifier.

A model Gasboard-3100 gas analyzer produced by Wuhan Cubic Optoelectronics Cp., Ltd. (Wuhan, China) is used to measure the composition of the producer gas. The resolution ratio and accuracy are 0.01% and 1%, respectively. Two sensors are used for gas composition measurement, which are a Non-Dispersive Infra-Red (DNIR) and a Thermal Conduction Detector (TCD). The volume concentrations of CO_2, CO, CH_4, H_2 and O_2 can be measured simultaneously. The measuring range for CO_2, CO, CH_4, H_2 and O_2 are 0%–50%, 0%–40%, 0%–25%, 0%–40% and 0%–25% respectively. Water vapor, dust and tar are removed before gas analysis. The gas analyzer is coupled with the tar sampling system.

The tar sampling system operates as follows: syngas is induced by a pump with a controlled flow rate of 1 L/min. The syngas is first cleaned by a dust filter, and then washed by four cold gas washing bottles, which are immersed in an ice-water bath. The sampling line and ash filter are electrically heated to 350 °C to avoid condensation of tar as noted in [7]. There might be some components that has a condensation temperature lower than 350 °C. These species cannot be collected in the sampling bottle. The first two bottles are filled with glass balls, in order to increase the contact area. Large molecule tars will be trapped in the first two bottles. Then, gas is washed by a solution of methanol and trichloromethane with a volume ratio of 1:4. After this process, tar can be trapped in four bottles. The clean gas is pumped into the gas analyzer for composition measurement. After gas sampling, all the pipes and bottles are washed with methanol and chloroform solution. The washing liquids are sealed and kept in a refrigerator for tar composition and mass analysis. After GC analysis, the solvent was evaporated in a RE3000A type rotary evaporator for 30 min, and the rest was placed in a drying oven with a temperature of 35 °C for 1 h in an inert atmosphere for the final evaporation and then weighed as gravimetric tar and sealed in a container. The total tar concentration is calculated by the mass of sampled tar minus the sampled gas volume.

Tar composition is analyzed by gas chromatography/mass spectrometry (GC/MS) using an Agilent 6890 N gas chromatograph and an Agilent 5975 C mass spectrometer (Agilent, Santa Clara, CA, USA). The GC split ratio is 10:1. The sample inlet temperature is set at 280 °C. A HP-5MS column (30 m × 0.25 mm, 0.25 μm) is used at a flow rate of 2.4 mL/min. The carrier gas is helium with a flow rate of 30 mL/min. The column temperature increases from 45 to 180 °C at 5 °C/min and then to 300 °C at 45 °C/min, where it is held for 10 min. The sample injection volume is 1 μL. The GC/MS was quantified by the external standard method with 2, 5, 10, 20, 40, 60, 80 and 100 ppm solutions of eight typical tar components: benzene, toluene, styrene, phenol, naphthalene, ethylbenzene, indene, and benzaldehyde. The chromatographic system has limitations for measuring higher tars as the heavier components are retained and do not elute through the GC column [7].

2.2.4. Biomass Feed Rate Calibration

Biomass is gravity feed by a loch hopper system to control the feed rate and exert a sealing function. In the gasifier, which is like a helical vane drum with 6 degree inclination, biomass could be transmitted by rotating the gasifier. The average biomass feed rate was plotted against the gasifier rotation speed. The determination coefficient was 0.9811, which was due to the non-uniform size of the wood chips. The loch hopper system runs intermittently to match the gasifier rotation speed in order to achieve a certain biomass feed rate. As the rotation speed varied from 1/6 r/min to 1 r/min, the biomass residence time inside the gasifier ranged from 114 min to 33 min. In the current study, the gasifier rotation speed was fixed at 5/9 r/min. The corresponding biomass feed rate and residence time were 300 kg/h and 55 min, respectively.

3. Results and Discussion

The aim of this study was to demonstrate the concept of a novel air-staged gasification process to produce tar-free fuel gas. A series of tests were conducted to investigate the impact of parameter variation such as air injection percentage (AIP) of the three regions and equivalent ratio (λ) on temperature distribution, gas composition, tar formation, gas production rate, cold gas efficiency, and carbon conversion rate.

The gasifier is designed for a nominal fuel power of 1.5 MW_{th}. This is equal to 300 kg/h biomass fuel fed into the gasifier with a low heat value of 17 MJ/kg.

The equivalent ratio (λ) is defined by Equation (1):

$$\text{Equivalent ratio } (\lambda) = \frac{\text{Air injected into the gasifier}}{\text{Stoichiometric air}} \tag{1}$$

The stoichiometric air V_0 is calculated by Equation (2):

$$V_0 = 0.0889C_{ar} + 0.265H_{ar} - 0.0333\,(O_{ar} - S_{ar}) \tag{2}$$

Here V_0 refers to the stoichiometric air required for fuel combustion (Nm^3/kg), Car, Har, Oar, Sar refer to the elemental mass compositions of C, H, O and S (wt %), which are available in Table 1.

Besides λ, AIP also plays an important role on the temperature profile inside the gasifier. According to the description of the conceptual design, the purpose of the three air injections is not identical. The first air is to supply the heat required for the pyrolysis reactions; the secondary air is used for partial oxidation of the pyrolytic gas, which results in a high temperature environment; the third air acts as an agent for char gasification. Hence, optimization of AIP is crucial for acquiring a desired temperature distribution.

A series of test runs were carried out with the 1.5 MW_{th} pilot scale gasifier to understand the behavior of this gasifier and improve its performance.

3.1. Optimization of AIP

We have made plenty of tests covering all the equivalence ratios (0.28 to 0.45). Three different AIP conditions were tested, which were condition 1 (1:7:2), condition 2 (2:6:2) and condition 3 (3:5:2). Results showed that an optimum ratio of (2:6:2) applied to all the equivalence ratios. In order to be concise, only the temperature curve when $\lambda = 0.36$ is drawn in Figure 5.

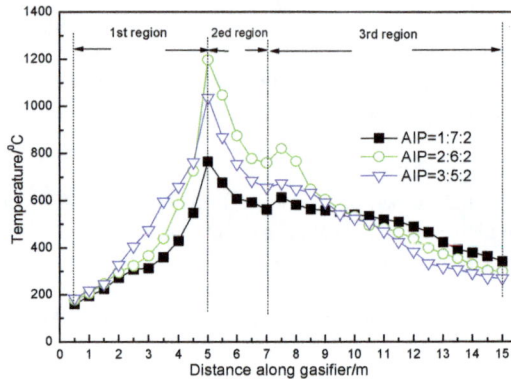

Figure 5. Temperature profile varied with AIP.

The temperature of the pyrolysis region rose up as the injected air increased. As described before, the heat required for the pyrolytic reactions is supplied by biomass and char oxidation. Hence, the more oxygen was injected, the higher the temperature rose. However, the second region behaved differently. In the second region, oxygen was injected into the gaseous phase, where it mixed with pyrolytic gas and resulted in a homogeneous partial oxidation reaction. Therefore, the environment temperature suddenly increased. In this experiment, although air injection in the second region decreased from 0.7 to 0.6, the peak temperature rose from 765 to 1196 °C. This indicated that more combustible gas was produced under condition 2. Under condition 1, the temperature in the first region was too low for the biomass decomposition reactions to take place. The pyrolytic gas produced was less than under condition 2. Therefore, even though the oxygen injection in the second region was higher; there was not enough gas for combustion, and as a result, the temperature decreased. Under condition 3, the temperature in the second region decreased. Although the pyrolysis reaction was enhanced due to the higher temperature, a large amount of injected oxygen may cause an oxidation of the pyrolytic products in the same time they are produced. In addition, a lack of oxygen in the second region restrains the secondary oxidation of the pyrolytic gas. For these reasons, the temperature in the second region was lower than under condition 2.

Heat required for char gasification is transferred by radiation and hot gas convection from the partial oxidation process. Therefore, the temperature profile in the third region was greatly affected by the second region, even though the air injection in the third region was identical for the three conditions.

Based on the above analysis, condition 2 is the most optimal. The highest temperature was achieved in the second region, which resulted in a maximum tar reduction efficiency and char gasification rate. Thus, in the following tests, AIP was fixed at 2:6:2 to obtain the best temperature distribution. Next the total air injected into the gasifier was varied to investigate the effect of λ on gasifier performance.

3.2. Temperature Profile vs. λ

We have conducted many experiments under different λ values (from 0.2 to 0.6). Results indicated that the auto thermal operation happens only if λ is larger than 0.28. Thus, we just present the results when $0.28 < \lambda < 0.45$. The temperature distribution along the axis of the gasifier is illustrated in Figure 6. It can be seen that the variation of temperature curves were alike for different λ values. In addition, the shape of the temperature curves was in accordance with the three regions inside the gasifier.

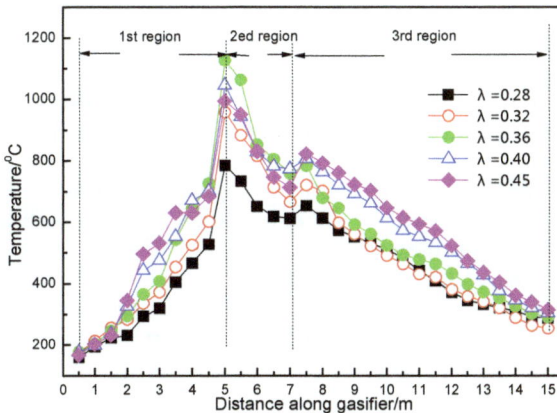

Figure 6. Temperature profile variation with λ.

When biomass fuel was fed into the gasifier, it was heated by radiation of the hot solid fuel and gas in the second region. The environment temperature at the inlet immediately rose up to around 200 °C. As the biomass was moved by rotating the gasifier, the temperature gradually increased. When the temperature reached 350 °C, the pyrolysis reactions started. It was reported that pyrolytic reactions could be enhanced under an oxidative environment [20]. At the end of the first region, the biomass decomposition reactions were almost finished. A slight increase of temperature could be observed due to hot gas radiation from the second region. Along the temperature curves in the first region, no obvious peaks were detected, which indicated that the oxidation reaction took place smoothly. Researchers have concluded that in such a temperature range (300–500 °C), the oxidation reactions almost take place between solid fuel and oxygen [21]. The combustion of pyrolytic gas requires a higher activation energy which cannot be reached within this temperature range [22,23]. Under the current temperature, a large amount of CO and CH_4 were produced, however they would not be burned in the first region and would experience partial oxidation in the next stage.

A step temperature increase appeared at the inlet of the second region. This is because of the combustion of pyrolytic gas. Oxygen injected at the first stage has all been consumed. The pyrolytic gas with high heat value was drawn to the partial oxidation area, where the gas met with oxygen and underwent an intense combustion reaction. Hence, the temperature exceeded 1000 °C.

At the inlet of the third region, another obvious peak was observed. The third injected air would cause combustion of the residual syngas that was not consumed in the second region. Besides, heterogeneous oxidation reactions would also occur when air contacts the hot biomass char. These homogeneous and heterogeneous oxidation reactions resulted in an increase of the environment temperature, while the heat released was instantly adsorbed by gasification reactions of biomass char. Hence, the temperature gradually decreased to around 300 °C at the outlet of the gasifier.

When λ increased, the temperature profile responded regularly. Firstly, the average temperature of the pyrolysis region increased. In the presented novel gasification process, pyrolysis reactions take place under an oxidative environment, and the oxygen concentration in the pyrolysis zone varied with λ. Since the heterogeneous oxidation reaction rate is slower than that of homogeneous oxidation, the heat release rate is slow and gentle, thus the temperature increased slightly. The oxidative reactions mainly happen on the surface of biomass or biomass semi-coke, where degradation of biomass large molecules occurs as well. The oxidation and degradation reactions almost take place at the same time and in the same location. Hence, it is an *"in-situ,"* oxidative degradation process [24]. In this way, the heat released from oxidation could be effectively adsorbed by reduction reactions. Therefore, the overall temperature won't rise too much, while a high decomposition rate of biomass fuel can be achieved. This is the major advantage of this air-staged gasifier.

The steep growth at the inlet of second region can be attributed to severe combustion of the pyrolytic gas. As shown in Figure 6, when λ increased from 0.28 to 0.36, the peak temperature rose from 965 to 1196 °C. It is obvious that a higher λ promoted partial oxidation reactions, and hence released more heat to this region. The temperature didn't vary linearly with λ. When λ increased from 0.36 to 0.45, the temperature decreased from 1196 to 1124 °C. This is because that under higher λ conditions, biomass pyrolysis is already finished. Increased air injection will dilute the pyrolytic gas since a large amount of N_2 is injected into the system. Therefore, the partial oxidation rate of the pyrolytic gas is slowed down and the temperature of the second region decreased accordingly.

In the third region, a distinct peak was observed at the inlet as well. However, it behaved differently with λ than in the second region. The peak temperature increased with λ in all tests. No inflection point was detected. This is because in the third region, the combustible gas concentration is low, therefore, heterogeneous oxidation reactions prevail. Oxygen was inclined to react with the biomass char rather than the gases. After the previous two stages, biomass devolatilization was completed. Char produced in the present gasification process would be more active, since it was produced by oxidative degradation. Researchers have reported that biomass char produced by pyrolysis in certain oxygen concentration environments has a highly porous structure, and the specific surface area could reach a value of 321 m^2/g [19]. Moreover, oxidative char is rich in oxygen-containing functional groups, which may act as an "activation center" during gasification and oxidation [25]. In conclusion, biomass char produced in an oxidative environment is more reactive. For this reason, an increase in λ results in a promotion of heterogeneous oxidation reactions and heat release and therefore, the average temperature of the third region rose with λ. At the outlet of the third region, the temperature dropped down to around 300 °C. Some components of the biomass tar would condense under 350 °C. Thus, a lower temperature would be helpful for trapping tar. In addition, a low temperature syngas outflow also indicates a high energy efficiency.

According to the above analysis, the temperature features of this gasifier can be summarized as follows: the moderate temperature and oxidative environment in the first region promoted the pyrolytic reaction and increased the production of gaseous products. A high temperature area is formed by homogeneous partial oxidation of the pyrolytic gas in the second region. This resulted in a high tar destruction rate. Biomass char produced by oxidative degradation had a high reaction activity. Hence, the gasification reaction in the third region was favored.

3.3. Syngas Quality

The fuel gas mass yield variation with λ is shown in Figures 7–9. The main components are CO, CO_2, H_2, CH_4 and trace amounts of C_xH_y. The unit of gas yield is kg/kg C, where C equals the mass of biomass plus the carbon mass content. The carbon mass content is listed in Table 1.

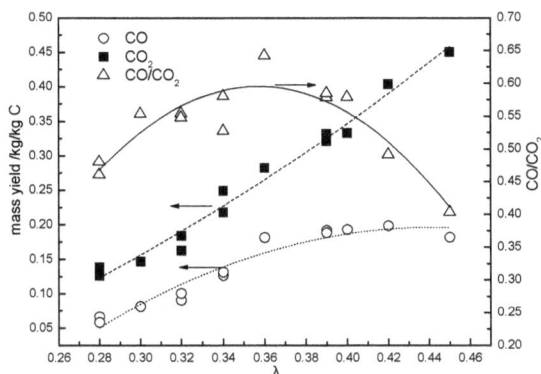

Figure 7. CO and CO_2 mass yield.

CO first increased then decreased as λ increased. As λ increased from 0.28 to 0.42, the yield of CO increased from 0.125 to 0.373 kg/kg C. After that, CO decreased to 0.341 kg/kg C. However, the mass yield of CO_2 continuously increases with λ. The mass of CO_2 is always larger than that of CO for all conditions.

During the biomass thermal chemical conversion process, the production of CO_2 is always accompanied by CO release. It seems that there is a rivalry between the formation of CO and CO_2. The proportion of CO and CO_2 is determined by the temperature and oxygen concentration in the environment. When the temperature was low, an increase of λ resulted in production of CO, while, CO_2 might be favored under a high temperature environment. Looking at the CO/CO_2 value in Figure 7, it is clear that when λ is below 0.36, the temperature is not high enough to initiate homogeneous oxidation. That is kinetically control reaction, thus CO keeps increasing as a result of the promoted pyrolysis reaction. Meanwhile, as λ surpasses 0.36, a significant increase of temperature in the second stage could be observed in Figure 6. Hence, CO would be oxidized into CO_2 and lead to a decrease of CO/CO_2. Under this condition, it seems that the diffusion rate dominates the oxidation reaction.

H_2 yield is shown in Figure 8. The peak value of 0.029 kg/kg C appears at λ = 0.36. The release of H_2 is mainly due to bonds between H radicals that are produced by the decomposition of larger molecules. Moreover, the polycondensation and dehydrocyclization of the biomass char structure would be another source of H_2 formation. Hence, H_2 production has a strong relationship with the environmental temperature. However, when λ exceeded 0.36, H_2 may be consumed through combustion.

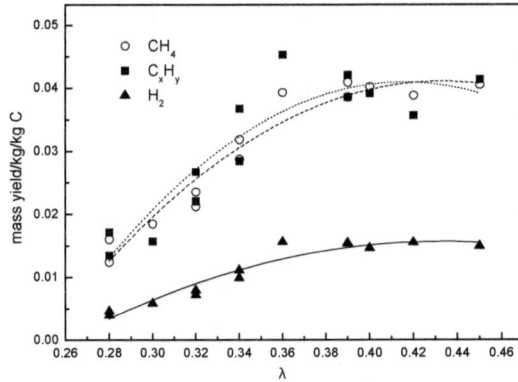

Figure 8. CH_4, C_xH_y and H_2 mass yield.

CH$_4$ and C$_x$H$_y$ are shown in Figure 8. The formation of CH$_4$ is always considered to be the product of demethylation reactions. Therefore, the amount of methyls in the biomass structure and the environmental temperature would be important for CH$_4$ production. The peak values of CH$_4$ and C$_x$H$_y$ were 0.074 kg/kg C and 0.085 kg/kg C at the point of λ = 0.36.

The syngas low heat value (LHV) is calculated by Equation (3):

$$LHV \ (kJ/Nm^3) = 126V_{CO}(\%) + 108V_{H2}(\%) + 359V_{CH4}(\%) + 595V_{CxHy}(\%) \tag{3}$$

where V_{CO}, V_{H2}, V_{CH4} and V_{CxHy} are the volume composition of CO, H$_2$, CH$_4$ and C$_x$H$_y$ in the syngas.

The variation of LHV is shown in Figure 9. A higher λ resulted in a rise of the temperature inside the gasifier. However, the quality of syngas was not always increasing. When λ increased from 0.28 to 0.36, LHV grew from 4214.66 kJ/Nm^3 to 5469.85 kJ/Nm^3, whereas, LHV decreased to 4065.50 kJ/Nm^3 when λ increased to 0.45. Thus, optimization of λ was critical in the present biomass gasification process.

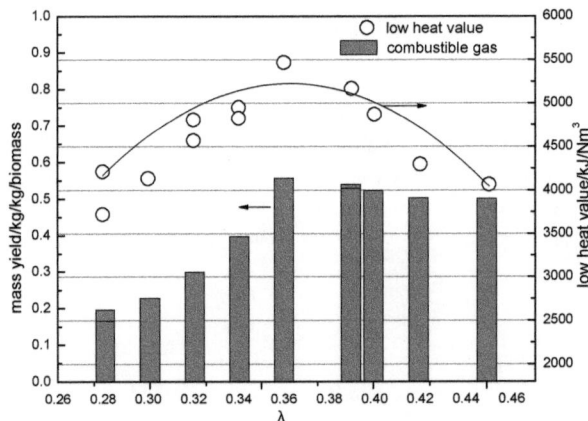

Figure 9. Mass of combustible gas and low heat value of syngas.

The total combustible gas and LHV in syngas reached a maximum value of 0.556 kg/kg biomass and 5469.85 kJ/Nm^3 at λ = 0.36 in the current study. Under these conditions, the temperature in the second region was comparatively high. Since the oxygen concentration was not too high, the combustible gases would not be consumed by combustion. It is also found that when λ exceeded 0.36, the environmental temperature rose up to a high level. Therefore, the oxidation of produced fuel gas would reduce the heat value of the syngas. It is clear that λ = 0.36 is a threshold point for improving or deteriorating syngas quality in this study. Several factors contributed to this transformation, such as the high decomposition rate of biomass in an oxidative environment, higher tar destruction efficiency, and more reactive char for gasification.

3.4. Tar Yield and Composition

Biomass tar sampled from sample port 1 (before conditioning) and 2 (after conditioning) were analyzed carefully. The total tar yield variation with λ is shown in Figure 10. It was observed that when λ increased from 0.28 to 0.45, the tar yield was reduced from 404 mg/(kg biomass) to 238 mg/(kg biomass). In the present gasifier, biomass tar was mainly formed at the pyrolysis stage. After that, tar decomposition and conversion happened in the second and third regions. During the tar conversion process, large molecular species not only cracked but polymerized to form high molecular weight aromatics. When the oxygen concentration in the first region was increased, tar production was suppressed. It is reported in the literature that under an oxidative environment, a large number of free radicals are formed [26]. These high reactive radicals will attack the weak bonds between tar molecules, hence pyrolytic tar would easily decompose. Increasing λ resulted in a temperature rise in the second and third regions as shown in Figure 6. A high temperature environment is beneficial for tar decomposition. Moreover, biomass char can act as an adsorbent and catalyst for tar conversion as described before. This effect would be enhanced under higher temperatures. Looking at the above factors, the total biomass tar production was reduced by an increase of λ.

Figure 10 also shows that tar yield obviously decreased after the gas conditioning process. Although the scrubbing efficiency slightly decreased as the input tar concentration was reduced, the average efficiency could be kept at around 65%. This indicated that the gas conditioning system has good load regulation ability.

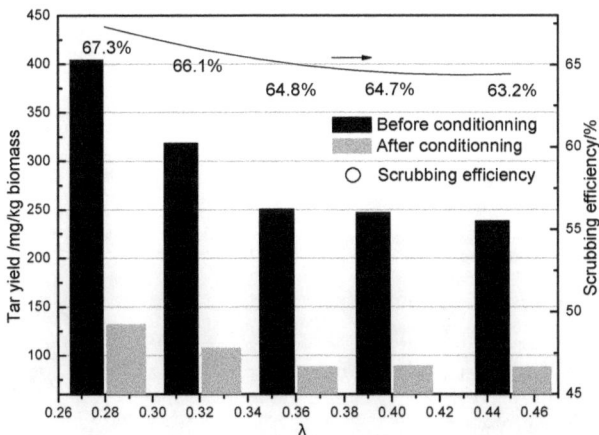

Figure 10. Tar concentration variation with λ.

Figure 11 shows a GC chromatogram illustrating the composition of a typical tar sample from sampling port 1. Each remarkable peak was marked with the corresponding species. The highest peak corresponded to benzene, followed by toluene, styrene, phenol and naphthalene. Benzene, toluene and xylene have lower boiling points. These components are not problematic for further usage. In addition, these species could increase the syngas lower heating value and anti-knocking properties during combustion in a gas engine. However, these tar components easily condense below 300 °C, which could cause fouling in the gas duct and other process facilities.

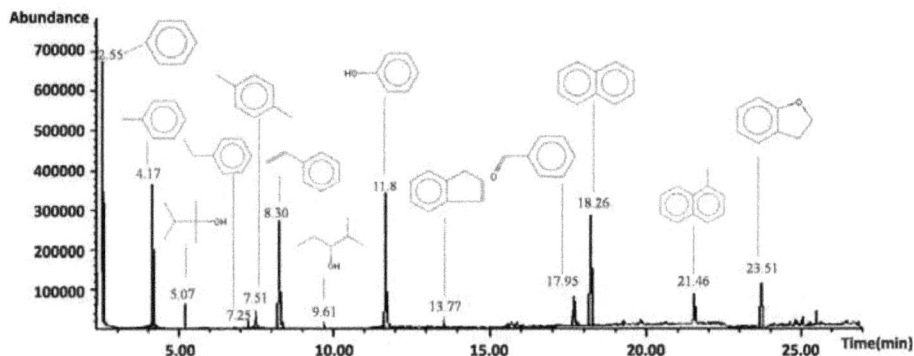

Figure 11. GC chromatogram of a typical sample from sampling port 1.

Tar composition and species relative concentrations are listed in Table 2. The relative concentrations are calculated by the corresponding peak area minus the sum of the areas of all the detected peaks. The listed 13 species represent about 85% of the total detected tar components. Among all the detected species, benzene was the most abundant one. Since benzene is the simplest aromatic with singe ring, any side chain breakage or decomposition of aromatics would produce benzene. It can be seen from Table 2 that when λ increased from 0.28 to 0.45, its relative concentration increased from 27.22% to 48.73%. This implies that tar decomposition and side chain breakage were favored under higher λ. Besides benzene, toluene and naphthalene are the other two most abundant components. When λ increased from 0.28 to 0.45, their concentration rose up from 9.76% and 6.75% to 15.32% and 13.52%, respectively. In comparison, toluene and naphthalene have more thermo-stable structures.

Table 2. Tar concentration and species relative concentration (%).

No.	Molecular Formula	Species Name	Equivalent Ratio (λ)					
			0.28	0.32	0.36		0.40	0.45
					Before Conditioning	After Conditioning		
1	C_6H_6	Benzene	37.22	39.21	43.25	41.35	45.64	48.73
2	C_7H_8	Toluene	9.76	9.86	12.46	12.06	14.63	15.32
3	$C_8H_{16}O_2$	2-Butanol, 2,3-dimethyl-	2.45	2.75	2.01	/	1.65	1.1
4	C_8H_{10}	Ethylbenzene	1.21	1.12	0.46	/	0.73	0.37
5	C_8H_{10}	*p*-Xylene	1.12	0.42	0.52	/	0.35	0.26
6	C_8H_8	Styrene	6.53	5.72	4.36	1.68	2.1	2.41
7	$C_6H_{14}O$	3-Pentanol, 2-methyl-	0.87	0.46	0.63	/	0.37	0.43
8	C_6H_6O	Phenol	9.24	6.26	2.68	1.72	2.74	2.35
9	C_9H_8	Indene	0.55	0.94	1.12	/	1.42	1.32
10	$C_7H_6O_3$	Benzaldehyde	2.13	0.96	0.86	/	0.87	0.91
11	C_8H_{10}	Naphthalene	6.75	10.14	12.56	31.32	13.22	13.52
12	$C_{11}H_{10}$	Naphthalene, 1-methyl-	2.31	2.41	2.33	3.17	2.42	2.51
13	C_8H_8O	Benzofuran, 2,3-dihydro-	4.31	4.14	0.85	/	0.67	0.72
	Percentage of above species (with benzene) from all detected species %		84.45	84.39	84.09	91.3	86.81	89.95
	Percentage of above species (without benzene) from all detected species %		47.23	45.18	40.84	49.95	41.17	41.22

The amounts of phenol and its derivatives were comparatively less. Since hydroxyls easily break away from benzene rings under high temperatures and in an oxidative environment, phenolic products could be converted to benzene easily. As seen from Table 2, as λ increased from 0.28 to 0.45, the phenol concentration decreased from 9.24% to 2.35%. Likewise, benzofuran decreased from 4.31% to 0.72%. Other oxygen-containing tar components presented a decreasing trend with increase of λ as well.

Tar composition variation before and after conditioning is illustrated in Table 2. It can be seen that after gas conditioning, the number of species was reduced remarkably. Six detected species accounted for up more than 90% of the total tar. The primary constituents are benzene, toluene, styrene, phenol, naphthalene and 1-methylnaphthalene.

3.5. Gas Production Rate and Carbon Conversion Rate

Gas production rate and carbon conversion rates are defined by Equations (4) and (5).
Gas production rate:

$$\text{gas production rate} = \frac{\text{syngas flow rate} \ \left(\text{Nm}^3/\text{h}\right)}{\text{biomass feeding rate} \ (\text{kg/h})} \tag{4}$$

Carbon conversion rate:

$$\text{carbon conversion rate} = \frac{\text{carbon element in syngas} \ (\text{kg})}{\text{carbon element in feedstock} \ (\text{kg})} \times 100\% \tag{5}$$

As shown in Figure 12, the gas production rate increased continuously with λ. This is because gas production rate takes no account of heat value. The increase of gas production rate after λ = 0.36 is mainly due to a rapid production of CO_2 (Figure 7). Similarly, the carbon conversion rate displays an increasing trend with λ as well. Under the conditions when λ was below 0.36, the carbon conversion rate increased rapidly. When λ was greater than 0.36, the rate of increased was slightly slowed down. As λ increased from 0.28 to 0.36, more air was injected; gasification and oxidation reactions were enhanced. As a result, a large number of combustible gases were produced. With further air being injected, the fuel gas would be combusted, since homogeneous oxidation reactions are faster than heterogeneous oxidation reactions under a high temperature [23]. However, homogeneous oxidation will not affect the carbon conversion rate, therefore, an increase of λ from 0.36 to 0.45 would have a weaker influence on carbon conversion rates than the condition when λ < 0.36.

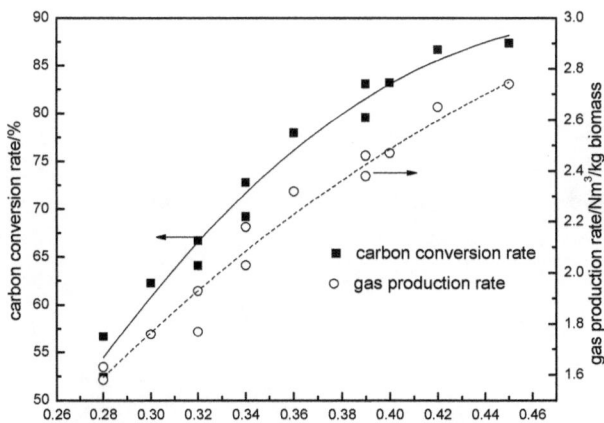

Figure 12. Carbon conversion rate and gas production rate.

3.6. Cold Gas Efficiency

Cold gas efficiency indicates the energy conversion efficiency from biomass feedstock to syngas. It is defined by Equation (6):

Cold gas efficiency:

$$\text{cold gas efficiency} = \frac{\text{syngas low heat value } (kJ/Nm^3) \times \text{ gas production rate } (Nm^3/kg)}{\text{biomass low heat value } (kJ/kg)} \times 100\% \quad (6)$$

A comparison of the theoretical and experimental cold gas efficiency results is plotted in Figure 13. The theoretical thermodynamic equilibrium calculations based on Gibbs free energy minimization is chosen to model the biomass gasification process. For the equilibrium calculations, the software program FactSage 5.4.1 has been used. FactSage is an integrated thermochemical databank system consisting of calculation modules and databases. It enables the user to access and manipulate pure substance and solution databases and perform thermochemical calculations such as multiphase chemical equilibria. It was seen that under equilibrium conditions there is a general increase in the cold gas efficiency with λ. The maximum value of 79% was obtained at λ = 0.32. This peak value is higher than the operational results and occurs at a lower λ. The highest cold gas efficiency of 75% can be achieved when λ was 0.36 under the operation conditions. Since the equilibrium model assumes an isothermal and long enough residence time to achieve equilibrium, the error may be attributed to the non-isothermal temperature inside the gasifier and the short residence time.

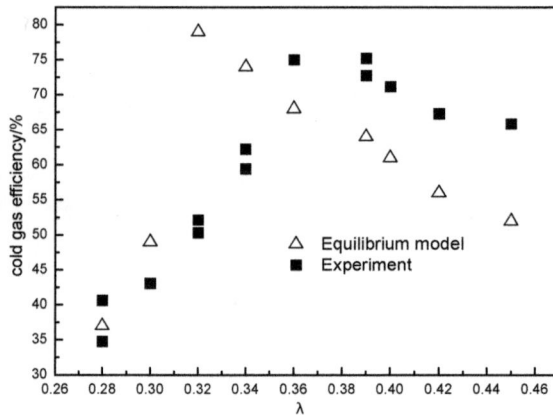

Figure 13. Cold gas efficiency of theoretical *vs.* experimental results.

3.7. Energy Balance Analysis

The total energy balance is calculated using Equations (7) to (10):

$$E_{input} = E_{biomass} + E_{air} + E_{motor} \quad (7)$$

$$E_{output} = E_{syngas} + E_{solid\ residue} + E_{condensation} \quad (8)$$

where E = mass × heat value:

$$\text{Energy conversion efficiency} = \frac{E_{syngas}}{E_{input}} \times 100\% \quad (9)$$

$$\text{Error} = \frac{E_{input} - E_{output}}{E_{input}} \times 100\% \quad (10)$$

Figure 14 shows the schematic diagram of the system with the energy input and output of products at each stage. Total energy input was determined by biomass feeding rate, total air injection and motor power. In the present test, the energy input was 1420.4 kW. The syngas energy was 1057.5 kW, so the energy conversion efficiency is 74.45%. This was due to heat losses from the system and also the heat reserved in the char residue. Besides, some of the energy is also lost in the gas conditioning system through the scrubbing and condensation processes. The characteristics of the solid residue are listed in Table 3.

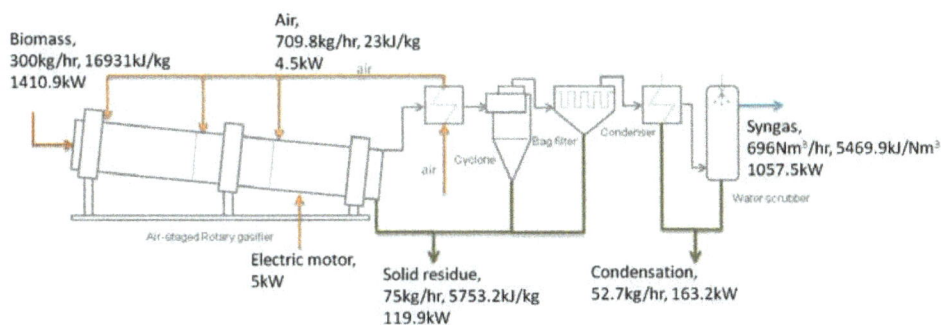

Figure 14. Energy input and output of the gasification system.

Table 3. Proximate and ultimate analysis of the solid residue.

Proximate Analysis [1] (wt %)				Ultimate Analysis [1] (wt %)					Low Heat Value (kJ/kg)
Water	Ash	Volatile	Fixed Carbon	C	H	O	N	S	
1.35	88.29	0	10.36	5.93	3.17	0.12	0.24	0.31	5753.2

[1] As received basis.

The heat value of the solid residue was determined by a bomb calorimeter test. The average heat value is 5753.2 kJ/kg. The solid residue ash production rate is 75 kg/h, which is measured by regular weighing of all the ash from the gasifier, cyclone and bag filter (shown in Figure 14). As syngas comes in contact with the cyclone internal walls, the temperature drops and tar components along with water vapor may condense. In this process, ash and char would contact with the condensates, causing the ash and char to become wet, decreasing its calorific value. The condensation heat is measured by monitoring the scrubbing and cooling water temperature variation as well as their flow rate. The heat inputs and outputs drawn in Figure 14 were not totally balanced. The error of 4.95% could be due to system heat dissipation and other possible heat losses that could not be measured.

3.8. Comparison of Performance of Biomass Gasifier

In Table 4 the experimental results are compared with those reported in the literature. Generally, fluidized bed gasifiers provide an intensive contact between gas and solid biomass which results in higher reaction rates and conversion efficiencies. Fixed bed gasifiers typically have a lower heat and mass transfer and often generate higher amounts of char. The peak temperature in a fixed bed system is also lways higher than in a fluidized bed gasifier, therefore, higher heat value syngas is produced. A recent review [14] reported that multi-stage gasifiers show an obvious advantage compared with other types. The separation of heating and drying, pyrolysis, oxidation and gasification processes makes it possible to control and optimize the different steps separately. Multi-stage gasifiers produce high purity gas with low levels of tar and high process efficiencies. In the present study, the rotary kiln gasifier combined both the advantages of fixed bed and fluidized bed gasifiers to form a new multi-stage gasifier concept, which has not been reported in any of the previous work. This gasifier provides a more intensive mixture of gas and solid fuel than a fixed bed one, meanwhile, its temperature is higher than that of a fluidized bed one. Table 4 shows that the present gasifier has a comparatively higher heat value of gas with low tar concentration.

Table 4. Comparison of the experimental results with published in literature.

Biomass Fuel	Low Heat Value of Syngas MJ/m^3	Tar Concentration (g/m^3)	Cold Gas Efficiency (%)	Gasifier Type	Reference
Biomass/waste	2.9–4.9	10–40/0.01–0.9/1~29.1–56.9/<70	fluidized bed	[14,27,28]
Agrol, willow, and DDGS, sawdust	2.7–4.3	5–12/1~30	50-70	Circulated fluidized bed	[14,28–30]
Sesame wood and rose wood/hazelnut shell	4.5–6.3/4~5	0.015–0.5/0.01~6	30–80.91	Downdraft fixed bed	[14,29,31–33]
Sewage sludge and woodpellets	4.9–5.3/5~6	30–150/10~150	20–60	Updraft fixed bed	[14,29,33–35]
Straw, manure fibres, sewage sludge	5.2~7/6.6	4.8/<0.015	87%~93%	Multi-stage	[14,36]
Waste wood	5.5	0.038	75	Rotary kiln	present study

4. Conclusions

A new gasification process concept was demonstrated in a pilot scale gasification system. This concept is based on air-staged injection with separated three thermo-chemical conversion regions inside the gasifier. Results showed that under optimum conditions, the gas production rate and syngas heat value could reach 2.32 Nm^3/kg and 5469.85 kJ/Nm^3, respectively. Tar concentration before and after gas conditioning could be reduced to 108 mg/Nm3 and 38 mg/Nm^3, respectively. Cold gas efficiency and carbon conversion efficiency could reach about 75% and 78%. A comparison of this study with literature data shows that a higher calorific value gas with lower tar concentration is produced in the present gasifier. The performance of this gasification system showed a considerable potential for implementing it in distributed electricity and heat supply projects.

Acknowledgments: This work was carried out with financial supports from National Natural Science Foundation of China (Grant No. 61171008 and No. 21103024).

Author Contributions: Huiyuan Shi and Wen Si conceived and designed the experiments; Huiyuan Shi performed the experiments; Wen Si and Xi Li analyzed the data; Huiyuan Shi wrote the paper.

Conflicts of Interest: The authors declare no conflict of interest.

References

1. Saxena, R.C.; Seal, D.; Kumar, S. Thermo-chemical routes for hydrogen rich gas from biomass: A Review. *Renew. Sustain. Energy. Rev.* **2008**, *12*, 1909–1927. [CrossRef]
2. Balat, M.; Balat, M.; Kırtay, E.; Balat, H. Main routes for the thermo-conversion of biomass into fuels and chemicals. Part 2: Gasification Systems. *Energy Convers. Manag.* **2009**, *50*, 3158–3168. [CrossRef]
3. Fredriksson, H.O.A.; Lancee, R.J.; Thüne, P.C.; Veringa, H.J.; Niemantsverdriet, J.W. Olivine as tar removal catalyst in biomass gasification: Catalyst dynamics under model conditions. *Appl. Catal. B Environ.* **2013**, *130–131*, 168–177. [CrossRef]
4. Virginie, M.; Adánez, J.; Courson, C.; de Diego, L.F.; García-Labiano, F.; Niznansky, D.; Kiennemann, A.; Gayán, P.; Abad, A. Effect of Fe–olivine on the tar content during biomass gasification in a dual fluidized bed. *Appl. Catal. B Environ.* **2012**, *121–122*, 214–222. [CrossRef]

5.   Barisano, D.; Freda, C.; Nanna, F.; Fanelli, E.; Villone, A. Biomass gasification and in-bed contaminants removal: Performance of iron enriched Olivine and bauxite in a process of steam/O$_2$ gasification. *Bioresour. Technol.* **2012**, *118*, 187–194. [CrossRef] [PubMed]
6.   Kirkels, A.F.; Verbong, G.P.J. Biomass gasification: Still promising? A 30-year global overview. *Renew. Sustain. Energy Rev.* **2011**, *15*, 471–481. [CrossRef]
7.   Milne, T.A.; Evans, R.J. *Biomass Gasifier"Tars": Their Nature, Formation, and Conversion*; National Renewable Energy Laboratory: Golden, CO, USA, 1998.
8.   Goyal, H.B.; Seal, D.; Saxena, R.C. Bio-fuels from thermochemical conversion of renewable resources: A Review. *Renew. Sustain. Energy Rev.* **2008**, *12*, 504–517. [CrossRef]
9.   Kirubakaran, V.; Sivaramakrishnan, V.; Nalini, R.; Sekar, T.; Premalatha, M.; Subramanian, P. A review on gasification of biomass. *Renew. Sustain. Energy Rev.* **2009**, *13*, 179–186. [CrossRef]
10. Evans, R.J.; Milne, T.A. Molecular characterization of the pyrolysis of biomass. 1. Fundamentals. *Energy Fuels* **1987**, *1*, 123–138. [CrossRef]
11. Wei, L.; Xu, S.; Zhang, L.; Zhang, H.; Liu, C.; Zhu, H.; Liu, S. Characteristics of fast pyrolysis of biomass in a free fall reactor. *Fuel Process. Technol.* **2006**, *87*, 863–871. [CrossRef]
12. Chen, Y.; Luo, Y.; Wu, W.; Su, Y. Experimental investigation on tar formation and destruction in a lab-scale two-stage reactor. *Energy Fuels* **2009**, *23*, 4659–4667. [CrossRef]
13. Houben, M.P. Analysis of tar removal in a partial oxidation burner. Ph.D. Thesis, Eindhoven University of Technology, Eindhoven, The Netherlands, 2004.
14. Heidenreich, S.; Foscolo, P.U. New concepts in biomass gasification. *Prog. Energy Combust. Sci.* **2015**, *46*, 72–95. [CrossRef]
15. Pan, Y.G.; Roca, X.; Velo, E.; Puigjaner, L. Removal of tar by secondary air injection in fluidized bed gasification of residual biomass and coal. *Fuel* **1999**, *78*, 1703–1709. [CrossRef]
16. Narv, P.I.; Orpso, A.; Aznar, M.P.; Corella, J. Biomass gasification with air in an atmospheric bubbling fluidized bed. Effect of six operational variables on the quality of produced raw gas. *Ind. Eng. Chem. Res.* **1996**, *35*, 2110–2120.
17. Yoshikawa, K. R&D (Research and Development) on distributed power generation from solid fuels. *Energy* **2006**, *31*, 1656–1665.
18. Henriksen, U.; Ahrenfeldt, J.; Jensen, T.K.; Gobel, B.; Bentzen, J.D.; Hindsgaul, C.; Sorensen, L.H. The design, construction and operation of a 75 kW two-stage gasifier. *Energy* **2006**, *31*, 1542–1553. [CrossRef]
19. El-Hendawy, A.-N.A. Surface and adsorptive properties of carbons prepared from biomass. *Appl. Surf. Sci.* **2005**, *252*, 287–295. [CrossRef]
20. Liu, X.; Li, W.; Xu, H.; Chen, Y. A comparative study of non-oxidative pyrolysis and oxidative cracking of cyclohexane to light alkenes. *Fuel Process. Technol.* **2004**, *86*, 151–167. [CrossRef]
21. Wang, H.; Dlugogorski, B.Z.; Kennedy, E.M. Coal oxidation at low temperatures: Oxygen consumption, oxidation products, reaction mechanism and kinetic modeling. *Prog. Energy Combust. Sci.* **2003**, *29*, 487–513. [CrossRef]
22. Putzeys, O.; Bar-Ilan, A.; Rein, G.; Fernandez-Pello, A.C.; Urban, D.L. The role of secondary char oxidation in the transition from smoldering to flaming. *Proc. Comb. Inst.* **2007**, *31*, 2669–2676. [CrossRef]
23. Senneca, O.; Chirone, R.; Salatino, P.; Nappi, L. Patterns and kinetics of pyrolysis of tobacco under inert and oxidative conditions. *J. Anal. Appl. Pyrolysis* **2007**, *79*, 227–233. [CrossRef]
24. Yi, S.U. Study on Product Properties and Reaction front Propagation Charateristics during Oxidative Pyrolysis of Biomass. Ph.D. Thesis, Shanghai Jiao Tong University, Shanghai, China, 2013.
25. Wang, H.; Dlugogorski, B.Z.; Kennedy, E.M. Analysis of the mechanism of the low-temperature oxidation of coal. *Comb. Flame* **2003**, *134*, 107–117. [CrossRef]
26. Hayes, C.J.; Merle, J.K.; Hadad, C.M. The chemistry of reactive radical intermediates in combustion and the atmosphere. In *Advances in Physical Orgnanic Chemistry*; Richard, J.P., Ed.; Academic Press: New York, NY, USA, 2009; pp. 79–134.
27. Gómez-Barea, A.; Leckner, B.; Perales, A.V.; Nilsson, S.; Cano, D.F. Improving the performance of fluidized bed biomass/waste gasifiers for distributed electricity: A New Three-Stage gasification System. *Appl. Therm. Eng.* **2013**, *50*, 1453–1462. [CrossRef]
28. Han, J.; Kim, H. The reduction and control technology of tar during biomass gasification/pyrolysis: An Overview. *Renew. Sustain. Energy Rev.* **2008**, *12*, 397–416. [CrossRef]

29. Meng, X.; de Jong, W.; Fu, N.; Verkooijen, A.H.M. Biomass gasification in a 100 kW$_{th}$ steam-oxygen blown circulating fluidized bed gasifier: Effects of operational conditions on product gas distribution and tar formation. *Biomass Bioenergy* **2011**, *35*, 2910–2924. [CrossRef]

30. Li, X.T.; Grace, J.R.; Lim, C.J.; Watkinson, A.P.; Chen, H.P.; Kim, J.R. Biomass gasification in a circulating fluidized bed. *Biomass Bioenergy* **2004**, *26*, 171–193. [CrossRef]

31. Sheth, P.N.; Babu, B.V. Experimental studies on producer gas generation from wood waste in a downdraft biomass gasifier. *Bioresour. Technol.* **2009**, *100*, 3127–3133. [CrossRef] [PubMed]

32. Dogru, M.; Howrath, C.R.; Akay, G.; Keskinler, B.; Malik, A.A. Gasification of hazelnut shells in a downdraft gasifier. *Energy* **2002**, *27*, 415–427. [CrossRef]

33. IEA (International Energy Agency). Task33: Small Scale Gasification: Gas Engine CHP for Biofuels, 2011. Swedish Energy Agency Report. Available online: http://www.ieatask33.org/app/ webroot/files/file/publications/new/Small%20Small_scale_gasification_overview.pdf (accessed on 29 December 2015).

34. Seggiani, M.; Vitolo, S.; PuccIni, M.; Bellini, A. Cogasification of sewage sludge in an updraft gasifier. *Fuel* **2012**, *93*, 486–491. [CrossRef]

35. Plis, P.; Wilk, R.K. Theoretical and experimental investigation of biomass gasification process in a fixed bed gasifier. *Energy* **2011**, *36*, 3838–3845. [CrossRef]

36. Zwart, R.; van der Heijden, S.; Emmen, R.; Bentzen, J.D.; Ahrenfeldt, J.; Stoholm, P.; Krogh, J. Tar Removal from Low-Temperature Gasifiers, 2010. ECN Report. Available online: http://www.ecn.nl/ docs/library/reort/2010/e10008.pdf (accessed on 29 December 2015).

energies

MDPI

Article

Computational Fluid Dynamic Analysis of Co-Firing of Palm Kernel Shell and Coal

Muhammad Aziz [1],*, Dwika Budianto [2],† and Takuya Oda [1],†

[1] Advanced Energy Systems for Sustainability, Tokyo Institute of Technology, Tokyo 152-8550, Japan; oda@ssr.titech.ac.jp

[2] Agency for the Assessment and Application of Technology (BPPT), Jakarta 10340, Indonesia; dwikabudianto@gmail.com

* Correspondence: aziz.m.aa@m.titech.ac.jp; Tel.: +81-3-5734-3809

† These authors contributed equally to this work.

Academic Editor: Tariq Al-Shemmeri
Received: 22 December 2015; Accepted: 18 February 2016; Published: 26 February 2016

Abstract: The increasing global demand for palm oil and its products has led to a significant growth in palm plantations and palm oil production. Unfortunately, these bring serious environmental problems, largely because of the large amounts of waste material produced, including palm kernel shell (PKS). In this study, we used computational fluid dynamics (CFD) to investigate the PKS co-firing of a 300 MWe pulverized coal-fired power plant in terms of thermal behavior of the plant and the CO_2, CO, O_2, NO_x, and SO_x produced. Five different PKS mass fractions were evaluated: 0%, 10%, 15%, 25%, and 50%. The results suggest that PKS co-firing is favorable in terms of both thermal behavior and exhaust gas emissions. A PKS mass fraction of 25% showed the best combustion characteristics in terms of temperature and the production of CO_2, CO, and SO_x. However, relatively large amounts of thermal NO_x were produced by high temperature oxidation. Considering all these factors, PKS mass fractions of 10%–15% emerged as the most appropriate co-firing condition. The PKS supply capacity of the palm mills surrounding the power plants is a further parameter to be considered when setting the fuel mix.

Keywords: co-firing; palm kernel shell (PKS); coal; computational fluid dynamics (CFD); mass fraction; temperature; exhaust gases

1. Introduction

The demand for palm oil and palm oil products has increased following rapid economic development in China, India, and South East Asian countries. Indonesia, Malaysia, and Thailand are the largest producers of palm oil and palm kernel oil, accounting for more than 80% of the total world production [1]. Indonesia has seen a considerable increase in palm oil production, with an annual growth of about 10% during 2002–2009 [2]. This reflects massive expansion in palm plantations, particularly in Sumatera and Kalimantan. The total plantation area is projected to reach approximately 13 million hectares by 2020 [3]. The export of palm oil, palm kernel oil, and oil products has become one of Indonesia's main sources of income. Palm oil and palm kernel oil are extracted from the mesocarp fiber and palm kernel, respectively. The former is rich in unsaturated acids (palmitic, oleic, and linoleic), while the latter is rich in saturated acids (lauric and myristic) [4].

Unfortunately, palm plantations and oil production have brought serious environmental problems including greenhouse gas emissions (GHGs), land conversion, and the production of huge amounts of agricultural waste. Basiron [5] reported that only about 10% of the whole palm tree is used for palm oil production. The solid waste includes the stripped fruit bunches, fiber, and palm kernel shell (PKS).

Fiber is often used as fuel to generate both the electricity and steam required for milling. The stripped fruit bunches are generally recycled back to the plantation as mulch to maintain the nutrient cycle and prevent soil erosion. Compared with other solid wastes, PKS has the advantage of a higher calorific value and lower moisture and ash contents [6]. It is recovered during the extraction of the palm kernel after the palm oil has been recovered from the mesocarp fiber. As the percentage of PKS is about 5%–7% of the fresh fruit bunch (FFB), the total potential of PKS in Indonesia has been estimated at about 54.87 GJ· y^{-1} [7].

Methods of utilizing biomass for power generation through combustion can be divided into dedicated firing and co-firing. Biomass co-firing has advantages compared with dedicated biomass burning when used in existing power plants, especially coal-fired power plants. These include lower capital costs and higher combustion efficiency [8]. The addition of biomass in a coal-fired power plant can significantly reduce GHG emissions and reduce slagging inside the combustor. Newly-built coal fired power plants in some European countries, Japan, and China are largely co-fired, with biomass accounting for 10%–20% of output on a calorie basis [9]. Studies of biomass combustion have also progressed significantly, using both experiments and numerical analysis. Karim and Naser [10] reviewed the combustion of biomass in packed beds, and identified emerging trends. Modeling of combustion technologies including coal and biomass co-firing has been described in detail in [11].

In Indonesia, electricity is mainly supplied by coal. Indonesia is one of the biggest coal exporters, and domestic coal reserves are estimated to be sufficient to meet future demand for about 80 years [12]. Most domestic coal is classified as low rank coal (LRC), with a heating value near that of biomass. LRC also has a high moisture content, a low sulfur content, high reactivity, and a low calorific value [13–15]. Co-firing of biomass can help improve domestic energy security as well as increasing the share of renewable biomass used in power plants initially designed for coal burning. Indonesia has released a vision for renewable energy, called Vision 25/25, in which renewable energy will meet 25% of the total energy demand in 2025 [16].

As PKS is widely available in Indonesia, the development of PKS co-firing in existing or planned coal-fired power plants has great potential. Unfortunately, almost no studies have been conducted to evaluate PKS co-firing. In this study, computational fluid dynamics (CFD) was used to analyze the thermal behavior of PKS co-fired coal-fired power plants, and the composition of the exhaust gases CO_2, CO, O_2, NO_x, and SO_x.

2. Computational Modeling of Palm Kernel Shell Co-Firing

Co-firing of biomass in a coal-fired combustor can be performed using different methods, including injection, co-milling, pre-gasification, and parallel co-firing. Injection co-firing, in which pre-milled biomass is mixed with the pulverized coal fed to the combustor, is considered to be the most feasible method, because of its relatively low capital cost and high co-firing ratio [17]. Injection co-firing was therefore chosen for investigation in this study. It was assumed that the PKS is injected with the pulverized coal into the combustor, after being dried and ground separately.

Figure 1 shows a basic schematic diagram of a PKS and coal system employing injection co-firing. It comprises three modules, one each for drying, combustion, and power generation. The raw wet coal is ground into small particles before being fed to the dryer. A fluidized bed or rotary dryer may be used, considering solid mixing, heat and mass transfer, and temperature uniformity [18,19]. The CO_2-rich exhaust gas from the power generation module is fed to the dryer as the drying medium. In a fluidized bed dryer, the dried coal particles rise to the top of the bed due to their lower density, resulting in overflow of the dry coal particles.

The coal and PKS are then blended and fed to the combustor using air as the feed gas. The heat from combustion is recovered by a superheater and economizer and used to generate steam for the turbine. Flue gas with a relatively low exergy rate is used to supply heat to the drying module and as the fluidizing gas in the bed.

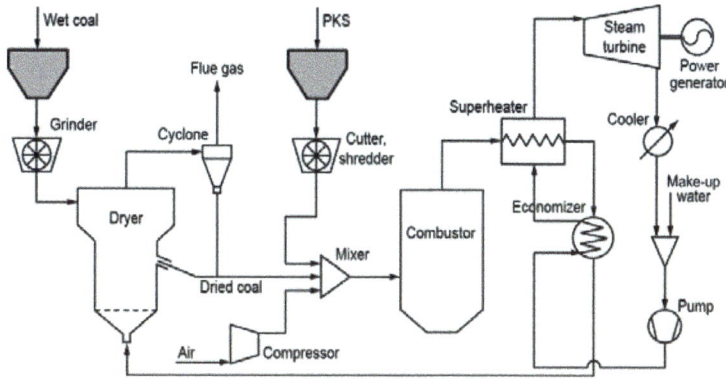

Figure 1. Basic schematic diagram of the integrated co-firing of palm kernel shell (PKS) and coal in the drying, combustion, and power generation modules.

Figure 2 shows a schematic view of the combustor used in this study, including its dimensions and the layout of the meshing and feeding inlets in cross section. The feed inlet was a coaxial dual tube in which the fuel was fed through the inner tube. The combustor was based on those of existing 300 MWe coal-fired power plants. The combustor had a height, width, and breadth of 45 m, 12 m, and 15 m, respectively. Commercial CFD software was used to create the simulation. ANSYS Design Modeler was used to build the combustor model in 3D, and ANSYS Fluent version 16.2 (ANSYS Inc., Concord, MA, USA) was used to analyze the co-firing behavior. The co-firing simulation took account of governing equations (mass, momentum, and enthalpy), turbulence, radiative heat transfer, and reactions in both the particle and gas phases.

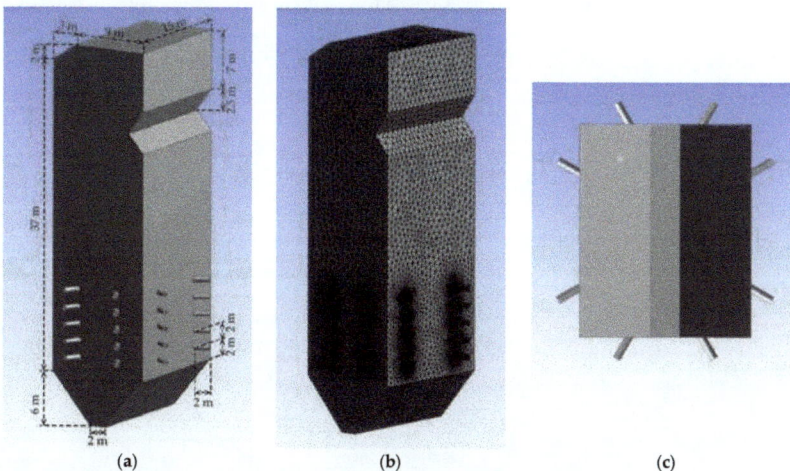

Figure 2. Schematic diagram of the combustor design used in this study: (a) combustor dimensions; (b) meshing layout; and (c) inlet feed distribution (cross section).

The study modeled the co-firing behavior of PKS and coal using the CFD method, as this is an effective tool for calculating fluid flows, heat and mass transfers, chemical reactions, and solid and fluid interactions [20,21]. CFD modeling is significantly more time and cost effective than physical investigation through experiments, and is also safe and easy to scale up. It is therefore often used as a precursor to experimental studies. When applied to co-firing, CFD analysis was expected to clarify the combustion process including the combustion temperature and the concentrations of the gases produced.

Coal and PKS particles entered the combustor through feeding inlets distributed at the lower side. Air was used as the feeding gas as well as supplying the oxygen required for combustion. There were two different flows: primary and secondary. The former acted as the feeding gas for the fuel and flowed through the inner tube, while the latter supplied the extra air required for combustion and flowed through the outer tube. Once the particles entered the combustor, continuous reactions occurred, including heating, drying, devolatilization, gas and char combustion, pollutant formation, and heat radiation [22]. The numerical calculation continued until the convergent condition was achieved (less than 10^{-4} for all residuals). The combination of velocity and pressure in the Navier-Stokes equations was calculated using the SIMPLE algorithm, which is a semi-implicit method for pressure linked equations. The Eulerian-Lagrangian approach was used to solve the gas-solid two-phase flow, and gas phase modeling was performed in the Eulerian domain by solving the steady-state Reynolds averaged Navier-Stokes (RANS) equations.

In this study, the initial coal flowrate $(\text{kg}\cdot\text{s}^{-1})$ required for combustion was stipulated as the baseline rate from which the flowrate of the PKS and coal was calculated using the designated PKS mass fraction. Five different PKS mass fractions were evaluated: 0% (100% coal), 10%, 15%, 25%, and 50%. The temperature distribution and the concentration of the gases produced during co-firing were observed.

2.1. Governing Equations

Co-firing is typically approximated and modeled as a dilute two-phase (solid and gas) flow using the Eulerian-Lagrangian approach, in which the gas phase is approximated through the Navier-Stokes model, while the solid phase is treated as a discrete phase. The trajectory of each particle was modeled by Newton's laws of motion and particle collision is treated as a sphere model [23]. The gas concentration and temperature of the particles were calculated using the energy and mass transfer equations. The interactions of mass, momentum and energy between the gas and solid particles were solved using the particle-in-cell (PIC) approach considering the particle state along the particle trajectories.

The mathematical calculations were essentially governed by the flow of fluid and the heat transfer. It was assumed that the fluid dynamics represented a viscous flow, so that each governing equation can be written as follows:

$$\frac{D\rho}{Dt} + \rho \nabla \cdot \vec{U} = 0 \tag{1}$$

$$\rho \frac{Du}{Dt} = -\frac{\partial p}{\partial x} + \frac{\partial \tau_{xx}}{\partial x} + \frac{\partial \tau_{yx}}{\partial y} + \frac{\partial \tau_{zx}}{\partial z} + \rho f_x \tag{2}$$

$$\rho \frac{Dv}{Dt} = -\frac{\partial p}{\partial y} + \frac{\partial \tau_{xy}}{\partial x} + \frac{\partial \tau_{yy}}{\partial y} + \frac{\partial \tau_{zy}}{\partial z} + \rho f_y \tag{3}$$

$$\rho \frac{Dw}{Dt} = -\frac{\partial p}{\partial z} + \frac{\partial \tau_{xz}}{\partial x} + \frac{\partial \tau_{yz}}{\partial y} + \frac{\partial \tau_{zz}}{\partial z} + \rho f_z \tag{4}$$

$$\begin{aligned}
\frac{D}{Dt}\left(e + \frac{U^2}{2}\right) = {} & \rho\dot{q} + \frac{\partial}{\partial x}\left(k\frac{\partial T}{\partial x}\right) + \frac{\partial}{\partial y}\left(k\frac{\partial T}{\partial y}\right) + \frac{\partial}{\partial z}\left(k\frac{\partial T}{\partial z}\right) - \frac{\partial(up)}{\partial x} - \frac{\partial(vp)}{\partial y} - \frac{\partial(wp)}{\partial z} \\
& + \frac{\partial(u\tau_{xx})}{\partial x} + \frac{\partial(u\tau_{yx})}{\partial y} + \frac{\partial(u\tau_{zx})}{\partial z} + \frac{\partial(v\tau_{xy})}{\partial x} + \frac{\partial(v\tau_{yy})}{\partial y} + \frac{\partial(v\tau_{zy})}{\partial z} \\
& + \frac{\partial(w\tau_{xz})}{\partial x} + \frac{\partial(w\tau_{yz})}{\partial y} + \frac{\partial(w\tau_{zz})}{\partial z} + \rho \vec{f}\vec{U}
\end{aligned} \tag{5}$$

where, ρ, $\dot{q}$, U, t, p, τ, f, and e are density, rate of volumetric heat addition per unit mass, velocity, time, pressure (Reynolds averaged), stress tensor (viscous and Reynolds stresses), the component of body force per unit mass, and the internal energy per unit mass, respectively. In addition, u, v, and w are the velocity in each corresponding x, y, and z direction. Equation (1) represents the mass (continuity equation) in non-conservation form. Equations (2)–(4) represent the momentum (Newton's second law) equation in non-conservation form in each component of the x, y, and z directions. Equation (5) is the energy (first law of thermodynamics) equation in non-conservation form [24,25].

2.2. Turbulence

The flow in the combustor during co-firing was turbulent due to fluid inertia, including time-dependent and convective acceleration, and characterized by fluctuations in velocity due to the complex geometry and high flow rates. Turbulence influences both heat and mass transfer and was therefore an important factor when modeling co-firing inside the combustor. In this study the k-ε turbulence model, which is one of the standard models used in Fluent, was adopted to solve the RANS equations used to model the co-firing. In CFD modeling, this k-ε turbulence model is widely used to calculate the swirling combustion flows. It is a robust and easy model to implement [19] and is used in many industrial applications [26]. Two main equations in this model relates to turbulent kinetic energy k and the turbulent dissipation rate ε. They can be expressed by Equations (6) and (7), respectively [27]:

$$\frac{\partial}{\partial t}(\rho\,k) + \frac{\partial}{\partial x_i}(\rho\,k\,u_i) = \frac{\partial}{\partial x_j}\left[\left(\mu + \frac{\mu_t}{\sigma_k}\right)\frac{\partial k}{\partial x_j}\right] + P_k + P_b - \rho\varepsilon - Y_M + S_k \qquad (6)$$

$$\frac{\partial}{\partial t}(\rho\,\varepsilon) + \frac{\partial}{\partial x_i}(\rho\,\varepsilon\,u_i) = \frac{\partial}{\partial x_j}\left[\left(\mu + \frac{\mu_t}{\sigma_\varepsilon}\right)\frac{\partial \varepsilon}{\partial x_j}\right] + C_{1\varepsilon}\frac{\varepsilon}{k}(P_k + C_{3\varepsilon}\,P_b) - \rho\,C_{2\varepsilon}\frac{\varepsilon^2}{k} + S_\varepsilon \qquad (7)$$

where μ_t, P_k, P_b, and Y_m represent turbulent viscosity, the production of the k term, the effect of the buoyancy term, and the contribution of dilatation fluctuation to the overall dissipation rate, respectively. Both S_k and S_ε are user-defined source terms while μ_t, P_k, and P_b were calculated using Equations (8)–(10):

$$\mu_t = \rho C_\mu \frac{k^2}{\varepsilon} \qquad (8)$$

$$P_k = -\overline{\rho u_i' u_j'}\frac{\partial u_j}{\partial x_i} \qquad (9)$$

$$P_b = \beta g_i \frac{\mu_t}{Pr_t}\frac{\partial T}{\partial x_i} \qquad (10)$$

$$\beta = -\frac{1}{\rho}\left(\frac{\partial \rho}{\partial T}\right)p \qquad (11)$$

where, g_i, T, and Pr_t are acceleration due to gravity in the ith direction, temperature, and the turbulent Prandtl number, respectively. Here, β is the coefficient of thermal expansion that can be derived by Equation (11). $C_{1\varepsilon}$, $C_{2\varepsilon}$, C_μ, σ_k, and σ_ε are constants used in this study, and their values were as follows: $C_{1\varepsilon}$ = 1.44; $C_{2\varepsilon}$ = 1.92; C_μ = 0.09; σ_k = 1.0; σ_ε = 1.3.

C_{3e} was calculated from the following equation:

$$C_{3\varepsilon} = \tanh\frac{|v_b|}{|u_b|} \qquad (12)$$

where, v_b and u_b are velocity components that are parallel and perpendicular to the gravitational vector, respectively.

2.3. Radiation

Heat transfer through radiation becomes dominant when the combustion temperature is relatively high. It governs both heat transfer and heat flux, especially during heating, drying, ignition, devolatilization, and char combustion. In this work, the P-1 radiation model was adopted to solve the radiative heat transfer, based on the expansion of radiation intensity to an orthogonal series of spherical harmonics [20]. The radiation model can be expressed as follows:

$$\nabla \cdot (\Gamma\,\nabla G) = (a + a_p)\,G - 4\pi\left(a\frac{\sigma\,T^4}{\pi} + E_p\right) \qquad (13)$$

$$\Gamma = 1/3 \left(a + a_p + \sigma_p \right) \tag{14}$$

$$G = 4 \, \sigma \, T^4 \tag{15}$$

$$a_p = \lim_{V \to 0} \sum_{n=1}^{N} \varepsilon_{pn} \frac{A_{pn}}{V} \tag{16}$$

$$E_p = \lim_{V \to 0} \sum_{n=1}^{N} \varepsilon_{pn} A_{pn} \frac{\sigma \, T_{pn}^4}{\pi \, V} \tag{17}$$

$$\sigma_p = \lim_{V \to 0} \sum_{n=1}^{N} \left(1 - f_{pn} \right) \left(1 - \varepsilon_{pn} \right) \frac{A_{pn}}{V} \tag{18}$$

where, G, σ, a, a_p, E_p, V, and σ_p are the incident radiation, the Stefan-Boltzmann constant, the absorption coefficient, the equivalent absorption coefficient due to the presence of particulates, the equivalent emission, the volume, and the equivalent particle scattering factor, respectively. In addition, ε_{pn}, A_{pn}, T_{pn}, and f_{pn} are emissivity, the projected area, the temperature, and a scattering factor associated with particle n.

2.4. Particle Phase Reaction Mechanisms

The mixture of PKS and coal can be considered as a typical gas-solid flow in which chemical reactions occur. The hydrodynamics can be approximated using the Eulerian-Lagrangian model. The particles were modeled separately as two discrete phase models. Some reactions occurred in the particle phase, especially during char combustion, which was assumed to be char oxidation producing CO, which was then released to the bulk gas in the combustor. It is important to note that char from biomass is generally more reactive than char from coal, and has a higher heating rate. A global 1-step heterogeneous reaction mechanism was adopted in this study to approximate char combustion under air. The combustion reactions for the coal char and PKS char can be expressed as follows:

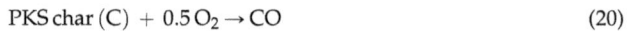

$$\text{coal char (C)} + 0.5\,O_2 \to CO \tag{19}$$

$$\text{PKS char (C)} + 0.5\,O_2 \to CO \tag{20}$$

2.5. Gas Phase Reaction Mechanisms

During devolatilization, the volatile matter from both PKS and coal are released and then react with oxygen, inducing further combustion. The oxidation reaction of volatile matter from both PKS and coal was approximated as a global 2-step reaction mechanism in which CO is an intermediate component [28]. PKS and coal were treated as different components, and both composition and enthalpy formation were derived from proximate and ultimate analyses of each material. The reaction mechanisms for the volatile matter in the gas phase can be expressed by the Westbrook-Dryer mechanism [29]:

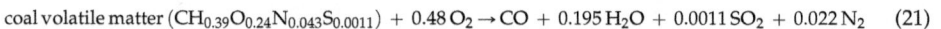

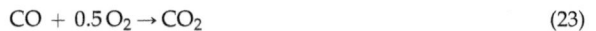

$$\text{coal volatile matter} \left(CH_{0.39}O_{0.24}N_{0.043}S_{0.0011} \right) + 0.48\,O_2 \to CO + 0.195\,H_2O + 0.0011\,SO_2 + 0.022\,N_2 \tag{21}$$

$$\text{PKS volatile matter} \left(CH_{0.8}O_{0.725}N_{0.0174} \right) + 0.3375\,O_2 \to CO + 0.4\,H_2O + 0.0087\,N_2 \tag{22}$$

$$CO + 0.5\,O_2 \to CO_2 \tag{23}$$

The kinetic reaction rate coefficient can be expressed as Equation (24), which is based on the Arrhenius equation for activation energy E_i. Table 1 shows the kinetic parameters used in the study for each corresponding reaction in both the particle and gas phases [22,30,31]. The coal was assumed to be lignite.

$$k = A \, e^{-\frac{E_i}{RT}} \tag{24}$$

Table 1. Kinetics parameters used in both particle and gas phases for each corresponding reaction [22,30,31].

Reaction Equation	A (s^{-1})	E_i (J·kmol^{-1}·K^{-1})
(19)	2.0×10^{11}	4.4×10^{7}
(20)	6.8×10^{15}	1.67×10^{8}
(21)	3.0×10^{8}	1.26×10^{8}
(22)	1.9×10^{15}	1.27×10^{8}
(23)	2.75×10^{9}	8.47×10^{7}

3. Calculation Conditions

The composition of the coal and PKS particles assumed in this study is shown in Table 2. The coal was sourced from Kalimantan, Indonesia, which is classified as LRC and has relatively high moisture content. The PKS was sourced from palm mills located in Sumatera, Indonesia. The as-received and as-used coal compositions represent the condition at the drying inlet when received from the mine, and at the combustion inlet after being dried. Initial drying was performed to reduce the moisture content of the coal from 48.76 wt % to 17.30 wt %. The PKS is as-used PKS because it was received from the palm mill and used for co-firing without any initial pre-treatment except crushing.

Table 2. Material composition of coal and PKS particles used in the study.

Component	Properties	Coal		PKS
		As-Received	As-Used	As-Used
Proximate analysis (wt %)	Fixed carbon	24.93	40.23	24.35
	Volatile matter	25.76	41.57	66.77
	Moisture	48.76	17.30	3.86
	Ash	0.56	0.90	5.02
Ultimate analysis (wt %)	Carbon	35.30	56.98	43.77
	Hydrogen	2.29	3.69	5.85
	Oxygen	11.23	18.13	42.32
	Nitrogen	1.75	2.83	0.89
	Sulfur	0.11	0.17	0.00
LHV (MJ·kg^{-1})	-		13.84	22.33	17.68

The simulated coal-fired power station had a power generation capacity of 300 MWe with a basic combustion efficiency of 30%. The flow rates of coal and air under ambient condition were 73 kg·s^{-1} and 630 kg·s^{-1}, respectively. Each particle was assumed to be a solid sphere, with sizes in the range 60–200 mesh (74–250 µm). The bulk densities of the coal and PKS were assumed to be 700 kg·m^{-3} and 600 kg·m^{-3}, respectively. The ambient temperature, combustor wall thickness, and external and internal emissivity coefficients were set at 300 K, 0.2 m, 0.9, and 0.6, respectively. The air was assumed to contain 79 mol % N$_2$ and 21 mol % O$_2$.

The total mesh of the 3D model used to represent the combustor was an approximately 1,805,305 tetrahedral cell unstructured grid. In the CFD modeling, the temperature distribution and concentration of the exhaust gases (CO$_2$, CO, O$_2$, NO$_x$, and SO$_x$) were observed, and their cross-sectional profiles were plotted on an *xy* plane at intervals of 6 m in height (the *z* axis), with the lowermost plot located 12 m above the base of the combustor. Six plot profiles were drawn, representing the distribution of each observed profile. Modeling was performed on a work station with a Quad-core Intel Core i7 2.9 GHz CPU and 16 GB of random-access memory (RAM).

4. Results and Discussion

4.1. Temperature Distribution

Figure 3 shows the temperature profile across the combustor for each corresponding PKS mass fraction. In general, higher PKS mass fractions produced higher flame temperatures inside the combustor as well as in the exhaust gases flowing to the superheater for energy recovery. The flame temperature became more uniform as the height in the combustor increased. Figure 4 shows the temperature distribution at the center of the combustor for each different PKS mass fraction. A higher PKS mass fraction resulted in a higher total volatile matter content during co-firing. The average temperature in the upper part of combustor, especially the freeboard, was lower than that of the lower part. This was probably a result of heat loss across the combustor.

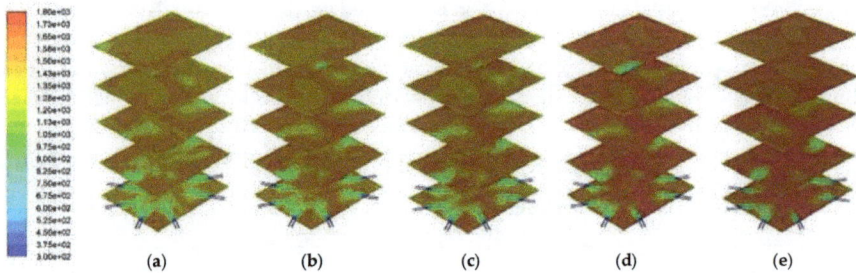

Figure 3. Cross-sectional temperature distribution profile across the combustor: (**a**) 0% PKS; (**b**) 10% PKS; (**c**) 15% PKS; (**d**) 25% PKS; and (**e**) 50% PKS.

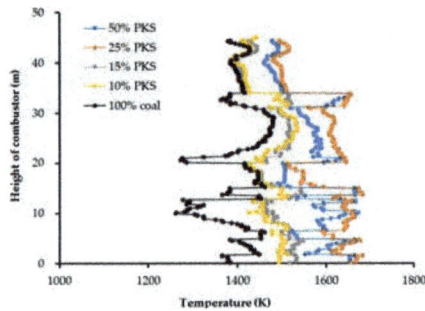

Figure 4. Temperature distribution at the center along the height of the combustor for each different PKS mass fraction.

As can be observed from Table 2, the PKS was higher in volatile matter but had a lower moisture content than the coal. As the coal used was LRC, it also had a relatively low calorific value. The lower moisture content of the PKS meant that devolatilization of the PKS particles was faster and earlier than devolatilization of the coal particles. The coal particles required a longer drying time before being devolatilized. As a result, at high PKS mass fractions, the combustion temperature was higher and more uniformly distributed in the lower part of the combustor where the feeding inlets were located. The lower moisture content of the PKS also meant that increasing the PKS mass fraction reduced the total moisture content of the mixed fuel. This affected the combustion temperature, as water has a relatively high heat capacity.

Co-firing with a PKS mass fraction of 25% produced a higher combustion temperature than a mass fraction of 50%, although the difference was insignificant. The highest center of combustor temperature was 1,683 K, recorded with a 25% PKS, and followed by 50% PKS (1675 K), 15% PKS (1548 K), 10% PKS (1536 K), and 0% PKS (1481 K). The average temperature at the combustor outlet for PKS mass fractions of 0%, 10%, 15%, 25%, and 50% were 1390 K, 1414 K, 1422 K, 1513 K, and 1494 K, respectively. In contrast to the moisture content, a higher PKS mass fraction reduced the total amount of fixed carbon (including char) in the mixed fuel, and the heat obtained from char combustion fell as the PKS mass fraction increased. The temperature distribution profile suggested that the optimum combustion performance of coal and PKS co-firing is achieved at a PKS mass fraction of 25%.

Large random temperature changes occurred at heights of less than 20 m, probably due to factors including the particle dynamics and random stages of combustion. The former is related both to the distribution of the feeding inlets, which ranged in height from 8 m to 16 m, and to the different physical properties of the coal and PKS particles. The latter is correlated with the different stages of combustion for different fuel particles. According to Williams *et al.* [32], combustion of coal and other solid fuels (including biomass) occurs in four stages: drying, devolatilization, char combustion, and ash formation. These stages of combustion are different for different fuels. This can be observed not only in the temperature distribution, but in other distributions discussed below.

4.2. Distribution of CO_2, CO, and O_2

Figures 5 and 6 represent the cross-sectional CO_2 distribution profile and CO_2 distribution at the center of the combustor. A higher PKS mass fraction resulted in a lower CO_2 concentration in the exhaust gas. Numerically, the average CO_2 mass fraction in the exhaust gas at the combustor outlet for PKS mass fractions of 0%, 10%, 15%, 25%, and 50% were 0.262, 0.256, 0.253, 0.246, and 0.220, respectively. In the reactions shown in (19–23), the carbon is mainly converted to CO_2. As coal has a higher carbon content than PKS, the total amount of carbon decreased as the PKS mass fraction increased.

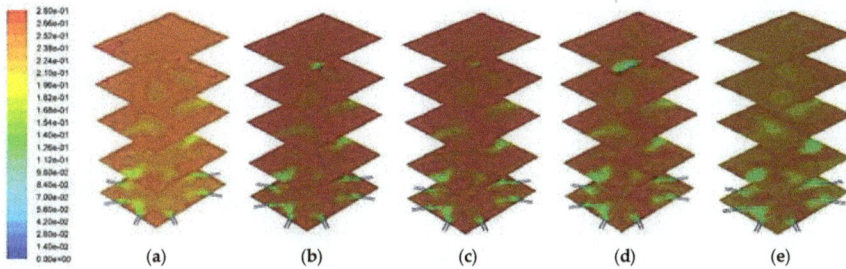

Figure 5. Cross-sectional CO_2 distribution profile across the combustor: (a) 0% PKS; (b) 10% PKS; (c) 15% PKS; (d) 25% PKS; and (e) 50% PKS.

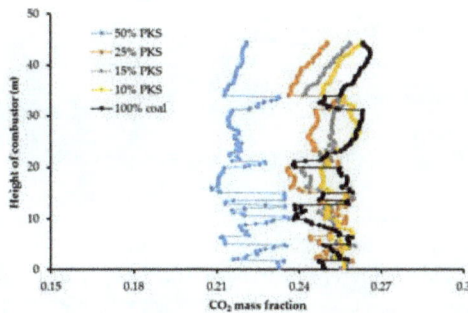

Figure 6. CO_2 distribution at the center along the height of the combustor at each different PKS mass fraction.

CO_2 also has a lower specific volume (higher density) than gases such as N_2 and O_2, leading to a higher volumetric heat capacity and a swirling tendency in the flame [33]. A higher mass fraction of CO_2 was found in the lower part of combustor. As a result, a higher PKS mass fraction produced less turbulence, and slightly increased the gas velocity in the combustor. The average gas velocities at PKS mass fractions of 0%, 10%, 15%, 25%, and 50% were 22.9 m·s^{-1}, 23.1 m·s^{-1}, 23.5 m·s^{-1}, 24.2 m·s^{-1}, and 25.1 m·s^{-1}, respectively.

Figures 7 and 8 show the cross-sectional CO distribution profile and its distribution at the center of combustor, respectively. As the PKS mass fraction increased, the mass fraction of CO decreased. The average CO mass fractions at the combustor outlet for PKS mass fractions of 0%, 10%, 15%, 25%, and 50% were 0.0041, 0.0017, 0.0014, 0.0008, and 0.0002, respectively. When co-firing with PKS, almost no CO remained in the upper part of the combustor, suggesting that all the CO produced during devolatilization was burned during combustion. Coal has a higher carbon content than PKS and produces a higher mass fraction of CO during devolatilization. In coal firing without PKS, this CO remains in the flue gas, although at very small concentrations. This causes environmental problems. Co-firing with PKS therefore has the additional advantage of eliminating the CO from the exhaust gas.

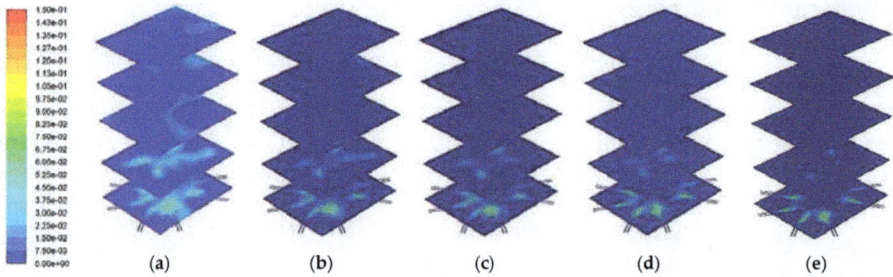

Figure 7. Cross-sectional CO distribution profile across the combustor: (a) 0% PKS; (b) 10% PKS; (c) 15% PKS; (d) 25% PKS; and (e) 50% PKS.

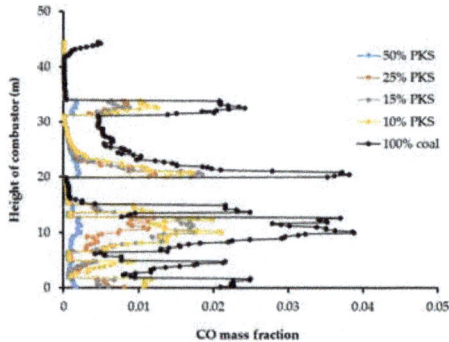

Figure 8. CO distribution at the center along the height of the combustor at each different PKS mass fraction.

Figures 9 and 10 show the cross-sectional oxygen distribution profile and the oxygen distribution at the center of the combustor at different PKS mass fractions. A higher PKS mass fraction led to a higher oxygen concentration. This is because PKS has a relatively high oxygen content, part of which persists and is exhausted together with the nitrogen and other flue gases. The averaged mass fraction of O_2 at the combustor outlet for PKS mass fractions of 0%, 10%, 15%, 25%, and 50% were 0.0017, 0.008, 0.011, 0.017, and 0.039, respectively.

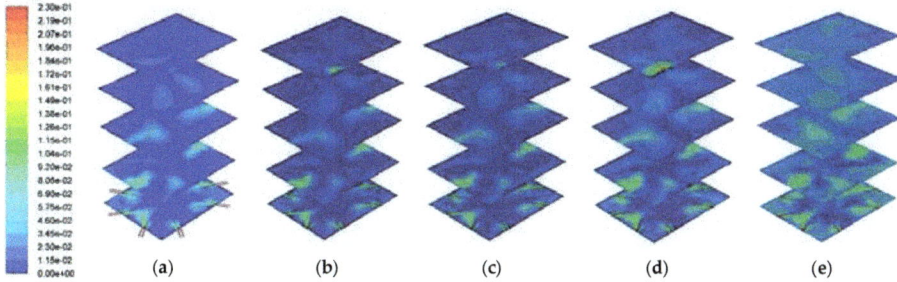

Figure 9. Cross-sectional O_2 distribution profile across the combustor: (**a**) 0% PKS; (**b**) 10% PKS; (**c**) 15% PKS; (**d**) 25% PKS; and (**e**) 50% PKS.

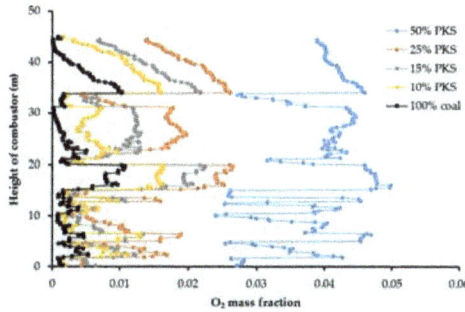

Figure 10. O_2 distribution at the center along the height of the combustor at each different PKS mass fraction.

4.3. Distribution of NO_x and SO_x

Figures 11 and 12 represent the cross sectional NO_x distribution profile and its distribution at the center of the combustor. A higher PKS mass fraction produced a higher NO_x concentration. Numerically, the average NO_x mass fractions at the combustor outlet for PKS mass fractions of 0%, 10%, 15%, 25%, and 50% were 1.2×10^{-9}, 5.5×10^{-9}, 1.2×10^{-8}, 2.7×10^{-7}, and 3.9×10^{-7}, respectively. Although the PKS contained less nitrogen than coal (Table 1), NO_x production increased in line with the PKS mass fraction, and was most pronounced at PKS mass fractions of 25% and 50%. It is known that there are three different primary sources of NO_x from combustion: thermal, fuel, and prompt NO_x. In this study, thermal NO_x was considered to account for most of the exhausted NO_x, and this was related to the PKS mass fraction.

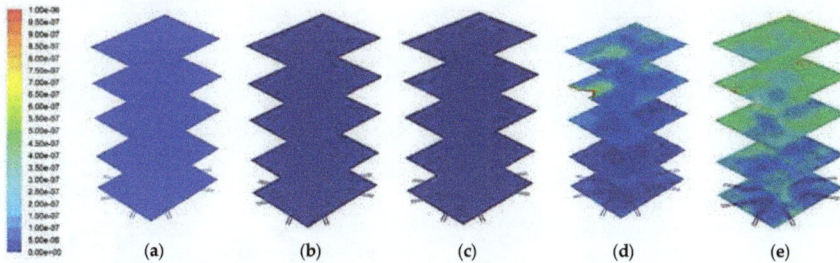

Figure 11. Cross-sectional NO_x distribution profile across the combustor: (**a**) 0% PKS; (**b**) 10% PKS; (**c**) 15% PKS; (**d**) 25% PKS; and (**e**) 50% PKS.

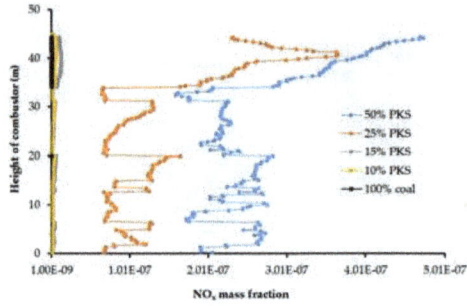

Figure 12. NO_x distribution at the center along the height of the combustor at each different PKS mass fraction.

The NO_x concentration increased significantly as the PKS mass fraction increased to 25%. According to the Zeldovich mechanism [34], thermal NO_x is generated at temperatures above 1600 K. The maximum combustion temperature at PKS mass fractions of 25% and 50% exceeded this value. This suggested that a lower PKS, of up to 15%, is the appropriate co-firing condition for limiting NO_x emissions.

Figures 13 and 14 show the cross sectional SO_2 distribution profile and its distribution at the center of the combustor at different PKS mass fractions. A higher PKS mass fraction was associated with a significant reduction in the SO_2 concentration. PKS has almost no sulfur content, in contrast with coal, so increasing the PKS mass fraction has the advantage of reducing the amount of SO_2 in the exhaust gas.

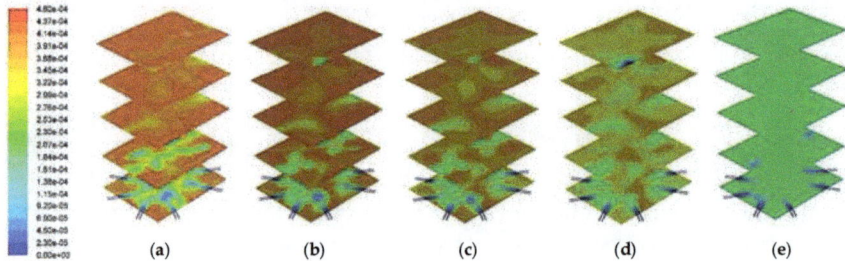

Figure 13. Cross-sectional SO_x distribution profile across the combustor: (**a**) 0% PKS; (**b**) 10% PKS; (**c**) 15% PKS; (**d**) 25% PKS; and (**e**) 50% PKS.

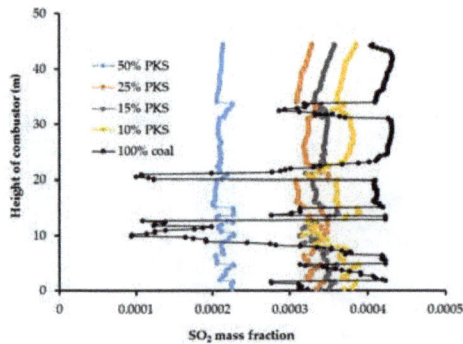

Figure 14. SO_x distribution at the center along the height of the combustor at each different PKS mass fraction.

5. Conclusions

Co-firing of PKS in a pulverized coal power plant was studied using CFD. We examined a range of co-firing behavior including the temperature distribution and the concentrations of CO_2, CO, O_2, NO_x, and SO_x in the exhaust gases. Inclusion of 25% PKS was shown to improve the combustion temperature and the concentrations of CO_2, CO, and SO_x in the flue gas. However, it produced relatively large amounts of NO_x, which is considered an atmospheric pollutant. Balancing these parameters, PKS mass fractions of 10% and 15% emerged as the appropriate co-firing conditions.

However, it is also very important to take account of the PKS supply from the surrounding palm mills. Given that the average palm mill capacity in Indonesia is 60 t-FFB h^{-1} and the ratio of produced PKS is 7%, PKS would need to be collected from about 10 palm mills to supply sufficient PKS for co-firing at 15% PKS and from 7 palm mills for co-firing at 10% PKS.

Co-firing of PKS and coal should be conducted at a pilot experimental scale before being applied in real power plants. For the future, we plan to undertake theoretical and experimental studies of the use of other solid wastes from palm mills, including empty fruit bunches.

Acknowledgments: The authors deeply thank New Energy Foundation (NEF), Japan, for the financial support through the Researcher Invitation Program 2015 through which this research has been conducted. The proximate and ultimate analyses of PKS and coal are provided by Agency for the Assessment and Application of Technology (BPPT), Indonesia.

Author Contributions: Dwika Budianto modeled and performed the CFD analysis for co-firing; Muhammad Aziz created the overall research plan and results analysis; Muhammad Aziz and Dwika Budianto wrote the manuscript; and Takuya Oda analyzed the results and checked the manuscript.

Conflicts of Interest: The authors declare no conflict of interest.

Abbreviations

The following abbreviations are used in this manuscript:

CFD	Computational fluid dynamics
FFB	Fresh fruit bunch
LRC	Low rank coal
PIC	Particle-in-cell
PKS	Palm kernel shell
RANS	Reynolds averaged Navier-Stokes
SIMPLE	Semi-implicit method for pressure linked equations
VM	Volatile matter

References

1. Aziz, M.; Oda, T.; Kashiwagi, T. Innovative steam drying of empty fruit bunch with high energy efficiency. *Dry. Technol.* **2015**, *33*, 395–405. [CrossRef]
2. Aziz, M.; Prawisudha, P.; Prabowo, B.; Budiman, B.A. Integration of energy-efficient empty fruit bunch drying with gasification/combined cycle systems. *Appl. Energy* **2015**, *139*, 188–195. [CrossRef]
3. Miettinen, J.; Hoiijer, A.; Tollenaar, D.; Page, S.; Malins, C.; Vernimmen, R.; Shi, C.; Liew, S.C. *Historical Analysis and Projection of Oil Palm Plantation Expansion on Peat Land in Southeast Asia*; ICCT White Paper No. 17; International Council on Clean Transportation: Washington, DC, USA, 2012.
4. Aziz, M.; Oda, T.; Kashiwagi, T. Design and analysis of energy-efficient integrated crude palm oil and palm kernel oil processes. *J. Jpn. Inst. Energy* **2015**, *94*, 143–150. [CrossRef]
5. Basiron, Y. Palm oil production through sustainable plantations. *Eur. J. Lipid Sci. Technol.* **2007**, *109*, 289–295. [CrossRef]
6. Bazargan, A.; Rough, S.L.; McKay, G. Compaction of palm kernel shell biochars for application as solid fuel. *Biomass Bioenergy* **2014**, *70*, 489–497. [CrossRef]

7. Febriansyah, H.; Setiawan, A.A.; Suryopratomo, K.; Setiawan, A. Gama stove: Biomass stove for palm kernel shell in Indonesia. *Energy Procedia* **2014**, *47*, 123–132. [CrossRef]
8. Griffin, W.M. Availability of biomass residues for co-firing in peninsular Malaysia: Implications for cost and GHG emissions in the electricity sector. *Energies* **2014**, *7*, 804–823. [CrossRef]
9. Livingston, W.R. Biomass ash and the mixed ashes from co-firing biomass with coal. In Proceedings of the IEA Clean Coal Workshop, Drax Power Station, London, UK, 25–26 January 2011.
10. Karim, M.R.; Naser, J. Progress in numerical modelling of packed bed biomass combustion. In Proceedings of the 19th Australasian Fluid Mechanics Conference, Melbourne, Australia, 8–11 December 2014.
11. Bhuiyan, A.B.; Karim, M.R.; Naser, J. Modeling of solid and bio-fuel combustion technologies. In *Thermofluid Modeling for Energy Efficiency Applications*; Khan, M.M.K., Hassan, N.M.S., Eds.; Elsevier: London, UK, 2016; pp. 259–309.
12. Indonesia Investments. Coal Mining in Indonesia. Available online: http://www.indonesia-investments.com/business/commodities/coal/item236 (accessed on 2 December 2015).
13. Aziz, M.; Kansha, Y.; Kishimoto, A.; Kotani, Y.; Liu, Y.; Tsutsumi, A. Advanced energy saving in low rank coal drying based on self-heat recuperation technology. *Fuel Process. Technol.* **2012**, *104*, 16–22. [CrossRef]
14. Aziz, M.; Oda, T.; Kashiwagi, T. Energy-efficient low rank coal drying based on enhanced vapor recompression technology. *Dry. Technol.* **2014**, *32*, 1621–1631. [CrossRef]
15. Liu, Y.; Aziz, M.; Kansha, Y.; Tsutsumi, A. A novel exergy recuperative drying module and its application for energy-saving drying with superheated steam. *Chem. Eng. Sci.* **2013**, *100*, 392–401. [CrossRef]
16. Saptoadi, H. The future of biomass energy in Indonesia. In Proceedings of the 2nd AUN/SEED-Net Regional Conference on Energy Engineering, Bangkok, Thailand, 13–14 November 2015.
17. Xu, W.; Niu, Y.; Tan, H.; Wang, D.; Du, W.; Hui, S. A new agro/forestry residues co-firing model in a large pulverized coal furnace: Technical and economic assessments. *Energies* **2013**, *6*, 4377–4393. [CrossRef]
18. Aziz, M.; Oda, T.; Kashiwagi, T. Advanced energy harvesting from algae—Innovative integration of drying, gasification and combined cycle. *Energies* **2014**, *7*, 8217–8235. [CrossRef]
19. Aziz, M.; Oda, T.; Kashiwagi, T. Clean hydrogen production from low rank coal: Novel integration of drying, gasification, chemical looping, and hydrogenation. *Chem. Eng. Trans.* **2015**, *45*, 613–618.
20. Ranade, V.V.; Gupta, D.F. *Computational Modeling of Pulverized Coal Fired Boilers*; CRC Press: Boca Raton, FL, USA, 2015; pp. 19–31.
21. Bhuiyan, A.A.; Naser, J. CFD modelling of co-firing of biomass with coal under oxy-fuel combustion in a large scale power plant. *Fuel* **2015**, *159*, 150–168. [CrossRef]
22. Tabet, F.; Gokalp, I. Review on CFD based models for co-firing coal and biomass. *Renew. Sustain. Energy Rev.* **2015**, *51*, 1101–1114. [CrossRef]
23. Oevermann, M.; Gerber, S.; Behrendt, F. Euler-Lagrange/DEM simulation of wood gasification in a bubbling fluidized bed reactor. *Particuology* **2009**, *7*, 307–316. [CrossRef]
24. Anderson, J.D. Governing Equations of Fluid Dynamics. In *Computational Fluid Dynamics: An Introduction*; Wendt, J., Ed.; Springer: Berlin, Germany, 2009; pp. 15–51.
25. Norton, T.; Sun, D.W.; Grant, J.; Fallon, R.; Dodd, V. Application of computational fluid dynamics (CFD) in the modelling and design of ventilation systems in the agricultural industry: A review. *Bioresour. Technol.* **2007**, *98*, 2386–2424. [CrossRef] [PubMed]
26. Bhuiyan, A.A.; Naser, J. Computational modelling of co-firing of biomass with coal under oxy-fuel condition in a small scale furnace. *Fuel* **2015**, *143*, 455–466. [CrossRef]
27. Launder, B.E.; Sharma, B.I. Application of the energy dissipation model of turbulence to the calculation of flow near a spinning disc. *Lett. Heat Mass Transf.* **1974**, *1*, 131–138. [CrossRef]
28. Yin, C.; Kaer, S.K.; Rosendahl, L.; Hvid, S.L. Modeling of pulverized coal and biomass co-firing in a 150 kW swirling stabilized burner and experimental validation. In Proceedings of the International Conference on Powder Engineering-09 (ICOPE-09), Kobe, Japan, 16–20 November 2009.
29. Westbrook, C.K.; Dryer, F.L. Simplified reaction-mechanisms for the oxidation of hydrocarbon fuels in flames. *Combust. Sci. Technol.* **1981**, *27*, 31–34. [CrossRef]
30. Sami, M.; Annamalai, K.; Wooldridge, M. Co-firing of coal and biomass fuel blends. *Prog. Energy Combust. Sci.* **2001**, *27*, 171–214. [CrossRef]

31. Ninduangdee, P.; Kuprianov, V.I.; Cha, E.Y.; Kaewrath, R.; Youngyuen, P.; Atthawethworawuth, W. Thermogravimetric studies of oil palm empty fruit bunch and palm kernel shell: TG/DTG analysis and modeling. *Energy Procedia* **2015**, *79*, 453–458. [CrossRef]

32. Williams, A.M.; Pourkashanian, M.; Jones, J.M. The combustion of coal and some other solid fuels. *Proc. Combust. Inst.* **2000**, *28*, 2141–2162. [CrossRef]

33. Khare, S.P.; Wall, T.F.; Farida, A.Z.; Liu, Y.; Moghtaderi, B.; Gupta, R.P. Factors influencing the ignition of flames from air-fired swirl pf burners retrofitted to oxy-fuel. *Fuel* **2008**, *87*, 1042–1049. [CrossRef]

34. Clean Air Technology Center (MD-12), Nitrogen Oxides (NO$_x$), Why and How They Are Controlled, Technical Bulletin of Environmental Protection Agency, EPA 456/F-99-006R, 1999, North Carolina, USA. Available online: http://www3.epa.gov/ttncatc1/cica/other7_e.html (accessed on 5 December 2015).

energies

MDPI

Article

Comparing the Bio-Hydrogen Production Potential of Pretreated Rice Straw Co-Digested with Seeded Sludge Using an Anaerobic Bioreactor under Mesophilic Thermophilic Conditions

Asma Sattar [1,†], Chaudhry Arslan [1,2,†], Changying Ji [1,*], Sumiyya Sattar [3], Irshad Ali Mari [4], Haroon Rashid [2] and Fariha Ilyas [5]

[1]   College of Engineering, Nanjing Agricultural University, Nanjing 210031, China;
      asma2005_2182@hotmail.com (A.S.); arslanakrampk@hotmail.com (C.A.)
[2]   Department of Structures and Environmental Engineering, University of Agriculture, Faisalabad 38000,
      Pakistan; haroon9a@gmail.com
[3]   Veterinary Research Institute, Lahore Cantt 54810, Pakistan; sumiyyasattar@hotmail.com
[4]   Khairpur College of Engineering and Technology Sindh Agriculture University, Khairpur Mir 66020,
      Pakistan; irshad_mari@hotmail.com
[5]   Department of Soil Science, Bahauddin Zakariya, University, Multan 60800, Pakistan;
      aafma12345@hotmail.com
*   Correspondence: chyji@njau.edu.cn; Tel.: +86-25-5860-6571
†   These authors contributed equally to this work.

Academic Editor: Tariq Al-Shemmeri
Received: 19 January 2016; Accepted: 8 March 2016; Published: 15 March 2016

Abstract: Three common pretreatments (mechanical, steam explosion and chemical) used to enhance the biodegradability of rice straw were compared on the basis of bio-hydrogen production potential while co-digesting rice straw with sludge under mesophilic (37 °C) and thermophilic (55 °C) temperatures. The results showed that the solid state NaOH pretreatment returned the highest experimental reduction of LCH (lignin, cellulose and hemi-cellulose) content and bio-hydrogen production from rice straw. The increase in incubation temperature from 37 °C to 55 °C increased the bio-hydrogen yield, and the highest experimental yield of 60.6 mL/g $VS_{removed}$ was obtained under chemical pretreatment at 55 °C. The time required for maximum bio-hydrogen production was found on the basis of kinetic parameters as 36 h–47 h of incubation, which can be used as a hydraulic retention time for continuous bio-hydrogen production from rice straw. The optimum pH range of bio-hydrogen production was observed to be 6.7 ± 0.1–5.8 ± 0.1 and 7.1 ± 0.1–5.8 ± 0.1 under mesophilic and thermophilic conditions, respectively. The increase in temperature was found useful for controlling the volatile fatty acids (VFA) under mechanical and steam explosion pretreatments. The comparison of pretreatment methods under the same set of experimental conditions in the present study provided a baseline for future research in order to select an appropriate pretreatment method.

Keywords: bio-hydrogen production; pretreatments; kinetic parameters; volatile fatty acids; response surface methodology

1. Introduction

Global energy demand is rising due to the industrialization and population growth. As fossil fuels are the dominant source of energy, the heavy reliance on fossil fuels is not only depleting them, but also contributing to climate change. In order to overcome this issue, efficient utilization of alternative energy sources, such as biomass, solar, wind and hydro, are getting more and more attention. Among all renewable sources, biomass is becoming an auspicious alternative due to near-carbon neutrality and ample availability [1].

In China, 0.75 billion tons of biomass energy resources were generated during the year 2010, out of which 52% was crop residue. One of the major shareholders in crop residue was rice straw, contributing 62% of total crop residue resources. About 1.35 tons of rice straw are produced for every ton of rice grain harvested, resulting in 1.9 million tons of rice straw production at 15% moisture content [2,3]. Although rice straw is used as a fuel for domestic purposes, a part of animal feed and in the paper making industry, still, a huge quantity of rice straw is left useless in the field. The burning of leftover straw in the open field causes serious environmental issues. Therefore, converting the rice straw into more valuable products, like methane, ethanol and bio-hydrogen, not only solves the issue of rice straw management, but also addresses the energy challenges faced by the world in recent times. Such conversion can be done by thermo-chemical means, like combustion, pyrolysis or liquefaction, which are not environmentally-friendly techniques. The other option of converting rice straw into a valuable energy resource is biological means, which include anaerobic fermentation [4]. No doubt, this biological technique is environmentally friendly, but it requires a variety of substrates for converting biomass into biofuels, which makes it the most promising option for treating lignocellulosic materials [5,6]. In this regard, bio-hydrogen production along with bio-ethanol and methane through anaerobic fermentation have great potential to develop a sustainable energy production system. Hawkes *et al.* [7] reported that bio-hydrogen production from such agricultural waste is more advantageous over other fuels, as hydrogen-producing microorganisms could consume a wide range of sugar hydrolysates as compared to other microbes. These sugar hydrolysates are available in rice straw in the form of cellulose and hemicellulose, entangled by the lignin moieties, which hinder the biological degradation of sugar content in rice straw [8]. To overcome this issue, pretreatment of rice straw is required to break the crystallinity of cellulose and the lignin seal [9]. Commination of lignocellulosic biomass is a traditional pretreatment method which changes the ultrastructure of rice straw. In this technique, a final particle size of 0.2–2 mm is achieved, which increases the surface area and reduces the cellulose crystallinity for better biodegradability [9,10]. Steam explosion is another widely-opted pretreatment method in which lignocellulosic biomass is exposed to 160–260 °C temperatures under 0.69–4.83 MPa for several seconds to a few minutes [11]. Under such conditions, hemicellulose is hydrolyzed into component sugars, and lignin is redistributed, which enhances the biodegradation process [12]. Although steam explosion and comminution are effective pretreatment techniques, still there is the need of a pretreatment technique that has less energy intake as compared to the techniques discussed above. In this regard, alkaline treatment is a simple and effective one, as it causes delignification, increases internal surface area and porosity, reduces crystallinity and the degree of polymerization and breaks down the links of polymers with lignin [13,14]. Apart from all these benefits, there are some environmental issues, like disposal and recycling of chemicals associated with alkaline treatment, which can be overcome by opting for solid state treatment instead of liquid state treatment [15]. Although NaOH, $Ca(OH)_2$, KOH and $NH_3 \cdot H_2O$ can be used for alkali treatment, NaOH is widely used for lignocellulosic biomass and especially for rice straw [13]. As a whole, much work has been done on different pretreatment methods, and every pretreatment method has its own merits and demerits under the tested conditions. It is difficult to compare the efficiency of these pretreatment methods from a review, as every study presented that the tested method is the optimum method for pretreating rice straw. Therefore, in order to address this issue, these methods need to be studied under similar conditions to compare the treatment efficiency, especially on the basis of bio-hydrogen production potential.

Bio-hydrogen production through anaerobic digestion cannot be done only with a pretreated rice straw, but also requires some source of microorganisms. In this regard, a mixed consortium of Clostridium is the best option, which is easily available in the form of sludge [16]. Although, sludge has some hydrogen consumers, like methanogens, along with hydrogen producers, which can be inactivated through heat treatment efficiently [17].

The following study was conducted to compare the effect of mechanical, thermal and chemical treatment on rice straw for bio-hydrogen production, co-digested with sludge under mesophilic and thermophilic conditions. The volatile solids, volatile fatty acids, soluble chemical oxygen demand and pH were also measured to observe different aspects of the fermentation process.

2. Material and Methods

2.1. Pretreatment of Rice

The rice straw was collected from Ba Bai Qiao experimental field of Nanjing Agricultural University, cut into short pieces and air dried. Later, three different pretreatments were performed on rice straw, *i.e.*, mechanical, chemical and thermal. In the mechanical treatment, straw was ground in a grinder (LH-08B Speed Grinder, CNC Instruments Inc.: Zhejieng, China), passed through a 2-mm sieve, and sieved straw was used for bio-hydrogen production.

In order to perform chemical treatment, the solid state NaOH pretreatment proposed by He *et al.* [15] was opted after some modification. The straw was first chopped by a specially-designed chopper and then ground into 5 mm–10 mm-sized particles. Later, 100 g of straw were mixed with 80 g of distilled water containing 6 g of NaOH and mixed thoroughly to make the resultant moisture content at 80% on a dry basis. Later, the straw was placed in a 1-L beaker for three weeks at room temperature. By the end of pretreatment, the straw was dried in an oven and stored in a refrigerator. The selected pretreatment has no environmental issues, and it does not require washing to remove leftover NaOH.

The steam explosion was done by chopping the straw into 3–4 cm-sized particles [18]. On the basis of initial moisture content, water was added to the straw, so that the total solids (TS) level could be maintained around 20% [19]. After adding water, straw was thoroughly mixed and left for 4–5 h, so that water is absorbed by the straw uniformly. Later, the straw was added into the steam explosion chamber (2 L) till it was half filled and sealed from the top. The saturated steam was added into the chamber till the temperature of the chamber reached 240 °C, after which, the timing of the reaction was started. After 240 s, the valve was opened, so that explosive depressurization could occur [20]. The resultant straw was collected and stored in bags.

2.2. Seeded Sludge

The sludge was obtained from a settling channel in Pokuo and was sieved and washed with tap water to remove dust and foreign materials [21]. Later, it was placed in a preheated oven at 100 °C for 30 min in order to deactivate hydrogenotrophic methanogens [22,23]. The volatile solids, volatile fatty acids, alkalinity and pH of the sludge were 2.87%, 13950 mg/L, 3700 mg/L and 7.1, respectively.

2.3. Anaerobic Bio-Reactor

In the present study, a 20-L stainless steel double jacket anaerobic bioreactor was developed in collaboration with Zhejiang Instruments Limited (Figure 1). The reactor was equipped with a proportional integral derivative (PID) controller (CAN-C700, Aivpen Instruments, Le Qing, China) to manage temperature with the help of a heating unit and a platinum resistance temperature sensor (PT-100). The flow through heating unit was controlled by a solenoid valve (D01-4104, YuYao Sanlixin Solenoid valve Co.: Shanghai, China) and water circulation pump (UP Basic, Grundfos: Sozhou, China) connected to a 100-L water reservoir. The pH was managed by another PID pH controller (PH900, Acitek Instrumentsm: Shanghai, China), pH sensor (Easyferm plus 120, Hamilton Bonaduz AG: Bonaduz, Switzerland) and a peristaltic pump to add the desired amount of chemical from a 500-mL glass bottle to maintain pH at specific points. Thorough mixing was done by a three-stage stirrer connected to a permanent magnet DC servo motor (ZSD05A, Shanghai ShuDong Motor Co., Ltd.: Shanghai, China). There was an inlet port (1-inch diameter) for feeding materials at the top, and an outlet port (1-inch diameter) along with a ball valve was at the bottom. A vacuum pump (FY-1H-N,

Zhejiang E & M Value Co.: Zhejiang, China) was also attached to the reactor to develop anaerobic conditions [24,25].

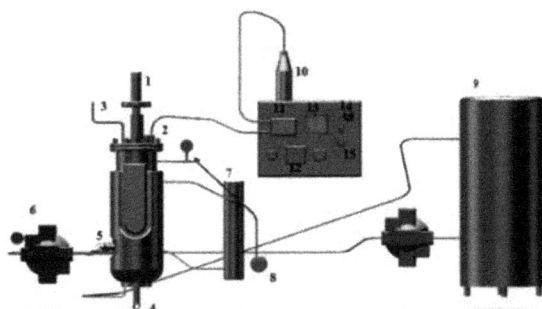

1- Stirring Motor, 2- Feeding Ports, 3- Gas Collecting Unit, 4- Outlet Port, 5- Temperature and pH Sensors
6- Water Circulation Pump, 7- Heating Unit, 8- Vacuum Pump, 9- Water Reservoir, 10- Chemical storage
11- Peristaltic Pump, 12- Temperature Controller, 13- pH Controller, 14- Stirrer Controller, 15- Power

Figure 1. Schematic diagram for the double jacket anaerobic bio-reactor.

2.4. Analytical and Assay Methods

The total solids (TS), volatile solids (VS), chemical oxygen demand (COD), volatile fatty acids (VFA) and alkalinity were measured by standard methods [26]. The volume of hydrogen gas was measured in the same way opted in our previous studies [24,25,27]. The compositional properties of straw were measured by the procedure opted by Ververis [28]. The bio-hydrogen production was modeled by the modified Gompertz equation for the determination of kinetic parameters [29]:

$$H = P\exp\left\{-\exp\left[\frac{R_m e}{P}(\lambda - t) + 1\right]\right\} \tag{1}$$

where H, t, P, R_m, λ and e represent cumulative bio-hydrogen production (mL), incubation time (h), bio-hydrogen production potential, maximum bio-hydrogen production rate (mL/h), lag phase duration (h) and 2.71828, respectively. The values of H, t, P and R_m were solved by using the curve fitting tool in MATLAB (Ver. 2010 a).

In order to develop 2D plots through the response surface methodology, the full quadratic model as shown below was used to model the bio-hydrogen production, pH and volatile fatty acids [30,31]:

$$Y = a_0 + \sum_{i=1}^{n} a_i X_i + \sum_{i=1}^{n} a_{ii} X_i^2 + \sum_{i=1}^{n}\sum_{i<j=2}^{n} a_{ij} X_i X_j \tag{2}$$

where X_i and X_j are the controlled parameters, which influence Y and a_0, a_{ii} and a_{ij}, are the offset term, linear and quadratic coefficients, respectively.

2.5. Batch Experiments

The plant was operated at 10% TS by adding pretreated straw and seeded sludge in equal proportions on a TS basis, and the initial pH was maintained to 7.5 with the help of pH controller using 3 M HCl or 3 M NaOH [18]. The co-digestion was done under mesophilic (37 °C) and thermophilic (55 °C) conditions, and all experiments were performed in duplicate [27]. The volume of bio-hydrogen and pH was measured after 12 h-intervals, and VFAs were measured on a daily basis. The incubation time was set to 7 days, after which TS and VS were analyzed [32].

3. Results and Discussion

3.1. Pretreatment Effect on Kinetic Parameters

The bio-hydrogen production rate (R_m) increased with an increase in temperature, and the intensity of increase was different under different pretreatments (Table 1). The maximum and minimum increase in the R_m value due to the increase in temperature from 37 °C to 55 °C was observed in steam explosion (12.56%) and chemical pretreatment (4.39%), respectively. On the other hand, the same increase in temperature increased the cumulative bio-hydrogen production (P) by 39.16%, 26.86% and 10.97% for chemical pretreatment, mechanical pretreatment and steam explosion, respectively. In the case of steam explosion, the increase in P and R_m due to the increase in temperature was close, *i.e.*, 10.96% and 12.56%, respectively. However, in the case of chemical pretreatment, the difference in P and R_m was much higher. The difference was observed due to the bio-hydrogen production period between two temperatures, *i.e.*, bio-hydrogen production started after 12 h of incubation and continued till 120 h at 37 °C, whereas it started before 12 h of incubation and continued till 144 h at 55 °C (Figure 2). As a whole, the increase in temperature from mesophilic to thermophilic increased the bio-hydrogen production, as reported by Alemehdi *et al.* [33] and Kim *et al.* [32]. The increase in hydrogen production might be due to the presence of *Thermoanaerobacterium thermosaccharolyticum* that grow at a higher temperature and produce more hydrogen [34]. On the other hand, due to the early start of bio-hydrogen production at 55 °C under chemical pretreatment, there was a 550-mL difference in the volume of gas produced between the two reactors under different temperatures that remained almost the same till 60 h of incubation. After 60 h, the difference in bio-hydrogen production kept on increasing. This means that the R_m for both temperatures was almost the same till 60 h of incubation, after which R_m increased, which increased the value of P for thermophilic conditions (Figure 2), whereas such an impact of temperature can be observed in the studies reported by Leilei *et al.* [19] and Chen *et al.* [35]. A similar impact of temperature on P and R_m was observed for mechanical pretreatment.

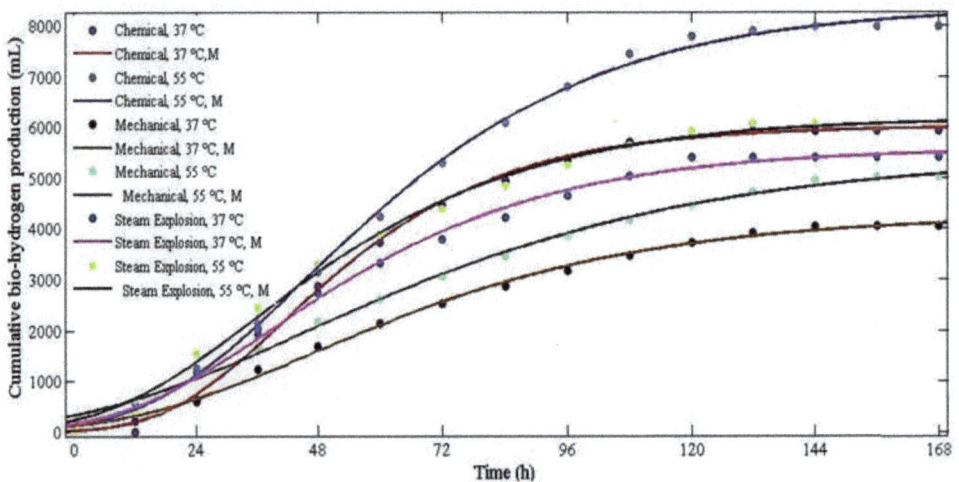

Figure 2. Cumulative bio-hydrogen production under tested treatments.

It was also observed that the P-values for thermophilic mechanical pretreatment and mesophilic steam explosion were close to each other, whereas the respective R_m values for both treatments were much different from each other. One of the main reasons for such a difference in P-values is the lag phase (λ), *i.e.*, the λ for thermophilic mechanical pretreatment (1.271 h) is smaller than the λ observed for mesophilic steam explosion (6.538 h), which represented the early production of bio-hydrogen in the thermophilic reactor under mechanical pretreatment. The early start of bio-hydrogen production in thermophilic reactor under mechanical pretreatment continued till 156 h of incubation, whereas the production was closed after 120 h in the mesophilic reactor under steam explosion. Because of these two facts, the P-values for both reactors were close to each other, even after having a huge difference in R_m values.

Table 1. Properties of rice straw under different pretreatments.

Properties (%)	Mechanical	Steam Explosion	Chemical
Holocellulose	66.79	56.92	54.53
LCH	76.51	70.64	61.51
Lignin	9.72	13.72	6.98
Ash	11.21	21.81	19.46
TS	92.05	20.53	90.49
VS	77.36	16.94	74.46

LCH = lignin, cellulose, hemicellulose; Ash = Solid remaining after ignition; TS, total solids; VS, volatile solids.

Kinetic parameters can be used to derive the time (t_{max}) required to attain the maximum value of R_m. This can be done by taking the first derivative with respect to time of Equation (1) and comparing the results with zero [36]. The resultant equation is:

$$t_{max} = \lambda + \frac{P}{e \cdot R_m} \tag{3}$$

By placing the kinetic parameters in Equation (1), the resultant t_{max} values are shown in Table 1. The increase in temperature decreased the t_{max} for mechanical pretreatment and steam explosion and increased it for chemical pretreatment. The t_{max} can be used as the hydraulic retention time (HRT) in continuous production of bio-hydrogen. As the t_{max} for steam explosion at 55 °C is the smallest in all treatments, steam explosion is more suitable for continuous production processes.

The response surface methodology was opted for a better representation of the bio-hydrogen production with time. In the first step, a quadratic model was fit to the bio-hydrogen production data using false values for incubation time (x_1: −3 = 24 h; 3 = 168 h), pretreatment method (x_2 : −1 = mechanical, 0 = steam explosion and 1 = chemical) and temperature (x_3: −1 = 37 °C and 1 = 55 °C). The following equation was obtained:

$$Y = 4202.5 + 705.42x_1 + 487.5x_2 + 228.75x_3 - 93.33x_1^2 - 332.5x_2^2 + 158.75x_1x_2 - 23.33x_1x_3 + 26.25x_2x_3 \tag{4}$$

$$R^2 = 0.912; F = 414.56$$

Here, Y is the modeled cumulative bio-hydrogen production. The quadratic model developed has a high F value and can significantly explain 91.2% of the variability. The 2D contour plots developed for modeled bio-hydrogen production are shown in Figure 3. It was revealed from modeling that the impact of temperature during the first 24 h was the highest under chemical pretreatment and the lowest under steam explosion. This can be visualized in the contours by observing the variation in the size of the triangle filled with dark blue color (Figure 3). The decrease in bio-hydrogen production was observed first under mechanical pretreatment after 60 h of incubation, which became more prominent under mesophilic temperature after 108 h. Such a change in bio-hydrogen production can be visualized by the area of the color distribution in surface plots or by observing the change in the width of contours. The decrease in modeled bio-hydrogen was observed after 72 h and 84 h, which also became more protuberant after 120 h and 144 h of incubation at 37 °C under steam explosion and chemical pretreatment, respectively. The decrease in actual bio-hydrogen production was also observed during the same span of time, but it was difficult to observe the noticeable change in bio-hydrogen production, as observed in contour and surface plots. As a whole, the 2D illustration of the modeled bio-hydrogen production provides a better representation, which is more helpful to understand the impact of pretreatment and temperature as compared to the line graphs developed.

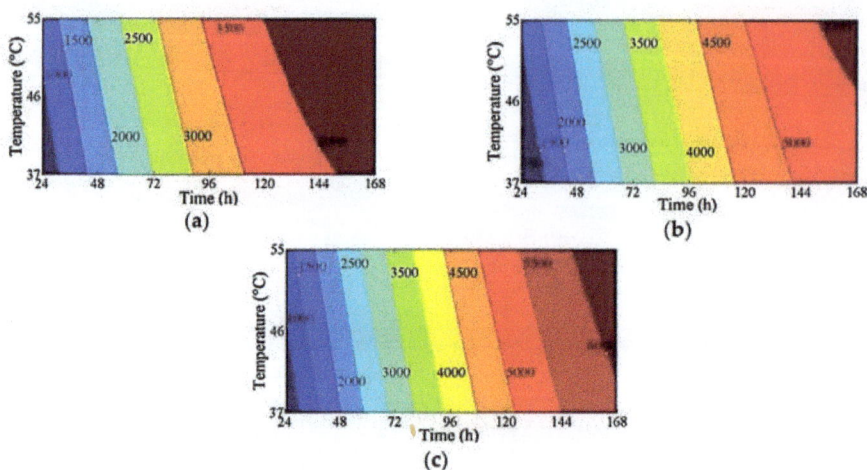

Figure 3. Modeled bio-hydrogen production: (**a**) mechanical pretreatment; (**b**) steam explosion; (**c**) chemical pretreatment.

3.2. Bio-Hydrogen Yield

The bio-hydrogen yield was calculated by dividing the P with initial volatile solid contents of the reactor (VS_{fed}). The impact of temperature was the same as observed on cumulative bio-hydrogen production, as the initial VS fed were the same for each treatment under both temperatures. The chemical treatment resulted in higher bio-hydrogen yield under both temperatures as compared to other treatments. It is clear from the Table 1 that the lignin content was decreased due to chemical treatment. The decrease in lignin content occurred in two steps. First, the lignin-carbohydrate complexes (LCCs) formed due chemical bonds among lignin and holocellulose being changed as the bonds were hydrolyzed by the saponification reaction and released holocellulose, which also increased the degradation of straw [37,38]. In the second stage, de-polymerization of lignin occurred and formed other substances that reduced the lignin content [15]. The cellulose released from LCCs was also affected by the NaOH used in the chemical treatment, which caused intramolecular and intermolecular changes [39]. Such changes resulted in decreased cellulose content, as well as made degradation easier for microbes. The same intramolecular and intermolecular degradation was also observed for hemicellulose because of the breakage and disruption of bonds during NaOH treatment, which ultimately reduced the hemicellulose (Table 1) [40]. On the other hand, steam explosion pretreatment increased the lignin content due to migration, re-condensation and re-localization of lignin onto outer surfaces, which also exposes the internal cellulose [41]. However, during this process, the crystalline structure of cellulose is preserved, while hemicellulose degraded and de-polymerized and resulted in higher holocellulose, as compared to chemical treatment [20]. This is the major difference in both pretreatments, which increased the bio-hydrogen yield from 10 to 15.30 mL/g VS_{fed} when the method was changed from steam explosion pretreatment to chemical pretreatment. Whereas, the mechanical treatment has no effect on the compositional properties, it reduced the crystalline nature of the cellulose, increased the active surface area, as well as increased the degree of polymerization [42]; which ultimately improved the hydrolysis of straw and the bio-hydrogen yield of 9.72 mL/VS_{fed}

obtained under thermophilic conditions. As a whole, the bio-hydrogen yields obtained in the present study are in agreement with the findings of Alimehdi [33].

The impact of the pretreatment method would be clearer if the yield were calculated on the removed fraction of volatile solids (VS_removed), as presented in Table 2. The increase in temperature from 37 °C to 55 °C increased the bio-hydrogen yield (VS_removed) by 21.48%, 6.03% and 9.34% and VS removal by 17.53%, 16.03% and 4.66% under chemical, steam explosion and mechanical pretreatment, respectively. The increase in temperature from 37 °C to 55 °C increased the degradation of cellulose by *Clostridium thermocellum*, which increased the bio-hydrogen yield, especially from rice straw under chemical treatment, as more cellulose was available under chemical treatment as compared to other tested treatments [43]. Apparently, the percentage increase in yield under mechanical pretreatment is higher than steam explosion, but if it is compared to the percentage of increase in VS removal, then steam explosion is more efficient than mechanical pretreatment. On the other hand, the mesophilic P-value (5570 mL) of steam explosion is 3.11% higher than the thermophilic P-value (5402 mL) of mechanical pretreatment, but the corresponding bio-hydrogen yield (VS_removed) of steam explosion is 19.37% higher than said mechanical pretreatment, which also makes steam explosion more efficient then mechanical pretreatment. Similarly, the thermophilic P-value (6181 mL) of steam explosion is 2.8% higher than the mesophilic P-value (6008 mL) of chemical pretreatment, but the bio-hydrogen yield of mesophilic chemical pretreatment is 0.98% higher than thermophilic steam explosion. As a whole, chemical pretreatment has a high efficiency in terms of cumulative bio-hydrogen production and yield, as well as in terms of VS removal. Apart from this, it has the lowest energy consumption compared to the other two methods, as well as zero pollutant emission, because there was no washing involved during pretreatment. Still, the bio-hydrogen yield obtained in the present study can be further improved by replacing the sludge source, as well as combining the pretreatment methods, like alkaline with chemical for more effective degradation of lignocellulosic biomass [19,32].

Table 2. Kinetic parameters and bio-hydrogen yield.

Pretreatment	Temperature	P (mL)	R_m (mL/h)	λ (h)	t_{max} (h)	R^2	Hydrogen Yield (mL/VS_fed)	(mL/VS_removed)
Mechanical	37 °C	4258	42.51	10.26	47.22	0.9973	7.66	36.62
	55 °C	5402	44.73	1.27	45.84	0.9955	9.72	40.04
Steam Explosion	37 °C	5570	66.78	6.53	37.32	0.9941	9.01	47.80
	55 °C	6181	75.17	5.88	36.23	0.9948	10.00	50.68
Chemical	37 °C	6008	92.44	17.65	41.63	0.9977	11.00	51.18
	55 °C	8361	96.5	14.36	46.33	0.9982	15.30	60.60

3.3. Change in pH

There was a sudden drop in pH during the first 12 h of incubation, especially under mesophilic reactors [44]. The increase in temperature from 37 °C to 55 °C decreased the initial drop in pH, which also increased the bio-hydrogen production during 12 h of incubation (Figures 2 and 4) [45]. The mesophilic reactor under mechanical pretreatment has a higher drop in pH during the first 12 h of incubation, but in next 12 h, the drop in pH was less compared to other mesophilic reactors. On the other hand, the drop in pH under steam explosion pretreatment was least effected by an increase in temperature from 37 °C to 55 °C till 60 h of incubation, after which, mesophilic pH was fluctuating between 5.7 and 5.8 till 120 h, and thermophilic pH decreased to 5.7 till 144 h of incubation; whereas, reactors under chemical pretreatment have the least pH drop during 12 h of incubation and the impact of an increase in temperature within the experimental range was higher than steam explosion treatment, but less than mechanical. The difference between pH under both temperatures kept on increasing till 72 h of incubation, after which, the mesophilic pH was stable at 5.9 till bio-hydrogen production was ceased at 120 h, and the thermophilic drop in pH continued till bio-hydrogen production was ceased at 144 h of incubation. Although the pH at which the bio-hydrogen production ceased for chemical pretreatment was the same under both temperatures (pH 5.9), the incubation time at which the bio-hydrogen production was ceased was different. On the other hand, the final pH at the end of incubation was also different, as the pH again started to decrease after bio-hydrogen production was ceased in the mesophilic reactor. The thermophilic pH at which the bio-hydrogen production ceased in other treatments was higher than the mesophilic pH. As a whole, pH from 6.7 ± 0.1 to 5.8 ± 0.1 and 7.1 ± 0.1 to 5.8 ± 0.1 was found suitable for co-digestion under mesophilic and thermophilic conditions, respectively.

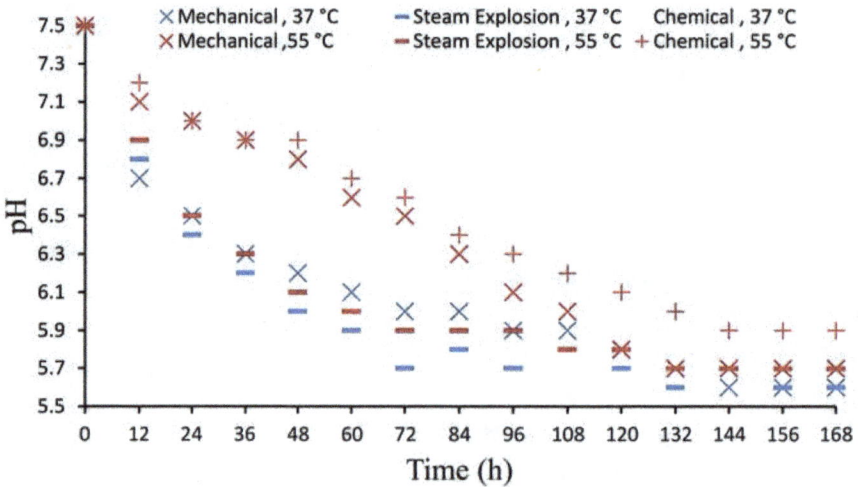

Figure 4. Drop in pH during incubation.

Quadratic modeling was done on pH data in the same way opted for bio-hydrogen production, and the following equation was obtained:

$$Y = 5.68 - 0.17x_1 + 0.07x_2 + 0.12x_3 + 0.03x_1^2 + 0.45x_2^2 - 0.13x_3^2 + 0.01x_1x_2 - 0.03x_1x_3 - 0.03x_2x_3 \quad (5)$$

$$R^2 = 0.8578; \ F = 241.37$$

Here, Y represents the pH within experimental conditions. The model explained the experimental results well for mechanical and chemical treatment, but not for steam explosion treatment, as the predicted values were 0.2–0.3 pH lesser than actual, but the trend was the same. Still, the R^2 value is acceptable in order to predict pH. Figure 5 represents the 2D plots by using Equation (4). The drop in pH represented a similar trend under mechanical and steam explosion till 96 h of incubation, whereas the variation was high under chemical treatment till 108 h of incubation. Mostly, the bio-hydrogen production was observed till 144 h of incubation, and the modeled values represented the same range of pH as observed experimentally. After 120 h, the modeled variation in pH was the least, which was in agreement with the actual results.

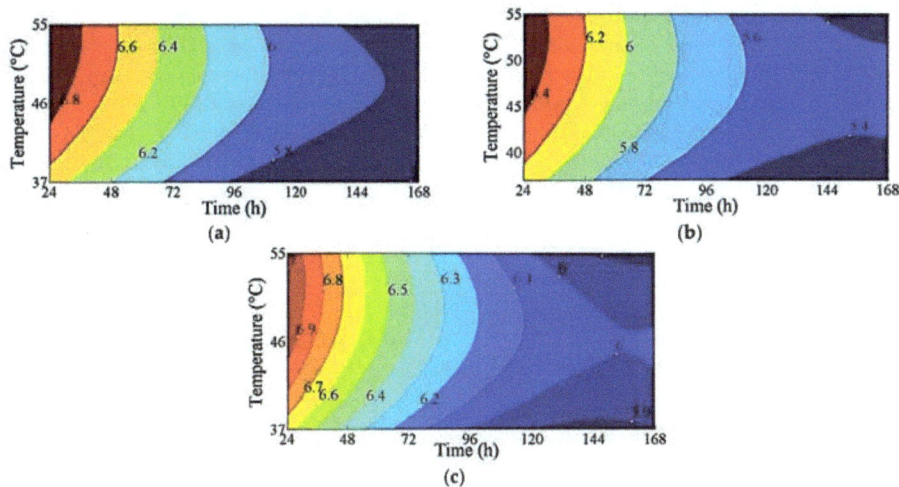

Figure 5. Modeled drop in pH: (**a**) mechanical pretreatment; (**b**) steam explosion; (**c**) chemical pretreatment.

3.4. VFA Production under Tested Pretreatment

In the present study, heat shocked sludge was used for co-digestion with pretreated straw, and it was observed that the VFA contents increased with an increase in time (168 h), which was in agreement with the finding of Kim [46]. The effect of pretreatment, temperature and time on VFA is represented in Figure 6. The highest experimental VFA at the end of incubation was observed in the mesophilic reactor under mechanical pretreatment. The increase in temperature decreased the VFA content for mechanical and steam explosion treatments, as also observed by Gadow [45]. The VFA contents were initially higher in the mesophilic rector under chemical pretreatment, but during 48 h–72 h of incubation, thermophilic VFA contents became higher. This was the same duration in which the bio-hydrogen production was increased dramatically in thermophilic reactors as compared to the mesophilic reactor under chemical pretreatment (Figure 2). As a whole, the VFA production rate was higher till 72 h, which was the same duration in which bio-hydrogen production was also higher, after which, VFA and the bio-hydrogen production rate both started to decrease. However, there was a sudden increase in VFA contents in the mesophilic reactor under mechanical pretreatment during 120 h–144 h of incubation. For the same duration, the VFA production was not so high in the thermophilic reactor under the same treatment, but bio-hydrogen production was much higher in the thermophilic reactor. The higher VFA contents can be inhibitory to the growth of bacteria, as they cause unfavorable physical changes in the cell and excessive energy is required to pump ions [25]. Such high energy is available at elevated temperatures, which increased the yield at elevated temperatures, as observed in the present study [47–49].

The following equation was obtained for VFA as a result of quadratic modeling:

$$Y = 2683.54 + 259.387x_1 - 166.82x_2 - 276.2x_2 - 11.81x_1^2 - 268.75x_2^2 + 10.99x_3^2 - 98.06x_1x_2 - 38.68x_1x_3 + 66.41x_2x_3 \quad (6)$$

$$R^2 = 0.8378; F = 206.58$$

The resultant 2D plots for VFA are shown in Figure 7. It is clear that the VFA contents were high at the low modeled temperature range (37 °C) at the end of incubation, as observed experimentally under mechanical pretreatment (Figure 7a). The variation in VFA contents decreased with an increase in temperature and incubation time under mechanical and steam explosion treatments, but the trend of variation was not the same under both treatments (Figure 7a,b). The VFA contents were almost the same at 96 h, 37 °C and 168 h, 55 °C under mechanical pretreatment and 60 h, 37 °C and 168 h, 55 °C under steam explosion, which represent higher variations of VFA in the case of steam explosion as compared to mechanical pretreatment. On the other hand, VFA production under chemical treatment was different than the other two tested treatments (Figure 7c). The VFA contents were higher under the mesophilic condition till 60 h of incubation, after which thermophilic VFA started to increase. After 120 h of incubation, the VFA variation increased with an increase in temperature, and VFA contents at 168 h, 37 °C and 144 h, 55 °C were almost same under chemical pretreatment. This represented the least effect of temperature under chemical pretreatment as compared to the other two tested treatments.

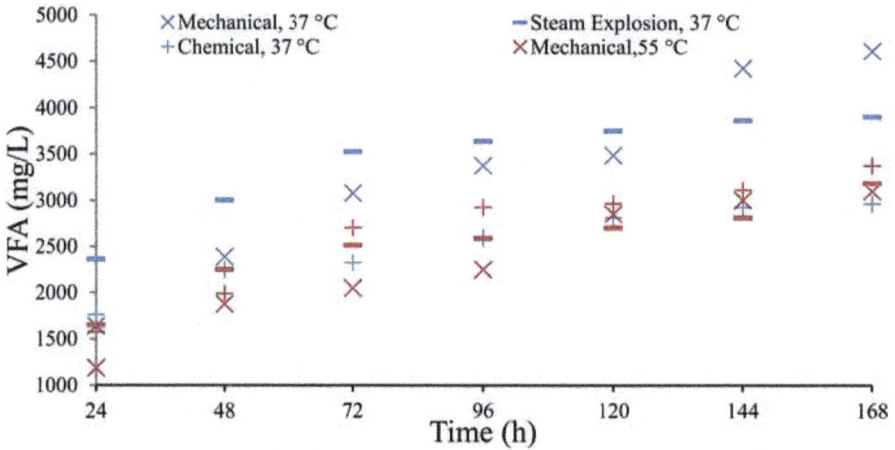

Figure 6. Volatile fatty acids (VFA) trend under various treatments.

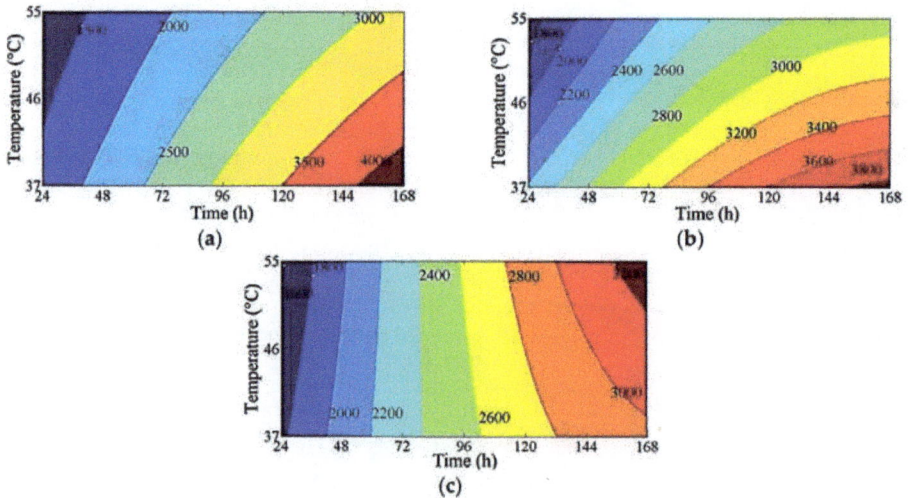

Figure 7. Modeled VFA: (**a**) mechanical pretreatment; (**b**) steam explosion; (**c**) chemical retreatment.

Energies **2016**, *9*, 198

4. Conclusions

The effect of mechanical, steam explosion and chemical pretreatments on the lignocellulosic properties of rice straw, as well as on the bio-hydrogen production potential of rice straw when co-digested with heat shocked sludge was studied under mesophilic and thermophilic conditions. The chemical pretreatment was observed to be the most effective way to reduce holocellulose contents, as well as decreasing the lignin contents of rice straw, which ultimately produced more bio-hydrogen under co-digestion as compared to the other two tested treatments. The increase in temperature from mesophilic to thermophilic conditions found an effective way to enhance bio-hydrogen production through co-digestion, and the highest experimental yield of 60.6 mL/g $VS_{removed}$ was obtained from chemical pretreatment under thermophilic conditions. The drop in pH was observed during the incubation pH from 6.7 ± 0.1 to 5.8 ± 0.1 and 7.1 ± 0.1 to 5.8 ± 0.1 to be suitable for bio-hydrogen production through co-digestion under mesophilic and thermophilic conditions, respectively. The 2D plots developed by the response surface methodology provided better representation of the experimental outcomes under various sets of treatments. As a whole, chemical pretreatment is recommended for rice straw, because of the higher bio-hydrogen yield through co-digestion, better pH environment and zero pollutant emission.

Supplementary Materials: The following are available online at www.mdpi.com/1996-1073/9/3/198/s1. Figure S1. Steps for mechanical treatment. Figure S2. Preparation for steam explosion treatment. Figure S3. Chemical pretreatment. Figure S4. Treated rice straw and sludge in the anaerobic bioreactor. Table S1. Average bio-hydrogen production.

Acknowledgments: We thank Chen Kunjie for providing lab facilities; Fang Himin, Huang Yu Ping and Kang Jin Bao for their help during lab work; Farman Ali Chandio and Fiaz Ahmad for helpful discussions and critical reviews. We also thank Uzma Sattar for improving the standard of the English. We extend our thanks to Higher Education Commission, Pakistan, the China Scholarship council and the College of Engineering, Nanjing Agricultural University, Nanjing, for supporting and providing research facilities for this study.

Author Contributions: Asma Sattar and Chaudhry Arslan designed the research and performed all of the lab work. Changying Ji provided the financial and technical support for designing and conducting the research, as well as supervised the whole research process. Sumiyya Sattar and Haroon Rashid developed and customized the 2D surface plots and assisted in the manuscript preparation. Irshad Ali Mari assisted the lab work and analysis. Fariha Ilyas performed the statistical analysis. Asma Sattar and Chaudhry Arslan wrote the manuscript with comments from all authors, and Asma Sattar finalized the manuscript under the supervision and guidelines of Changying Ji.

Conflicts of Interest: The authors declare no conflict of interest.

References

1. Ni, M.; Leung, D.Y.; Leung, M.K.; Sumathy, K. An overview of hydrogen production from biomass. *Fuel Process. Technol.* **2006**, *87*, 461–472. [CrossRef]
2. Yuan, Z.; Wu, C.; Huang, H.; Lin, G. Research and development on biomass energy in china. *Int. J. Energy Technol. Policy* **2002**, *1*, 108–144. [CrossRef]
3. Kadam, K.L.; Forrest, L.H.; Jacobson, W.A. Rice straw as a lignocellulosic resource: Collection, processing, transportation, and environmental aspects. *Biomass Bioenergy* **2000**, *18*, 369–389. [CrossRef]
4. Singh, A.; Sevda, S.; Abu Reesh, I.M.; Vanbroekhoven, K.; Rathore, D.; Pant, D. Biohydrogen production from lignocellulosic biomass: Technology and sustainability. *Energies* **2015**, *8*, 13062–13080. [CrossRef]
5. Claassen, P.; Van Lier, J.; Contreras, A.L.; Van Niel, E.; Sijtsma, L.; Stams, A.; De Vries, S.; Weusthuis, R. Utilisation of biomass for the supply of energy carriers. *Appl. Microbiol. Biotechnol.* **1999**, *52*, 741–755. [CrossRef]
6. Nath, K.; Das, D. Improvement of fermentative hydrogen production: Various approaches. *Appl. Microbiol. Biotechnol.* **2004**, *65*, 520–529. [CrossRef] [PubMed]
7. Hawkes, F.; Dinsdale, R.; Hawkes, D.; Hussy, I. Sustainable fermentative hydrogen production: Challenges for process optimisation. *Int. J. Hydrog. Energy* **2002**, *27*, 1339–1347. [CrossRef]
8. Lynd, L.R.; Van Zyl, W.H.; McBride, J.E.; Laser, M. Consolidated bioprocessing of cellulosic biomass: An update. *Curr. Opin. Biotechnol.* **2005**, *16*, 577–583. [CrossRef] [PubMed]

9. Mosier, N.; Wyman, C.; Dale, B.; Elander, R.; Lee, Y.; Holtzapple, M.; Ladisch, M. Features of promising technologies for pretreatment of lignocellulosic biomass. *Bioresour. Technol.* **2005**, *96*, 673–686. [CrossRef] [PubMed]
10. Kratky, L.; Jirout, T. Biomass size reduction machines for enhancing biogas production. *Chem. Eng. Technol.* **2011**, *34*, 391–399. [CrossRef]
11. Sun, Y.; Cheng, J. Hydrolysis of lignocellulosic materials for ethanol production: A review. *Bioresour. Technol.* **2002**, *83*, 1–11. [CrossRef]
12. Weil, J.; Sarikaya, A.; Rau, S.-L.; Goetz, J.; Ladisch, C.M.; Brewer, M.; Hendrickson, R.; Ladisch, M.R. Pretreatment of yellow poplar sawdust by pressure cooking in water. *Appl. Biochem. Biotechnol.* **1997**, *68*, 21–40. [CrossRef]
13. Zheng, Y.; Zhao, J.; Xu, F.; Li, Y. Pretreatment of lignocellulosic biomass for enhanced biogas production. *PrECS* **2014**, *42*, 35–53. [CrossRef]
14. Yasuda, M.; Kurogi, R.; Tsumagari, H.; Shiragami, T.; Matsumoto, T. New approach to fuelization of herbaceous lignocelluloses through simultaneous saccharification and fermentation followed by photocatalytic reforming. *Energies* **2014**, *7*, 4087–4097. [CrossRef]
15. He, Y.; Pang, Y.; Liu, Y.; Li, X.; Wang, K. Physicochemical characterization of rice straw pretreated with sodium hydroxide in the solid state for enhancing biogas production. *Energy Fuels* **2008**, *22*, 2775–2781. [CrossRef]
16. Fang, H.H.; Li, C.; Zhang, T. Acidophilic biohydrogen production from rice slurry. *Int. J. Hydrog. Energy* **2006**, *31*, 683–692. [CrossRef]
17. Oh, S.-E.; Van Ginkel, S.; Logan, B.E. The relative effectiveness of pH control and heat treatment for enhancing biohydrogen gas production. *Environ. Sci. Technol.* **2003**, *37*, 5186–5190. [CrossRef] [PubMed]
18. Li, D.; Chen, H. Biological hydrogen production from steam-exploded straw by simultaneous saccharification and fermentation. *Int. J. Hydrog. Energy* **2007**, *32*, 1742–1748. [CrossRef]
19. He, L.; Huang, H.; Lei, Z.; Liu, C.; Zhang, Z. Enhanced hydrogen production from anaerobic fermentation of rice straw pretreated by hydrothermal technology. *Bioresour. Technol.* **2014**, *171*, 145–151. [CrossRef] [PubMed]
20. Ibrahim, M.M.; El-Zawawy, W.K.; Abdel-Fattah, Y.R.; Soliman, N.A.; Agblevor, F.A. Comparison of alkaline pulping with steam explosion for glucose production from rice straw. *Carbohydr. Polym.* **2011**, *83*, 720–726. [CrossRef]
21. Nathao, C.; Sirisukpoka, U.; Pisutpaisal, N. Production of hydrogen and methane by one and two stage fermentation of food waste. *Int. J. Hydrog. Energy* **2013**, *38*, 15764–15769. [CrossRef]
22. Li, C.; Fang, H.H. Fermentative hydrogen production from wastewater and solid wastes by mixed cultures. *Crit. Rev. Environ. Sci. Technol.* **2007**, *37*, 1–39. [CrossRef]
23. Reungsang, A.; Sreela-or, C. Bio-hydrogen production from pineapple waste extract by anaerobic mixed cultures. *Energies* **2013**, *6*, 2175–2190. [CrossRef]
24. Sattar, A.; Arslan, C.; Ji, C.; Chen, K.; Nasir, A.; Fang, H.; Umair, M. Optimizing the physical parameters for bio-hydrogen production from food waste co-digested with mixed consortia of clostridium. *J. Renew. Sustain. Energy* **2016**, *8*, 013107. [CrossRef]
25. Arslan, C.; Sattar, A.; Ji, C.; Sattar, S.; Yousaf, K.; Hashim, S. Optimizing the impact of temperature on bio-hydrogen production from food waste and its derivatives under no pH control using statistical modelling. *BGeo* **2015**, *12*, 6503–6514. [CrossRef]
26. APHA. *Standard Methods for the Examination of Water and Wastewater*, 25th ed.; American Public Health Association: Washington, DC, USA, 2005; pp. 94–100.
27. Arslan, C.; Sattar, A.; Changying, J.; Nasir, A.; Ali Mari, I.; Zia Bakht, M. Impact of pH management interval on biohydrogen production from organic fraction of municipal solid wastes by mesophilic thermophilic anaerobic codigestion. *BioMed. Res. Int.* **2015**, *2015*. [CrossRef] [PubMed]
28. Ververis, C.; Georghiou, K.; Danielidis, D.; Hatzinikolaou, D.; Santas, P.; Santas, R.; Corleti, V. Cellulose, hemicelluloses, lignin and ash content of some organic materials and their suitability for use as paper pulp supplements. *Bioresour. Technol.* **2007**, *98*, 296–301. [CrossRef] [PubMed]
29. Ramos, C.; Buitrón, G.; Moreno-Andrade, I.; Chamy, R. Effect of the initial total solids concentration and initial pH on the bio-hydrogen production from cafeteria food waste. *Int. J. Hydrog. Energy* **2012**, *37*, 13288–13295. [CrossRef]

30. Kim, S.-H.; Han, S.-K.; Shin, H.-S. Optimization of continuous hydrogen fermentation of food waste as a function of solids retention time independent of hydraulic retention time. *Process. Biochem.* **2008**, *43*, 213–218. [CrossRef]

31. Jo, J.H.; Lee, D.S.; Park, D.; Choe, W.-S.; Park, J.M. Optimization of key process variables for enhanced hydrogen production by enterobacter aerogenes using statistical methods. *Bioresour. Technol.* **2008**, *99*, 2061–2066. [CrossRef] [PubMed]

32. Kim, M.; Liu, C.; Noh, J.-W.; Yang, Y.; Oh, S.; Shimizu, K.; Lee, D.-Y.; Zhang, Z. Hydrogen and methane production from untreated rice straw and raw sewage sludge under thermophilic anaerobic conditions. *Int. J. Hydrog. Energy* **2013**, *38*, 8648–8656. [CrossRef]

33. Alemahdi, N.; Man, H.C.; Nasirian, N.; Yang, Y. Enhanced mesophilic bio-hydrogen production of raw rice straw and activated sewage sludge by co-digestion. *Int. J. Hydrog. Energy* **2015**, *40*, 16033–16044. [CrossRef]

34. Shin, H.-S.; Youn, J.-H.; Kim, S.-H. Hydrogen production from food waste in anaerobic mesophilic and thermophilic acidogenesis. *Int. J. Hydrog. Energy* **2004**, *29*, 1355–1363. [CrossRef]

35. Chen, C.-C.; Chuang, Y.-S.; Lin, C.-Y.; Lay, C.-H.; Sen, B. Thermophilic dark fermentation of untreated rice straw using mixed cultures for hydrogen production. *Int. J. Hydrog. Energy* **2012**, *37*, 15540–15546. [CrossRef]

36. Chang, A.C.; Tu, Y.-H.; Huang, M.-H.; Lay, C.-H.; Lin, C.-Y. Hydrogen production by the anaerobic fermentation from acid hydrolyzed rice straw hydrolysate. *Int. J. Hydrog. Energy* **2011**, *36*, 14280–14288. [CrossRef]

37. Yang, S. *Plant Fiber Chemistry*; China Light Industry Press: Beijing, China, 2001; pp. 176–182. (In Chinese)

38. Durot, N.; Gaudard, F.; Kurek, B. The unmasking of lignin structures in wheat straw by alkali. *Phytochemistry* **2003**, *63*, 617–623. [CrossRef]

39. Lu, J.; Shi, S.; Yang, R.; Niu, M.; Song, W. Modification of reed cellulose microstructure and it change in enzymatic hydrolysis of reed pulp. *Trans. China Pulp. Pap.* **2005**, *20*, 85–90. (In Chinese)

40. Sun, X.F.; Sun, R.; Tomkinson, J.; Baird, M. Preparation of sugarcane bagasse hemicellulosic succinates using nbs as a catalyst. *Carbohydr. Polym.* **2003**, *53*, 483–495. [CrossRef]

41. Selig, M.J.; Viamajala, S.; Decker, S.R.; Tucker, M.P.; Himmel, M.E.; Vinzant, T.B. Deposition of lignin droplets produced during dilute acid pretreatment of maize stems retards enzymatic hydrolysis of cellulose. *Biotechnol. Prog.* **2007**, *23*, 1333–1339. [CrossRef] [PubMed]

42. Hendriks, A.; Zeeman, G. Pretreatments to enhance the digestibility of lignocellulosic biomass. *Bioresour. Technol.* **2009**, *100*, 10–18. [CrossRef] [PubMed]

43. Levin DB, I.R.; Cicek, N.; Sparling, R. Hydrogen production by clostridium thermocellum 27,405 from cellulosic biomass substrates. *Int. J. Hydrog. Energy* **2006**, *31*, 1496–1503. [CrossRef]

44. Li, Y.; Zhang, R.; He, Y.; Liu, X.; Chen, C.; Liu, G. Thermophilic solid-state anaerobic digestion of alkaline-pretreated corn stover. *Energy Fuels* **2014**, *28*, 3759–3765. [CrossRef]

45. Gadow, S.; Li, Y.-Y.; Liu, Y. Effect of temperature on continuous hydrogen production of cellulose. *Int. J. Hydrog. Energy* **2012**, *37*, 15465–15472. [CrossRef]

46. Kim, M.; Yang, Y.; Morikawa-Sakura, M.S.; Wang, Q.; Lee, M.V.; Lee, D.-Y.; Feng, C.; Zhou, Y.; Zhang, Z. Hydrogen production by anaerobic co-digestion of rice straw and sewage sludge. *Int. J. Hydrog. Energy* **2012**, *37*, 3142–3149. [CrossRef]

47. Gottschalk, G. *Bacterial Metabolism*, 2nd ed.; Springer: New York, NY, USA, 1986.

48. Zoetemeyer, R.J.; Cohen, A.; Boelhouwer, C. Product inhibition in the acid forming stage of the anaerobic digestion process. *Water Res.* **1982**, *16*, 633–639. [CrossRef]

49. Switzenbaum, G.-G.E.; Hickey, R.F. Monitoring of the anaerobic methane fermentation process. Enzyme microbial technology. *Enzyme Microb. Technol.* **1990**, *12*, 722–730. [CrossRef]

energies

Article

Catalytic Intermediate Pyrolysis of Napier Grass in a Fixed Bed Reactor with ZSM-5, HZSM-5 and Zinc-Exchanged Zeolite-A as the Catalyst

Isah Yakub Mohammed [1,6], Feroz Kabir Kazi [2], Suzana Yusup [3], Peter Adeniyi Alaba [5], Yahaya Muhammad Sani [5] and Yousif Abdalla Abakr [4,*]

[1] Department of Chemical and Environmental Engineering, the University of Nottingham Malaysia Campus, Jalan Broga, Semenyih 43500, Darul Ehsan, Malaysia; kebx3iye@nottingham.edu.my
[2] Department of Engineering and Mathematics, Sheffield Hallam University, City Campus, Howard Street, Sheffield S1 1WB, UK; f.kabir@shu.ac.uk
[3] Department of Chemical Engineering, Universiti Teknology Petronas (UTP) Bandar Seri Iskandar, Tronoh 31750, Malaysia; drsuzana_yusup@petronas.com.my
[4] Department of Mechanical, Manufacturing and Material Engineering, the University of Nottingham Malaysia Campus, Jalan Broga, Semenyih 43500, Darul Ehsan, Malaysia
[5] Department of Chemical Engineering, University of Malaya, Kuala Lumpur 50603, Malaysia; adeniyipee@live.com (P.A.A.); ymsani@siswa.um.edu.my (Y.M.S.)
[6] Crops for the Future (CFF), the University of Nottingham Malaysia Campus, Jalan Broga, Semenyih 43500, Darul Ehsan, Malaysia
* Correspondence: yousif.abakr@nottingham.edu.my; Tel.: +60-132-321-232

Academic Editor: Tariq Al-Shemmeri
Received: 28 January 2016; Accepted: 21 March 2016; Published: 29 March 2016

Abstract: The environmental impact from the use of fossil fuel cum depletion of the known fossil oil reserves has led to increasing interest in liquid biofuels made from renewable biomass. This study presents the first experimental report on the catalytic pyrolysis of Napier grass, an underutilized biomass source, using ZSM-5, 0.3HZSM-5 and zinc exchanged zeolite-A catalyst. Pyrolysis was conducted in fixed bed reactor at 600 °C, 30 °C/min and 7 L/min nitrogen flow rate. The effect of catalyst-biomass ratio was evaluated with respect to pyrolysis oil yield and composition. Increasing the catalyst loading from 0.5 to 1.0 wt % showed no significant decrease in the bio-oil yield, particularly, the organic phase and thereafter decreased at catalyst loadings of 2.0 and 3.0 wt %. Standard analytical methods were used to establish the composition of the pyrolysis oil, which was made up of various aliphatic hydrocarbons, aromatics and other valuable chemicals and varied greatly with the surface acidity and pore characteristics of the individual catalysts. This study has demonstrated that pyrolysis oil with high fuel quality and value added chemicals can be produced from pyrolysis of Napier grass over acidic zeolite based catalysts.

Keywords: Napier grass; intermediate pyrolysis; catalytic deoxygenation; zeolite; bio-oil characterization

1. Introduction

Fossil fuels remain the main global energy supply source despite the environmental impacts cum sociopolitical concerns which are well documented in the literature [1–4]. The fear of energy insecurity in the near future, in addition to the need for reduction of greenhouse gases, has led to the development of energy from alternative renewable sources such as biomass, wind, solar and mini-hydro [5–8]. Among these renewable resources, biomass is the only renewable resource that has carbon in its building blocks which can be processed into liquid fuel. Lignocellulosic biomass (non-food materials) such as forest residues, agro-wastes, energy grasses, aquatic plants and algae, *etc.*, have been seen as ideal raw materials in this direction as they avoid the initial public perception of food insecurity associated with first generation biofuels which were produced from food materials [9,10]. In addition, they have low levels of sulfur and nitrogen contents which make them relatively environmental friendly. Napier grass (*Pennisetum purpureum*) is one of the perennial grasses with potential high biomass yield, typically in the range of 25–35 oven dry tones per hectare annually, which correspond to 100 barrels of oil energy equivalent per hectare compared to other herbaceous plants [11,12]. Other advantages of Napier grass includes compatibility with conventional farming practices, and the fact that it outcompetes weeds, needs very little or no supplementary nutrients and therefore requires lower establishment cost. It can be harvested up to four times a year with a ratio of energy output to energy input of around 25:1 which makes it one of the highest potential energy crops for development of efficient and economic bioenergy systems [11,12].

Pyrolysis is currently one of the most promising thermochemical processes for converting biomass materials into products with high energy potential. Bio-oil, bio-char and non-condensable gas products are generally obtained in different proportions in any pyrolysis process. The distribution of pyrolysis products depends heavily on how the process parameters such as pyrolysis temperature, heating rate and vapor residence time are manipulated. Generally, there are different types of pyrolysis namely; slow, intermediate and fast pyrolysis. Slow pyrolysis is also referred to as carbonization. It is carried out at a temperature up to 400 °C, for 60 min to days, with a typical product distribution of about 35% bio-char, 30% bio-oil, and 35% non-condensable gas. Fast pyrolysis can produce up to 80% bio-oil, 12% bio-char, and 13% non-condensable gas at temperature around 500 °C, with high heating rates, a short vapor residence time of about 1 s, and rapid cooling of volatiles [13,14]. For intermediate pyrolysis, the operating conditions are 500–650 °C and the vapor residence time is approximately 10 to 30 s. About 40%–60% of the total product yield is usually bio-oil, 15%–25% bio-char and 20%–30% non-condensable gas. In addition, unlike fast pyrolysis, intermediate pyrolysis produce bio-oil with less reactive tar which can be used directly as a fuel in engines and boilers, and dry char suitable for both agricultural and energy applications [15–17].

The pyrolysis reactor represents the core unit of the entire pyrolysis process. It plays a very important role in the product distribution and accounts for about 10%–15% of the total capital cost [14]. A range of reactor designs are available, which include bubbling fluidized-bed reactors, circulating fluidized-bed reactors, fixed-bed reactors, auger reactors, ablative reactors, rotating cone reactors, *etc.* These reactors have been studied extensively to improve the efficiency of pyrolysis processes and the quality of bio-oil production. However, each reactor type has specific characteristics, pros and cons. In general, a good reactor design should exhibit high heating and heat transfer rates and should have an excellent temperature control capability [14].

Bio-oil from biomass pyrolysis is a complex mixture consisting predominantly oxygenated organic compounds, phenolics, light hydrocarbons and traces of nitrogen- and sulfur-containing compounds depending on the nature of the source biomass. The high level of oxygenated compounds in bio-oil is responsible for the oil's poor physicochemical characteristics such as low pH, low chemical stability, lower energy content and therefore render it unsuitable for direct application as fuel or refinery ready feedstock for the production of quality fuels and other consumer products. In order to meet the target of having alternative fuel sources and reduce the challenges of fossil fuel on our environment, there is need to develop methods for reducing the oxygen content of bio-oil to a minimum level. Several deoxygenation methods are being developed in this direction, one of which is *in situ* deoxygenation. *In situ* upgrading involves the use of catalytic material to reduce oxygenated volatiles generated during pyrolysis prior to condensation through a series of chemical reactions such as decarboxylation, dehydration, and decarbonylation where oxygen is removed in the form of CO_2, H_2O and CO, respectively. The process can be organized either by mixing the biomass with catalyst (in bed mixing) follow by pyrolysis, or by passing the pyrolysis vapor through a bed or beds of catalyst [18–23]. The most commonly used catalyst in this application are zeolite-based materials, particularly ZSM-5, due to their high acidity, shape selective pore structure, low affinity for coke formation owing to bulky molecules and high selectivity towards aromatic hydrocarbons [23–25]. Process parameters governing the yield and quality of bio-oil produced via the *in situ* upgrading include pyrolysis temperature, heating rate and catalyst- phthalenes decreased with an increasbiomass ratio (CBR) [26–29]. Liu *et al.* [29] studied the catalytic pyrolysis of duckweed with HZSM-5 revealed that the pyrolysis temperature and CBR affect the distribution of organic component in the product bio-oil. A high temperature was shown to favor the production of total monocyclic aromatic hydrocarbons such as benzene, toluene and xylene (BTX) while polyaromatic hydrocarbons (PAH) such as indenes and nae in pyrolysis temperature. This trend was attributed to the exothermic nature of the oligomerization reactions. Similar observations have been reported by Kim *et al.* [30] during the *in situ* upgrading of pyrolysis vapor from unshiu citrus peel over HZSM-5. Studies by Ojha and Vinu [31] on resource recovery from polystyrene through fast catalytic pyrolysis using a zeolite-based catalyst also followed a similar trend. In terms of CBR, Liu *et al.* [29] stated that the increase in CBR promoted formation of BTX, while a downward trend was observed for PAH. This was contrary to the observation made by Ojha and Vinu [31]. They reported that an increase in CBR favored production of benzene among the monoaromatics while PAH yield increased with CBR. This difference in the observed trend could be linked to the characteristics of the respective catalysts used during those studies.

The impact of different zeolite-based catalysts on the production of aromatic hydrocarbons from pyrolysis of biomass materials such rice husk, *Miscanthus*, rice stalks, wood, corncobs, algae, *etc.*, have been reported the literature. To the best of our knowledge, catalytic pyrolysis of Napier grass with zeolite catalysts has not been carried out. In addition, studies involving lower catalyst biomass ratios with potential practical applications are rarely carried out. Most literature studies employed the pyroprobe technique with high CBR and, in most cases the amount of catalyst used is equal or greater than the feedstock biomass. For large scale processes using in bed mixing technique, a large amount of catalyst material requirement may not be practicable technically and economically. The objective of this study was to investigate effect of the zeolites ZSM-5, and HZSM-5 and zinc-exchanged zeolite-A catalyst loading on the yield and characteristics of Napier grass pyrolysis bio-oil.

2. Experimental

2.1. Materials and Characterization

Napier grass stem (NGS) with a particle size of around 2 mm was used in this study and the biomass has volatile matter, fixed carbon, ash content and higher heating value of 81.51 wt %, 16.75 wt %, 1.75 wt % and 18.11 MJ/kg respectively. Ultimate analysis revealed that the biomass has 48.61 wt % carbon, 6.01 wt % hydrogen, 0.99 wt % nitrogen, 0.32 wt % sulfur and 44.07 wt % oxygen. Other details of its characteristics can be found in Mohammed *et al.* [12]. ZSM-5 and Zeolite A (zinc-exchanged: ZEOA) catalysts were purchased from Fisher Scientific Sdn. Bhd. (Selangor, Malaysia) and Sigma-Aldrich Sdn. Bhd. (Selangor, Malaysia) respectively. HZSM-5 was obtained from desilication of ZSM-5 with NaOH solution. ZSM-5 (30 g) was mixed with aqueous solution of 0.3 M NaOH (300 mL) for 2 h at 70 °C. The solid was filtered using vacuum filtration with the aid of a polyamide filter and thereafter oven dried at 100 °C. The dried sample was transformed to hydronium form with 0.2 M NH_4NO_3 solution at 80 °C for 24 h, followed by overnight drying at 100 °C and calcination at 550 °C for 5 h and the final solid was designated as 0.3HZSM-5. All the catalysts were characterized according to standard procedures. X-ray diffraction (XRD; PANalytical X′pertPro, DSKH Technology Sdn. Bhd.: Selangor, Malaysia; CuK_α radiation, λ = 0.1541 nm;) was used to examine the nature of the crystalline system at 2θ angles between 10° and 60°, 25 mA, 45 kV, step size of 0.025°, and 1.0 s scan rate. Scanning electron microscopy (SEM, FEI Quanta 400 FE-SEM, Hillsboro, OR, USA) was used to evaluate the surface and structural characteristics. Specific surface area and pore properties were determined using an ASAP 2020 physisorption analyzer (Micrometrics: Norcross, GA, USA). Acidity of the catalyst was determined via ammonia-temperature programmed desorption (TPD) using a ChemiSorb 2720 Pulse Chemisorption system (Micrometrics).

2.2. Catalytic Pyrolysis and Pyrolysis Oil Characterization

Our intermediate pyrolysis study was carried out in a fixed bed pyrolysis system as shown in Figure 1. The system consists of a fixed bed reactor made of stainless steel (115 cm length, 6 cm inner diameter), a distribution plate with 1.0 mm hole diameter which sits at 25 cm from the bottom of the tube, two nitrogen preheating sections, a cyclone, a water chiller operating at 3 °C attached to a coil condenser, oil collector and gas scrubbers. 200 g of NGS (bone dry, 2.5 mm particle size) mixed with catalyst was placed on the distribution plate inside the reactor tube and pyrolysis was conducted under nitrogen atmosphere at 7 L/min. A pyrolysis temperature of 600 °C was used and the reactor was heated electrically at the rate of 30 °C/min. The reaction temperature was monitored with a K-type thermocouple connected to a computer through a data logger. The reaction time was kept at 15 min (±2 min) or until no significant amount of non-condensable gas was observed after the reaction temperature reaches 600 °C. Effect of catalyst loading was first evaluated with 0.5, 1, 2 and 3 wt % ZSM-5. The optimum oil yield conditions were then used with 0.3HZSM-5 and ZEOA catalyst. Non-catalytic pyrolysis product yield was used as a control. Characterization of bio-oil collected was carried out according to standard procedures. A WalkLAB microcomputer pH meter TI9000 (Trans Instruments, Singapore) was used to determine the pH. Water content in the bio-oil was determined using a Karl Fischer V20 volumetric titrator (Mettler Toledo: Columbus, OH, USA) [32,33]. Higher heating value was determined using an oxygen bomb calorimeter (Parr 6100, Parr Instruments: Molin, IL, USA) [33,34]. Density and viscosity were determined using an Anton Paar density meter (DMA 4500 M, Ashland, VA, USA) and Brookfield (Hamilton, NJ, USA) DV-E viscometer, respectively. Bio-oil elemental compositions were determined using a Perkin Elmer 2400 Series II CHNS/O analyzer (Perkin Elmer Sdn Bhd.: Selangor, Malaysia). Chemical functional groups in the bio-oil were determined with FTIR (Spectrum RXI, PerkinElmer: Selangor, Malaysia) using a pair of circular demountable potassium bromide (KBr) cell windows (25 mm diameter and 4 mm thickness). Spectra were recorded with the Spectrum V5.3.1 software within the wavenumber range of 400–4000 cm^{-1} at 32 scans and a resolution of 4 cm^{-1}. Details of the chemical composition of the bio-oil was determined using a gas chromatograph-mass spectrometer (GC-MS) system (PerkinElmer Clarus® SQ 8: Akron, OH, USA) with a quadruple detector and PerkinElmer-Elite™-5ms column (30 m × 0.25 mm × 0.25 μm) (Akron, OH, USA). The oven was programmed at an initial temperature of 40 °C, ramp at 5 °C/min to 280 °C and held there for 20 min. The injection temperature, volume, and split ratio were 250 °C, 1 μL, and 50:1 respectively. Helium was used as carrier gas at a flow rate of 1 mL/min. The bio-oil samples in chloroform (10%, *w/v*) were prepared and used for the analysis. MS ion source at 250 °C with 70 eV ionization energy was used. Peaks of the chromatogram were identified by comparing with standard spectra of compounds in the National Institute of Standards and Technology (NIST: Gaithersburg, MD, USA) library.

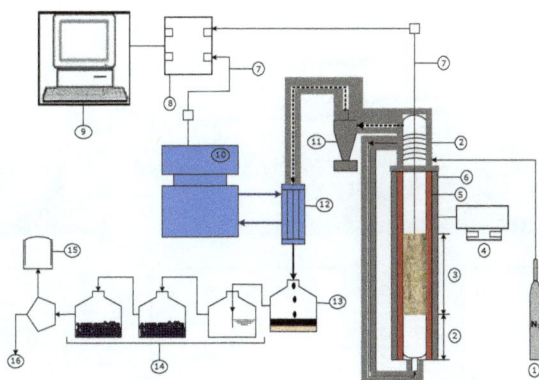

Figure 1. Experimental set-up. (1) Nitrogen cylinder; (2) nitrogen preheating sections; (3) pyrolysis section; (4) furnace controller; (5) heater; (6) insulator; (7) thermocouples; (8) data logger; (9) computer; (10) water chiller; (11) cyclone; (12) condenser; (13) bio-oil collector; (14) gas scrubber; (15) gas sampling bag; (16) gas venting.

Samples of the non-condensable pyrolysis product were collected in a gas SKC polypropylene fitted gas sampling bag for each experiment and its composition analyzed using a gas chromatography PerkinElmer Clarus 500 (Akron, OH, USA) equipped with a stainless steel column (Porapak R 80/100) and thermal conductivity detector (TCD). Helium was used as a carrier gas and the GC was programed at 60 °C, 80 °C and 200 °C for oven, injector and TCD temperature, respectively.

3. Results and Discussion

3.1. Catalyst Characteristics

XRD patterns of the zeolites are presented in Figure 2. The parent ZSM-5 exhibited main peaks at around 2θ between 20° and 25°, which are typical characteristic peaks for ZSM-5. The decrease in the intensity of the peaks observed in the 0.3HZSM-5 indicates a loss of crystallinity as a result of desilication [35] which may also be related to the formation of mesoporous structures in the material [36–41]. The peaks observed in ZEOA at around 10.1°, 16.1°, 21.4°, 27.1°, 29.9° and 35° are distinctive characteristics peaks of zeolite A [42–44]. SEM images showing the morphology of the zeolites are presented in Figure 3.

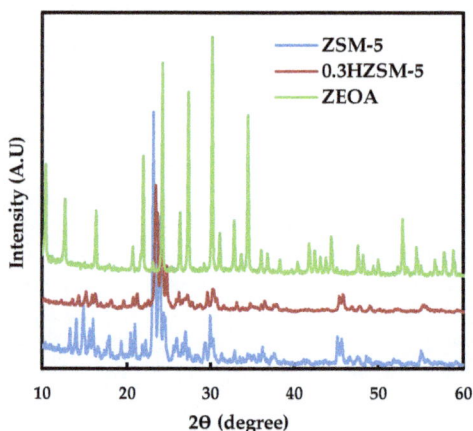

Figure 2. Diffractograms of ZSM-5, 0.3HZSM-5 and ZEOA.

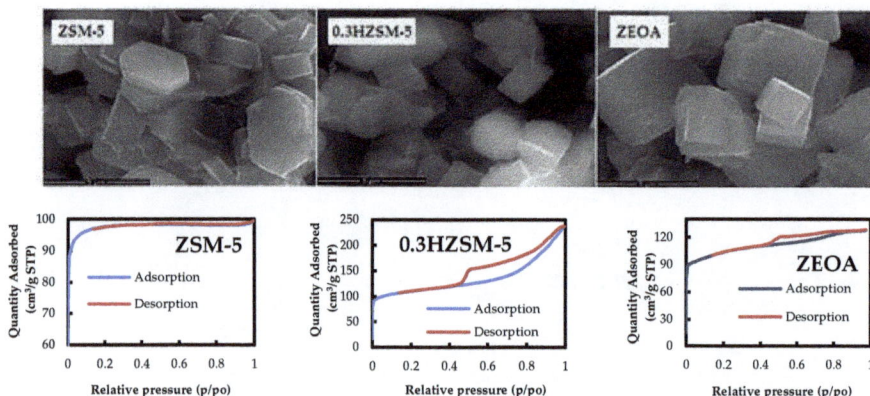

Figure 3. *Cont.*

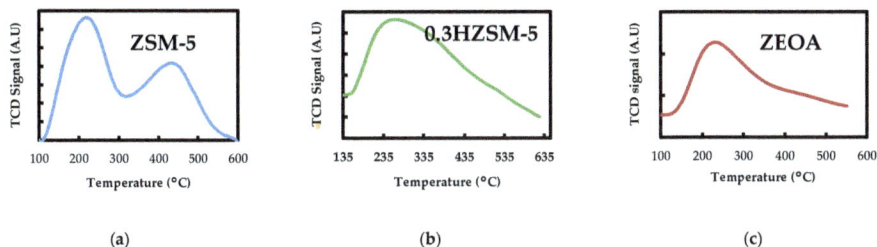

Figure 3. Characteristics (SEM, BET and TPD) of (**a**) ZSM-5, (**b**) 0.3HZSM-5 and (**c**) ZEOA.

The SEM micrographs indicated that ZSM-5 is highly crystalline, with hexagonal prismatic morphology and different particle size of less than 500 nm. 0.3HZSM-5 showed morphological characteristic similar to the parent ZSM-5 indicting that desilication does not affect the morphological integrity of the catalyst. ZEOA exhibited an extremely crystalline system with cubical structure having smooth ages, which is a typical characteristic of zeolite A.

From the result of the physisorption analysis (Figure 3), ZSM-5 displayed a type I isotherm according to the IUPAC classification. The isotherm showed a very strong adsorption in the initial region and a plateau at high relative pressure (>0.9). This pattern indicates that ZSM-5 is a microporous material [44]. 0.3HZSM-5 displayed a combination of type I and IV isotherms with a low slope region at the middle which indicates the presence of few multilayers and a hysteresis loop at relative pressures above 0.4 that could be attributed to capillary condensation in a mesoporous material [45,46]. This observation is in good agreement with the XRD results. ZEOA also displayed an isotherm similar to that of 0.3HZSM-5 with a visible but less pronounced hysteresis loop. This indicates that ZEOA is made up of some mesoporous structures. Other characteristics and physisorption analysis are summarized in Table 1. Comparing ZSM-5 and 0.3HZSM-5, Brunauer Emmet Teller (BET) specific surface area (S_{BET}) and S_{micro} decreased after desilication. A decrease in the micropore volume of the 0.3HZSM-5 was also observed after desilication compared to the parent ZSM-5, which suggest that the conversion of the microporous structure during the desilication contributed to the resultant mesoporosity in the 0.3HZSM-5 [45]. The results of NH_3-TPD analysis (Figure 3) showed two peaks at temperatures around 219 and 435 °C for ZSM-5, while single peaks around 258 °C and 229 °C was observed for 0.3HZSM-5 and ZEOA, respectively. The high temperature peak represents the desorption of NH_3 from strong acid sites while those at temperatures between 219 and 258 °C are attributed to desorption of NH_3 from weak acid sites [31,35,36,46]. Disappearance of the strong acid sites in the 0.3HZSM-5 is attributed to desilication. Similar observations have been reported in the literature [45,46]. The area under each peak was evaluated and the corresponding total surface acidity was 3.8085 mmol/g for ZSM-5, while 2.9635 and 1.21 mmol/g were recorded for 0.3HZSM-5 and ZEOA, respectively.

Table 1. Textural characteristics of ZSM-5, 0.3HZSM-5 and ZEOA.

Property	ZSM-5	0.3HZSM-5	ZEOA
Si/Al ratio	20.7600	12.5100	1.0000
S_{BET} (m^2/g)	385.2000	374.8800	367.0000
S_{micro} (m^2/g)	356.5400	240.2300	315.7200
S_{meso} (m^2/g)	28.6600	134.6500	93.3100
V_{micro} (m^3/g)	0.1383	0.1114	0.1240
Total acidity (mmol/g)	3.8085	2.9635	1.2100

3.2. Pyrolysis Product Distribution

The effect of ZMS-5 loading with respect to pyrolysis product distribution compared to the non-catalytic pyrolysis (NCP) is shown in Figure 4a. Bio-oil here constituted the total liquid product (organic: OR and aqueous: AQ phase); bio-char is the total solids, including coke. The bio-oil recorded from the NCP (raw) was 41.91 wt % (12.34 wt % OR; 29.57 wt % AQ) with corresponding bio-char and non-condensable gas values of 29.24 and 28.85 wt %. For the catalytic process, the bio-oil yield decreased to 40.07 wt % (11.25 wt % OR; 28.82 wt % AQ) at a catalyst loading of 0.5 wt %. Increasing the catalyst loading from 0.5 to 1.0 wt % showed no significant decrease in the bio-oil yield, particularly for the organic phase, and 38.88 wt % oil (11.15 wt % OR; 27.74 wt % AQ) was recorded. Thereafter, the oil yield decreased to 33.29 wt % (9.43 wt % OR; 23.86 wt % AQ) and 30.35 wt % (7.48 wt % OR; 22.87 wt % AQ) at catalyst loadings of 2.0 and 3.0 wt %, respectively. Comparing with the existing literature, most researchers employed high catalyst to biomass ratios which generally lead to less liquid yield and more gas production [21,47–49]. Studies involving catalyst loadings similar to the ones used in this study, particularly between 0.5 and 1.0 wt %, are seldom carried out. Research conducted by Park *et al.* [50] on catalytic pyrolysis of *Miscanthus* with ZSM-5 using a catalyst to biomass ratio of 0.1 and a reaction temperature of 450 °C in a fixed bed reactor resulted in a high yield of organic phase (21.5 wt %). Similarly, the work of Elordi *et al.* [51] on catalytic pyrolysis of polyethylene with ZSM-5 using a catalyst/biomass ratio of 0.03 at 500 °C in a spouted bed reactor generated about 25 wt % organic product. Furthermore, under low ZSM-5 loading conditions (0.5 and 1.0 wt %), a high yield of organic phase was recorded compared to 2.0 and 3.0 wt % ZSM-5 loading. This could be attributed to the generation of less reactive pyrolysis vapor via simultaneous dehydration, decarboxylation, and decarbonylation reactions. The non-condensable gas yield under this catalytic condition was higher compared to the non-catalytic pyrolysis and also increased with catalyst loading, suggesting a high degree of decarboxylation and decarbonylation reactions. The bio-char yield during the catalytic process with ZSM-5 was 29.24, 29.79, 30.12 and 30.22 wt % at 0.5, 1.0, 2.0 and 3.0 wt catalyst. These values are similar to that of bio-char yield obtained with the NCP (29.24 wt %). The small increment, particularly at ZSM-5 loadings from 1.0 to 3.0 wt %, could be attributed to coke deposits. This observation is in good agreement with the literature [48,52]. Also, comparing the impact of ZSM-5 with 0.3HZSM-5 and ZEOA, the bio-oil yield obeyed the following order: ZEOA > ZSM-5 > 0.3HZSM-5. The bio-oil from ZEOA constitutes a large percentage of the AQ phase (34.89 wt %) which is largely water resulting from deoxygenation reactions. The low yield of the OR phase with the ZEOA catalyst is an indication of secondary reactions in which oxygen is removed in form of water as the main reaction product. The non-condensable gas yield recorded with ZEOA was lower than that 0.3HZSM-5 and ZSM-5, while the bio-char yield was comparable to those obtained with the other catalysts. The overall impact of ZEOA on the product distribution could be attributed to a combined effect of the zinc cation in the catalyst, its acidity and porosity [53]. Therefore the non-condensable fraction may also consist of substantial amounts of CO_2. A similar trend of pyrolysis product distribution related to zinc cation in the catalyst and low SAR has been reported in the literature [54,55].

Bio-oil yield from 0.3HZSM-5 was 31.14 wt % (10.03 wt % OR and 25.11 wt % AQ phase) compared to 40.07 wt % (11.25 wt % OR; 28.82 wt % AQ) ZSM-5 oil yield. The reduction in the organic phase recorded with 0.3HZSM-5 may be attributed to the improved pore structures which perhaps led to cracking of large organic molecules. This can also be backed up by the higher non-condensable gas amount of 34.29 wt % recorded, compared to 30.69 wt % from ZSM-5. This shows that improvement in the pore structure reduces the diffusion resistance of large oxygenates which will otherwise be deposited on the catalyst surface as coke which is generally encountered with microporous materials [45].

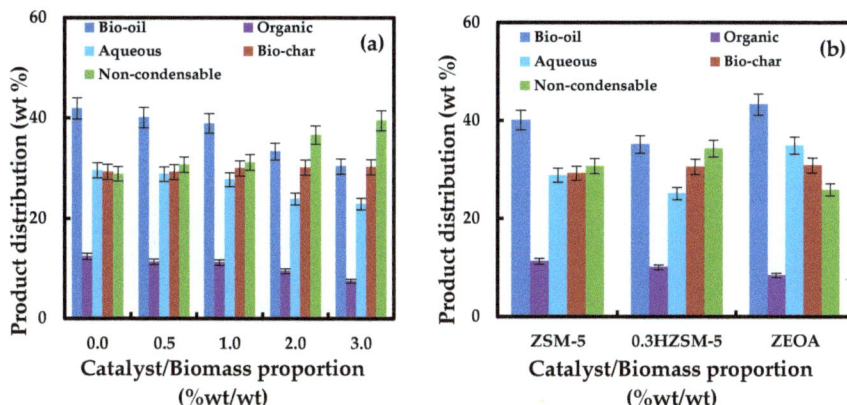

Figure 4. Pyrolysis products distribution (a) Effect of ZSM-5 loading; (b) impact of 0.3HZSM-5 and ZEOA at 0.5 wt % loading compared to ZSM-5.

3.3. Physicochemical Properties of Organic Phase Product

The properties of the organic phase bio-oil collected are summarized in Table 2. The pH values of the oil from all the ZSM-5 loadings and HZSM-5 were between 2.79 and 2.98, while the oil from ZEOA had a pH value of 3.57. Comparing the oil from the catalytic process with that from the non-catalytic process (pH of 2.71), some level of improvement in the acidity of the oil was recorded which can be attributed to a reduction of phenolic compounds through decarbonylation reactions. The water content of the catalytically produced oil was between 9.1 and 10.8 wt % compared to 8.64 wt % recorded for the oil from the non-catalytic process. Decreases in the density and viscosity of the oil from all the catalysts was observed compared to the oil from the non-catalytic process. This indicates that the catalysts increased the production of small molecules, which contributed to a lower viscosity [39]. A high content of reaction generally accounts for the lower density and viscosity of bio-oils [56–58]. From the ultimate analysis result (Table 2), the impact of catalyst on the elemental composition of the bio-oil was more pronounced on its carbon and oxygen contents, which also directly affected the heating value.

Table 2. Physicochemical properties of the organic phase bio-oils.

Bio-Oil		ZSM-5 (wt %)				0.3HZSM-5	ZEOA
Property	Raw	0.5	1.0	2.0	3.0	0.5	0.5
Proximate analysis							
pH	2.71 ± 0.01	2.83 ± 0.01	2.93 ± 0.01	2.98 ± 0.01	2.79 ± 0.01	2.91 ± 0.01	3.57 ± 0.01
H_2O (wt %)	8.64 ± 0.23	9.50 ± 0.25	9.60 ± 0.23	10.00 ± 0.26	10.20 ± 0.23	9.60 ± 0.24	10.80 ± 0.26
Viscosity (cP) *	2.82 ± 0.14	2.80 ± 0.13	2.80 ± 0.13	2.78 ± 0.13	2.74 ± 0.14	2.81 ± 0.15	2.70 ± 0.13
Density (g/cm³) *	1.082 ± 0.0	1.059 ± 0.0	1.056 ± 0.0	1.040 ± 0.0	1.002 ± 0.0	1.051 ± 0.0	0.998 ± 0.0
Ultimate analysis (wt % w.b)							
Carbon (C)	49.97 ± 1.50	59.92 ± 1.79	61.65 ± 1.66	63.84 ± 1.84	64.69 ± 1.82	65.61 ± 1.80	63.98 ± 1.79
Hydrogen (H)	6.79 ± 0.07	6.82 ± 0.07	6.63 ± 0.06	7.47 ± 0.07	6.6 ± 0.06	6.46 ± 0.07	6.34 ± 0.07
Nitrogen (N)	1.35 ± 0.03	0.95 ± 0.02	0.85 ± 0.02	0.53 ± 0.02	0.97 ± 0.02	0.89 ± 0.02	0.57 ± 0.02
Sulfur (S)	0.6 ± 0.01	0.51 ± 0.01	0.46 ± 0.01	0.43 ± 0.01	0.4 ± 0.01	0.44 ± 0.01	0.41 ± 0.01
Oxygen (O) **	41.29 ± 1.07	31.8 ± 0.86	30.41 ± 0.88	27.73 ± 0.83	27.34 ± 0.79	26.6 ± 0.88	27.34 ± 0.88
HHV (MJ/kg)	26.23 ± 0.10	28.29 ± 0.10	27.87 ± 0.10	27.57 ± 0.10	28.23 ± 0.10	28.24 ± 0.10	27.29 ± 0.10
Ultimate analysis (wt % d.b)							
C	54.7	66.21	68.2	70.93	71.17	72.58	71.73
H	6.38	6.37	6.15	7.07	6.15	5.97	5.76
O	36.79	25.81	24.2	20.93	21.18	19.99	21.41
HHV (MJ/kg)	28.92	31.5	31.07	30.88	31.29	31.48	30.87
DOD (%)	0.00	29.85	34.22	43.11	42.43	45.66	41.80

* Measured at 20 °C; ** by difference $[O = 100 - (C + H + N + S)]$.

All the catalytically produced oils have higher carbon and lower oxygen contents, which resulted in higher energy content compared to the oil produced without catalyst. Increased ZSM-5 loading and the improved pore characteristics of 0.3HZSM-5 and ZEOA produced oil with higher carbon and lower oxygen content. The dry basis elemental composition and relative degree of deoxygenation (DOD) was computed according to Equations (1), (2), (3) [59] and (4), respectively:

$$C_{dry} = \frac{C_{wet}}{\left[1 - \left(\frac{H_2O}{100}\right)\right]} \tag{1}$$

$$H_{dry} = \frac{[H_{wet}] - \left[H_2O \times \left(\frac{(2 \times MWH)}{(2MWH + MWO)}\right)\right]}{\left[1 - \left(\frac{H_2O}{100}\right)\right]} \tag{2}$$

$$O_{dry} = \frac{[O_{wet}] - \left[H_2O \times \left(\frac{(MWO)}{(2MWH + MWO)}\right)\right]}{\left[1 - \left(\frac{H_2O}{100}\right)\right]} \tag{3}$$

$$DOD(\%) = \left[1 - \left(\frac{O_{cat}}{O_{Ncat}}\right)\right] \times 100 \tag{4}$$

where C, H and O is carbon, hydrogen and oxygen content in wt %; H_2O is the water content of the oil (wt %); MWH and MWO is atomic weight of hydrogen and oxygen; O_{cat} and O_{Ncat} are the oxygen contents (wt %) of oil from the catalytic and non-catalytic process. The results show that better quality bio-oil can be produced with high catalyst loading, but at the expense of quantity.

3.4. Thermogravimetric Analysis of the Organic Phase Bio-Oil

The use of thermogravimetric method for analyzing bio-oil provides insights into the type or groups of organic compound present in the oil based on thermal characteristics such as evaporation and reactivity with respect to temperature. The bio-oil collected in this study was subjected to a thermogravimetric study (TGA) using a Perkin Elmer simultaneous thermal analyzer thermogravimetric analyzer (STA 6000) in a nitrogen atmosphere, flow rate 20 mL/min at temperatures between ambient to 500 °C and a heating rate of 10 °C/min. Approximately 10 mg of sample was used in each run and the results are shown in Figure 5.

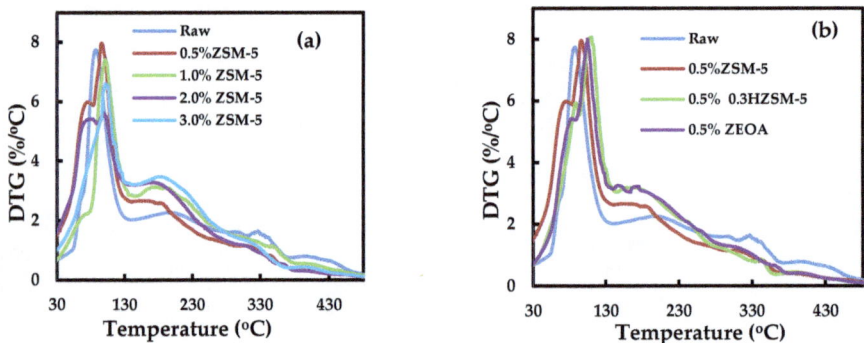

Figure 5. *Cont.*

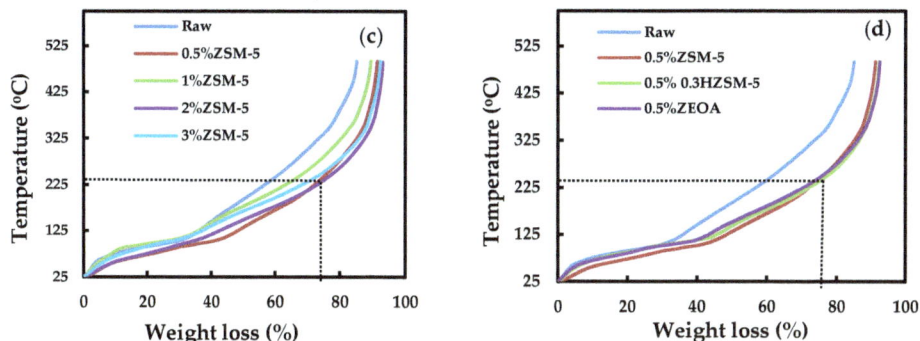

Figure 5. DTG (a,b) and TG (c,d) curves of organic bio-oil produced with and without catalysts.

The bio-oil fractions have been grouped into light, medium and heavy fractions. The light fractions consist of volatile organic compounds such as acids, alcohols and other compounds with boiling points close to the boiling point of water. Phenolics, furans and simple sugars such as levoglucosan constitute the medium fractions, while oligomers derived from hemicellulose, cellulose and lignin made up the heavy fractions [60]. The study by Garcia-Perez *et al.* [61] categorized bio-oil fractions into macro-groups such as volatile non-polar and polar compounds, monolignols, polar compounds with moderate volatility, sugars, extractive derived compounds, and heavy polar and non-polar compounds. From the results obtained in this present study (Figure 5), the DTG curves exhibited four (4) characteristic peaks at temperatures between of 45–100 °C, 140–230 °C, 275–365 °C and 370–440 °C. The peaks between 45 and 100 °C could be due to evaporation of the light fractions. ZSM-5 loading led to formation of a shoulder around 70–80 °C (Figure 5a). This represents a generation of more volatile organic compounds during pyrolysis with catalyst loading which may be ascribed to alkanes, alcohols and alkenols. Alkenols are generally intermediate products of the decarboxylation of acetic acid at elevated temperature [62]. The peak around 83 °C for the oil from the non-catalytic process shifted to 93, 96, 98 and 99 °C for ZSM-5 loadings of 0.5, 1.0, 2.0 and 3.0 wt %, respectively. These may be attributed to vaporization of volatile non-polar compounds such as aromatic hydrocarbons [61]. The peaks observed around 140–230 °C may be related to vaporization of phenolics and hydroxybenzenes. The peak in this temperature range was more pronounced with increased ZSM-5 loading which can be attributed to the presence of more hydroxybenzenes in the bio-oil produced with catalyst. Peaks observed at 275–365 °C and 370–440 °C may be attributed to extractive-derived compounds such as oligomers from the structural carbohydrates [60,62]. The peaks level off with ZSM-5 loading which are indications of a reduction of oligomers by ZSM-5 during the pyrolysis. A similar trend was also observed with 0.3HZSM-5 and ZEOA as presented in Figure 5b. The TG curves (Figure 5c,d) give the percentage of weight lost with respect to the temperature. It provides information about the volatile fraction of bio-oils. At 250 °C, the total weight loss of the raw bio-oil was about 60 wt %, which corresponds to the amount of volatiles present in it. This fraction increased to about 72 wt % in the oil produced by the catalytic process. The weight loss observed above 250 °C could be attributed to the thermal degradation of oligomers. Furthermore, this result suggests that only 60 wt % and 72 wt % of the bio-oil from the non-catalytic and catalytic pyrolysis can be analyzed with GC-MS. Hence, the injector temperature of 250 °C chosen in Section 2.2 above for the GC-MS analysis is adequate.

3.5. Functional Group Analysis of Bio-Oils

The FTIR spectra of chemical compounds in the bio-oil samples are shown in Figure 6a,b. The broad peak around 3420 cm^{-1} in all bio-oil samples is an indication that samples contain chemical compounds with hydroxyl groups (–OH) such as water, alcohols and phenols [63,64]. The peak became wider (Figure 6a) in the oil produced by the catalytic process with increasing ZSM-5 loading, which can be ascribed to the increased moisture level in the oil samples. The peak at a frequency around 2920 cm^{-1} is due to the C–H stretching vibrations of methyl and methylene groups which are common to all the bio-oil samples, indicating the presence of saturated hydrocarbons while the broad peak at frequency around 2091 cm^{-1} is attributed to $C \equiv C$ functional groups which denote the presence of alkynes [63,64]. Sharp vibrations observed around 1707 cm^{-1} in the oil produced by the non-catalytic process are ascribed to $C = O$ which signifies the presence of aldehydes, ketones or carboxylic acids. This peak diminished in all the oils from the ZSM-5 process with increasing catalyst loading. Also, the stretching vibration observed in the former around 1625 cm^{-1} due to the $C = O$ functional group of ketones disappeared completely in the latter. These observations confirm the extent of deoxygenation in the oil by the ZSM-5 catalyst. The vibrations around 1462 and 1384 cm^{-1} common to all the samples are ascribed $C = H$ and $C - H$, indicating the presence of alkenes/aromatic hydrocarbons and alkanes, respectively [65–67]. The sharp band around 1220 is due to $C - O$ vibrations indicating the presence of alcohols and esters. The fingerprint region bands between 900 and 620 cm^{-1} are ascribed to aromatic $C - H$ bending vibrations while the ones at around 550 cm^{-1} are due to alkyl halides [63–68]. Similar spectral characteristics were observed with bio-oil produced with HZSM-5 and ZEOA (Figure 6b) with respect to the oil from the non-catalytic process. However, comparing the spectra of oils from HZSM-5 and ZEOA with ZSM-5, improvements in the peaks around 1462 and 1384 cm^{-1}, and the fingerprint region between 900 and 620 cm^{-1} were also noted, which implies more alkenes in the oil. This may be connected to the improved pore characteristics of the respective catalysts which promote the deoxygenation of large oxygen-containing organic molecules to smaller hydrocarbons [45].

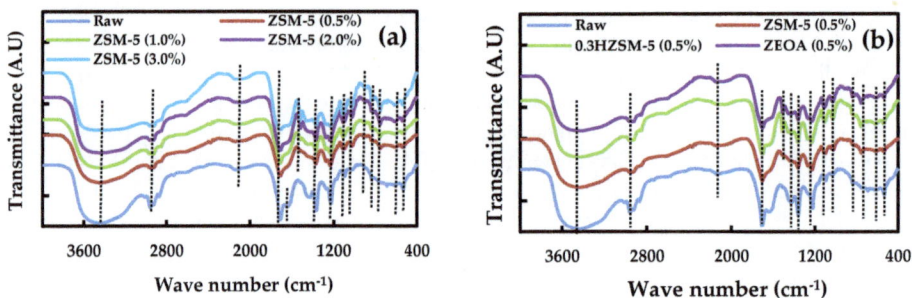

Figure 6. Averaged FTIR spectra (**a**) auto-smoothed and (**b**) auto-baseline corrected) of bio-oil samples.

3.6. GC-MS Analysis of the Organic Phase Bio-Oil

Identification of chemical compounds in the bio-oil samples was carried out by GC-MS. A library search using the MS NIST Library 2011 revealed that the oil was made up of various hydrocarbons, aromatics, phenols, furans, acids, ketones and alcohols . These organic compounds were further categorized into hydrocarbons, aromatics, phenolics, alcohols and other oxygenates (Figure 7) in order to evaluate the effect of ZSM-5 catalyst loading on the distribution of chemical compounds in the oil. The hydrocarbons consist of alkanes, alkenes and alkynes which account for 23.2% in the oil from the non-catalytic process while 29.52%, 23.20%, 28.89% and 26.95% was recorded in the oil produced with 0.5, 1.0, 2.0 and 3.0 wt % catalyst loading, respectively. The proportion of olefins in the total hydrocarbons in the oil from catalytic process (12.4%–29.53%) was higher than that obtained from the non-catalytic process (1.55%). This observation can be attributed to the acidity of the ZSM-5 catalyst which is known for the selective production of olefins through cracking of oxygenated compounds at higher temperatures similar to the temperature (600 °C) used in this study [29,69,70]. The aromatic hydrocarbons were detected only in the oil from the catalytic process which consist of 4a-Methyl-1,2,4a,5,8,8a-hexahydro-naphthalene, 1,1′-ethylidenebisbenzene, and bis (methylthio) methylbenzene. These compounds are produced via a series of complex chemical reactions such as cracking, oligomerization, dehydrogenation, and aromatization promoted by the Brønsted acid cites of the ZSM-5. With increasing catalyst loading, the total aromatic yield increased. 12.18% and 15.19% naphthalene was recorded with 0.5 and 1.0 wt % ZSM-5 loading, respectively, which is mainly the product of condensed fragments from the surface active cites of the ZSM-5 while 15.69% and 18.88% of benzene was observed with 2.0 and 3.0 wt % catalyst loading. Similar observations have been reported in the literature [23,31,71]. The phenolics, mainly phenols and other oxygenates observed in the bio-oil decreased with catalyst loading as a result of dehydration, decarbonylation and decarboxylation which transformed them to smaller molecular units such as benzene [23,31,71]. Comparing the product distribution of ZSM-5 with 0.3HZSM-5 and ZEOA, the hydrocarbon yield from 0.3HZSM-5 and ZEOA was 23.51% and 30.69% (Figure 7b), respectively, compared to ZSM-5 which had 29.52%.

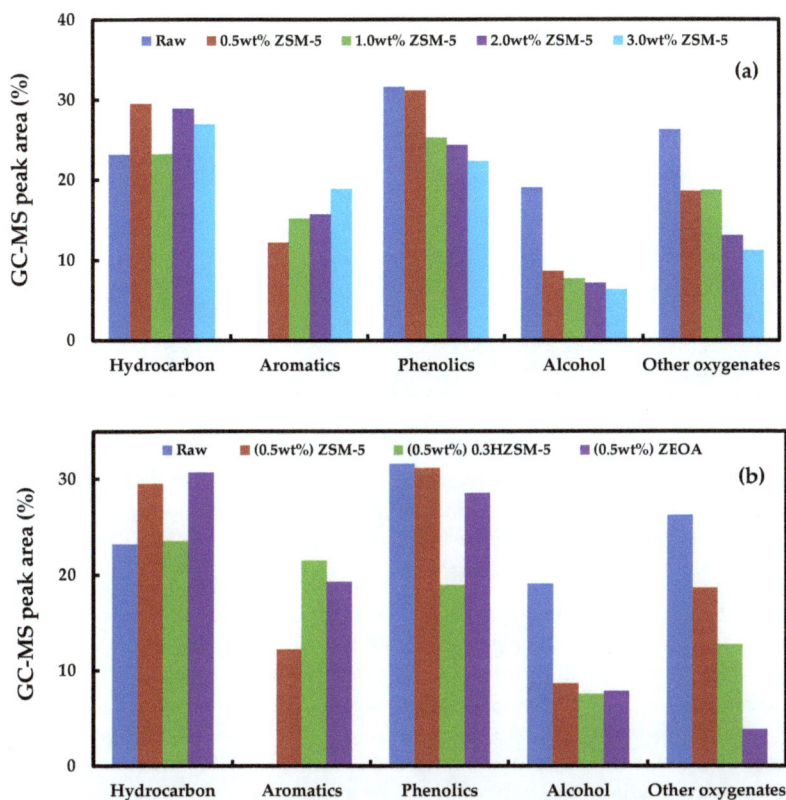

Figure 7. Groups of organic compounds detected in the bio-oil samples using GC-MS (**a**) ZSM-5 loading (0.5–3.0 wt %); (**b**) 0.5 wt % loading of ZSM-5, 0.3HZSM-5 and ZEOA.

The aromatic production was 21.45% and 19.22% with 0.3HZSM-5 and ZEOA while that recorded with ZSM-5 was 12.18%. Phenolics recorded with ZSM-5 were 31.14%, but decreased to 28.53% and 18.89% with ZEOA and 0.3HZSM-5. Also, other oxygenates decreased with ZEOA (3.77%) and 0.3HZSM-5 (12.69%) while 18.59% oxygenate was observed with ZSM-5. The increase in the aromatics, hydrocarbons and reduction in the phenolics and other oxygenates in the oil produced with ZEOA and 0.3HZSM-5 can be linked to the mesoporousity of the respective catalyst which perhaps reduces the steric hindrance of large organic molecules associated with the microstructure of ZSM-5 [45,69,72]. Furthermore, the composition of aromatics from 0.3HZSM-5 and ZEOA was mainly benzene, compared to the aromatics from ZSM-5 which mainly consisted of naphthalene (PAH). The increase in the benzene content and reduction of phenols could also be attributed to the differences in the amount of acid sites. Similar observations have been reported in the literature [45,72].

3.7. GC Analysis of the Non-Condensable Gas

The composition of the non-condensable gas analyzed by GC is summarized in Table 3. High levels of methane in the non-condensable gas from the non-catalytic pyrolysis implies thermal cracking mechanisms that produce small organic molecules during the pyrolysis. The proportion of methane in the gas decreased in the catalytic pyrolysis. With increasing ZSM-5 loading, a significant drop in the methane yield was observed, which suggests that the catalysts promoted the conversion of methane precursors to form aromatic hydrocarbons [73]. Higher composition of CO and CO_2 in the non-condensables from the catalytic process was recorded, compared to the non-catalytic pyrolysis which is an indication of decarbonylation and decarboxylation [27,49,66–74].

Table 3. Composition of non-condensable gases.

Catalyst Type	Catalyst Loading (wt %)	Gas Composition (vol %) N_2-Free Basis			
		CH_4	H_2	CO	CO_2
Raw	0.00	2.72	0.56	14.04	25.32
ZSM-5	0.50	2.40	0.32	20.87	29.94
ZSM-5	1.00	2.26	0.31	22.97	33.32
ZSM-5	2.00	2.01	0.34	24.67	36.07
ZSM-5	3.00	1.94	0.36	26.14	37.95
ZEOA	0.50	2.19	0.33	26.16	31.65
0.3HZSM-5	0.50	2. 32	0.29	29.23	31.80

4. Conclusions

This study gives a background on the catalytic pyrolysis of Napier grass. It dwells on the operating parameters that affect the product distribution and the quality of the resulting bio-oil and discusses experimental results using three zeolite-based catalysts: ZSM-5, 0.3HZSM-5 and zinc exchanged Zeolite A. The results summary can be summarized as follows:

- ZSM-5 catalyst loading between 0.5 and 1.0 wt % had no significant impact on the oil yield compared to higher catalyst loadings at 2.0 and 3.0 wt %. The yield of non-condensable gas increased with catalyst loading. Impact of ZSM-5 on the yield of bio-char was minimal.
- Organic compounds in the bio-oil produced with ZSM-5 were made up of mainly hydrocarbons, aromatics and phenols. Catalyst loadings between 0.5 and 1.0 wt % promoted the yield of polyaromatic hydrocarbon (naphthalene) while benzene dominated the aromatics when 2.0 and 3.0 wt % catalyst loading were employed.
- Desilication of ZSM-5 with NaOH produced a mesoporous 0.3HZSM-5. Bio-oil yield decreased with HZSM-5 and increased with ZEOA compared to ZSM-5 at 0.5 wt % loading. The organic phase composition of the bio-oil from 0.3HZSM-5 and ZEOA were lower than that from of ZSM-5. Higher hydrocarbon yield was recorded, particularly with ZEOA, and the aromatics were mainly benzenes. Reduction in the phenolic content and other oxygenated compounds were also recorded. This observation was attributed to the improved pore structure and the acid sites of the catalysts. Higher composition of CO and CO_2 was observed in the non-condensable gas from the catalytic pyrolysis compared to the non-catalytic pyrolysis and was attributed to decarbonylation and decarboxylation reactions.
- This study has demonstrated that bio-oil with high fuel quality and other value added chemicals can be produced from pyrolysis of Napier grass over acidic zeolite-based catalysts.

Acknowledgments: This project was supported by the Crops for the Future (CFF) and the University of Nottingham under the grant BioP1-005.

Author Contributions: Isah Yakub Mohammed, Yousif Abdalla Abakr, Feroz Kabir Kazi and Suzana Yusup conceived and designed the experiments; Yahaya Muhammad Sani and Peter Adeniyi Alaba synthesized and characterized the catalysts; Isah Yakub Mohammed and Yousif Abdalla Abakr produced and characterized the bio-oil; Yahaya Muhammad Sani and Peter Adeniyi Alaba analyzed the gas composition; Feroz Kabir Kazi and Suzana Susup analyzed the data; Isah Yakub Mohammed, Yousif Abdalla Abakr, Feroz Kabir Kazi and Suzana Yusup wrote the manuscript.

Conflicts of Interest: The authors declare no conflict of interest.

References

1. Yakub, M.I.; Mohamed, S.; Danladi, S.U. Technical and Economic Considerations of Post-combustion Carbon Capture in a Coal Fired Power Plant. *Int. J. Adv. Eng. Technol.* **2014**, *7*, 1549–1581.

2. Mohammed, I.Y. Optimization and Sensitivity Analysis of Post-combustion Carbon Capture Using DEA Solvent in a Coal Fired Power Plant. *Int. J. Adv. Eng. Technol.* **2015**, *7*, 1681–1690.

3. Mohammed, I.Y.; Samah, M.; Mohamed, A.; Sabina, G. Comparison of Selexol™ and Rectisol® Technologies in an Integrated Gasification Combined Cycle (IGCC) Plant for Clean Energy Production. *Int. J. Eng. Res.* **2014**, *3*, 742–744. [CrossRef]

4. Yakub, M.I.; Abdalla, A.Y.; Feroz, K.K.; Suzana, Y.; Ibraheem, A.; Chin, S.A. Pyrolysis of Oil Palm Residues in a Fixed Bed Tubular Reactor. *J. Power Energy Eng.* **2015**, *3*, 185–193. [CrossRef]

5. Gebreslassie, B.H.; Slivinsky, M.; Wang, B.; You, F. Life cycle optimization for sustainable design and operations of hydrocarbon biorefinery via fast pyrolysis, hydrotreating and hydrocracking. *Comput. Chem. Eng.* **2013**, *50*, 71–91. [CrossRef]

6. Liew, W.H.; Hassim, M.H.; Ng, D.S.K. Review of evolution, technology and sustainability assessments of biofuel production. *J. Clean. Prod.* **2014**, *71*, 11–29. [CrossRef]

7. Park, S.R.; Pandey, A.K.; Tyagi, V.V.; Tyagi, S.K. Energy and exergy analysis of typical renewable energy systems. *Renew. Sustain. Energy Rev.* **2014**, *30*, 105–123. [CrossRef]

8. Ming, Z.; Ximei, L.; Yulong, L.; Lilin, P. Review of renewable energy investment and financing in China: Status, mode, issues and countermeasures. *Renew. Sustain. Energy Rev.* **2014**, *31*, 23–37. [CrossRef]

9. Nigam, P.S.; Singh, A. Production of liquid biofuels from renewable resources. *Prog. Energy Combust. Sci.* **2011**, *37*, 52–68. [CrossRef]

10. Srirangan, K.; Akawi, L.; Moo-Young, M.; Chou, C.P. Towards sustainable production of clean energy carriers from biomass resources. *Appl. Energy* **2012**, *100*, 172–186. [CrossRef]

11. Samson, R.; Mani, S.; Boddey, R.; Sokhansanj, S.; Quesada, D.; Urquiaga, S.; Reis, V.; Ho Lem, C. The Potential of C4 Perennial Grasses for Developing a Global BIOHEAT Industry. *Crit. Rev. Plant Sci.* **2005**, *24*, 461–495. [CrossRef]

12. Mohammed, I.Y.; Abakr, Y.A.; Kazi, F.K.; Yusup, S.; Alshareef, I.; Chin, S.A. Comprehensive Characterization of Napier Grass as a Feedstock for Thermochemical Conversion. *Energies* **2015**, *8*, 3403–3417. [CrossRef]

13. Mohammed, I.Y.; Abakr, Y.A.; Kazi, F.K.; Yusup, S.; Alshareef, I.; Chin, S.A. Pyrolysis of Napier Grass in a Fixed Bed Reactor: Effect of Operating Conditions on Product Yields and Characteristics. *BioResources* **2015**, *10*, 6457–6478. [CrossRef]

14. Bridgwater, A.V. Review of Fast Pyrolysis of biomass and product upgrading. *Biomass Bioenergy* **2012**, *38*, 68–94. [CrossRef]

15. Kebelmann, K.; Hornung, A.; Karsten, U.; Griffiths, G. Intermediate pyrolysis and product identification by TGA and Py-GC/MS of green microalgae and their extracted protein and lipid components. *Biomass Bioenergy* **2013**, *49*, 38–48. [CrossRef]

16. Mahmood, A.S.N.; Brammer, J.G.; Hornung, A.; Steele, A.; Poulston, S. The inter-mediate pyrolysis and catalytic steam reforming of Brewers spent grain. *J. Anal. Appl. Pyrolysis* **2013**, *103*, 328–342. [CrossRef]

17. Tripathi, M.; Sahu, J.N.; Ganesan, P. Effect of process parameters on production of biochar from biomass waste through pyrolysis: A review. *Renew. Sustain. Energy Rev.* **2016**, *55*, 467–481. [CrossRef]

18. Iliopoulou, E.F.; Stefanidis, S.D.; Kalogiannis, K.G.; Delimitis, A.; Lappas1, A.A.; Triantafyllidis, K.S. Catalytic upgrading of biomass pyrolysis vapors using transition metal-modified ZSM-5 zeolite. *Appl. Catal. B Environ.* **2010**, *127*, 281–290. [CrossRef]

19. Carlson, T.R.; Cheng, Y.; Jae, J.; Huber, G.W. Production of green aromatics and olefins by catalytic fast pyrolysis of wood sawdust. *Energy Environ. Sci.* **2011**, *4*, 145–161. [CrossRef]
20. Compton, D.L.; Jackson, M.A.; Mihalcik, D.J.; Mullen, C.A.; Boateng, A.A. Catalytic pyrolysis of oak via pyroprobe and bench scale, packed bed pyrolysis reactors. *J. Anal. Appl. Pyrolysis* **2011**, *90*, 174–181. [CrossRef]
21. Du, S.; Sun, Y.; Gamliel, D.P.; Valla, J.A.; Bollas, G.M. Catalytic pyrolysis of miscanthus × giganteus in a spouted bed reactor. *Bioresour. Technol.* **2014**, *169*, 188–197. [CrossRef] [PubMed]
22. Choi, Y.S.; Lee, K.-H.; Zhang, J.; Brown, R.C.; Shanks, B.H. Manipulation of chemical species in bio-oil using *in situ* catalytic fast pyrolysis in both a bench-scale fluidized bed pyrolyzer and micropyrolyzer. *Biomass Bioenergy* **2015**, *81*, 256–264. [CrossRef]
23. Park, Y.-K.; Yoo, M.L.; Jin, S.H.; Park, S.H. Catalytic fast pyrolysis of waste pepper stems over HZSM-5. *Renew. Energy* **2015**, *79*, 20–27. [CrossRef]
24. Jae, J.; Tompsett, G.A.; Foster, A.J.; Hammond, K.D.; Auerbach, S.M.; Lobo, R.F.; Huber, G.W. Investigation into the shape selectivity of zeolite catalysts for biomass conversion. *J. Catal.* **2011**, *279*, 257–268. [CrossRef]
25. Zhang, B.; Zhong, Z.-P.; Wang, X.-B.; Ding, K.; Song, Z.-W. Catalytic upgrading of fast pyrolysis biomass vapors over fresh, spent and regenerated ZSM-5 zeolites. *Fuel Process. Technol.* **2015**, *138*, 430–434. [CrossRef]
26. Carlson, T.R.; Vispute, T.P.; Huber, G.W. Green gasoline by catalytic fast pyrolysis of solid biomass derived compounds. *ChemSusChem* **2008**, *1*, 397–400. [CrossRef] [PubMed]
27. Gamliel, D.P.; Du, S.; Bollas, G.M.; Valla, J.A. Investigation of *in situ* and ex situ catalytic pyrolysis of miscanthus-giganteus using a PyGC–MS microsystem and comparison with a bench-scale spouted-bed reactor. *Bioresour. Technol.* **2015**, *191*, 187–196. [CrossRef] [PubMed]
28. Yildiz, G.; Ronsse, F.; Venderbosch, R.; Van Duren, R.; Kersten, S.R.A.; Prins, W. Effect of biomass ash in catalytic fast pyrolysis of pine wood. *Appl. Catal. B Environ.* **2015**, *168–169*, 203–211. [CrossRef]
29. Liu, G.; Wright, M.M.; Zhao, Q.; Brown, R.C. Catalytic fast pyrolysis of duckweed: Effects of pyrolysis parameters and optimization of aromatic production. *J. Anal. Appl. Pyrolysis* **2015**, *112*, 29–36. [CrossRef]
30. Kim, B.-S.; Kim, Y.-M.; Jae, J.; Watanabe, C.; Kim, S.; Jung, S.-C.; Kim, S.C.; Park, Y.-K. Pyrolysis and catalytic upgrading of Citrus unshiu peel. *Bioresour. Technol.* **2015**, *194*, 312–319. [CrossRef] [PubMed]
31. Ojha, D.K.; Vinu, R. Resource recovery via catalytic fast pyrolysis of polystyrene using zeolites. *J. Anal. Appl. Pyrolysis* **2015**, *113*, 349–359. [CrossRef]
32. *Standard Test Method for Water Using Volumetric Karl Fischer Titration*; ASTM International: West Conshohocken, PA, USA, 2001.
33. Mohammed, I.Y.; Kazi, F.K.; Abakr, Y.A.; Yusuf, S.; Razzaque, M.A. Novel Method for the Determination of Water Content and Higher Heating Value of Pyrolysis Oil. *BioResources* **2015**, *10*, 2681–2690. [CrossRef]
34. *Standard Test Method for Heat of Combustion of Liquid Hydrocarbon Fuels by Bomb Calorimeter*; ASTM International: West Conshohocken, PA, USA, 2009.
35. Alaba, P.A.; Sani, Y.M.; Mohammed, I.Y.; Abakr, Y.A.; Daud, W.M.A.W. Synthesis and application of hierarchical mesoporous HZSM-5 for biodiesel production from shea butter. *J. Taiwan Inst. Chem. Eng.* **2016**, *59*, 405–412. [CrossRef]
36. Shirazi, L.; Jamshidi, E.; Ghasemi, M.R. Effect of Si/Al ratio of ZSM-5 zeolite on its morphology, acidity and crystal size. *Cryst. Res. Technol.* **2008**, *43*, 1300–1306. [CrossRef]
37. Gong, F.; Yang, Z.; Hong, C.; Huang, W.; Ning, S.; Zhang, Z.; Xu, Y.; Li, Q. Selective conversion of bio-oil to light olefins: Controlling catalytic cracking for maximum olefins. *Bioresour. Technol.* **2011**, *102*, 9247–9254. [CrossRef] [PubMed]
38. Suyitno; Purbaningrum, P.S.; Danardono, D.; Salem, A.E.; Mansur, F.A. Synthesis of zeolite Socony mobil from blue silica gel and rice husk ash as catalysts for hydrothermal liquefaction. *J. Eng. Sci. Technol.* **2015**, *10*, 982–993.
39. Shaikh, I.R.; Shaikh, R.A.; Shaikh, A.A.; War, J.A.; Hangirgekar, S.P.; Shaikh, A.L.; Shaikh, P.R.; Shaikh, R.R. H-ZSM-5 Zeolite Synthesis by Sourcing Silica from the Wheat Husk Ash: Characterization and Application as a Versatile Heterogeneous Catalyst in Organic Transformations including Some Multicomponent Reactions. *J. Catal.* **2015**, *2015*, 805714. [CrossRef]
40. Huang, Y.; Wei, L.; Julson, J.; Gao, Y.; Zhao, X. Converting pine sawdust to advanced biofuel over HZSM-5 using a two-stage catalytic pyrolysis reactor. *J. Anal. Appl. Pyrolysis* **2015**, *111*, 148–155. [CrossRef]

41. Wang, Y.; Fan, S.; Zhang, J.; Zhao, T.-S. Effect of Synthesis Conditions on the yields and properties of HZSM-5. *Cryst. Res. Technol.* **2015**, *50*, 522–527. [CrossRef]

42. Ayele, L.; Perez-Pariente, J.; Chebude, Y.; Díaz, I. Synthesis of zeolite A from Ethiopian kaolin. *Microporous Mesoporous Mater.* **2015**, *215*, 29–36. [CrossRef]

43. Bieseki, L.; Penha, F.G.; Perghera, S.B.C. Zeolite A Synthesis Employing a Brazilian Coal Ash as the Silicon and Aluminum Sourceand its Applications in Adsorption and Pigment Formulation. *Mater. Res.* **2013**, *16*, 38–43. [CrossRef]

44. Dyballa, M.; Obenaus, U.; Lang, S.; Gehring, B.; Traa, Y.; Koller, H.; Hunger, M. Brønsted sites and structural stabilization effect of acidic low-silica zeolite A prepared by partial ammonium exchange. *Microporous Mesoporous Mater.* **2015**, *212*, 110–116. [CrossRef]

45. Sing, K.S.W.; Everett, D.H.; Haul, R.A.W.; Moscou, L.; Poerotto, R.A.; Rouquerol, J.; Siemieniewska, T. Reporting physisorption data for gas/solid system with special reference to the determination of surface area and porosity. *Pure Appl. Chem.* **1985**, *57*, 603–619. [CrossRef]

46. Li, J.; Li, X.; Zhou, G.; Wang, W.; Wang, C.; Komarneni, S.; Wang, Y. Catalytic fast pyrolysis of biomass with mesoporous ZSM-5 zeolites prepared by desilication with NaOH solutions. *Appl. Catal. A Gen.* **2014**, *470*, 115–122. [CrossRef]

47. You, S.J.; Park, E.D. Effects of dealumination and desilication of H-ZSM-5 on xylose dehydration. *Microporous Mesoporous Mater.* **2014**, *186*, 121–129. [CrossRef]

48. Wang, L.; Lei, H.; Ren, S.; Bu, Q.; Liang, J.; Wei, Y.; Liu, Y.; Lee, G.-S.J.; Chen, S.; Tang, J.; *et al.* Aromatics and phenols from catalytic pyrolysis of Douglas fir pellets in microwave with ZSM-5 as a catalyst. *J. Anal. Appl. Pyrolysis* **2012**, *98*, 194–200. [CrossRef]

49. Jae, J.; Coolman, R.; Mountziaris, T.J.; Huber, G.W. Catalytic fast pyrolysis of lignocellulosic biomass in a process development unit with continual catalyst addition and removal. *Chem. Eng. Sci.* **2014**, *108*, 33–46. [CrossRef]

50. Naqvi, S.R.; Uemura, Y.; Yusup, S.B. Catalytic pyrolysis of paddy husk in a drop type pyrolyzer for bio-oil production: The role of temperature and catalyst. *J. Anal. Appl. Pyrolysis* **2014**, *106*, 57–62. [CrossRef]

51. Park, H.J.; Park, K.-H.; Jeon, J.-K.; Kim, J.; Ryoo, R.; Jeong, K.-E.; Park, S.H.; Park, Y.-K. Production of phenolics and aromatics by pyrolysis of Miscanthus. *Fuel* **2012**, *97*, 379–384. [CrossRef]

52. Elordi, G.; Olazar, M.; Lopez, G.; Castaño, P.; Bilbao, J. Role of pore structure in the deactivation of zeolites (HZSM-5, Hb and HY) by coke in the pyrolysis of polyethylene in a conical spouted bed reactor. *Appl. Catal. B Environ.* **2011**, *102*, 224–231. [CrossRef]

53. Lappas, A.A.; Samolada, M.C.; Iatridis, D.K.; Voutetakis, S.S.; Vasalos, I.A. Biomass pyrolysis in a circulating fluid bed reactor for the production of fuels and chemicals. *Fuel* **2002**, *81*, 2087–2095. [CrossRef]

54. Veses, A.; Puértolas, B.; López, M.J.; Callén, M.S.; Solsona, B.; García, T. Promoting Deoxygenation of Bio-Oil by Metal-Loaded Hierarchical ZSM-5 Zeolites. *ACS Sustain. Chem. Eng.* **2016**, *4*, 1653–1660. [CrossRef]

55. Nilsen, M.H.; Antonakou, E.; Bouzga, A.; Lappas, A.; Mathisen, K.; Stöcker, M. Investigation of the effect of metal sites in Me-Al-MCM-41 (Me = Fe, Cu or Zn) on the catalytic behavior during the pyrolysis of wooden based biomass. *Microporous Mesoporous Mater.* **2007**, *105*, 189–203. [CrossRef]

56. Kelkar, S.; Saffron, C.M.; Andreassi, K.; Li, Z.; Murkute, A.; Miller, D.J.; Pinnavaia, T.J.; Kriegel, R.M. A survey of catalysts for aromatics from fast pyrolysis of biomass. *Appl. Catal. B Environ.* **2015**, *174–175*, 85–95. [CrossRef]

57. Imam, T.; Capareda, S. Characterization of bio-oil, syn-gas and bio-char from switchgrass pyrolysis at various temperatures. *J. Anal. Appl. Pyrolysis* **2012**, *93*, 170–177. [CrossRef]

58. Abu Bakar, M.S.; Titiloye, J.O. Catalytic pyrolysis of rice husk for bio-oil production. *J. Anal. Appl. Pyrolysis* **2013**, *103*, 362–368. [CrossRef]

59. Fan, Y.; Cai, Y.; Li, X.; Yin, H.; Yu, N.; Zhang, R.; Zhao, W. Rape straw as a source of bio-oil via vacuum pyrolysis: Optimization of bio-oil yield using orthogonal design method and characterization of bio-oil. *J. Anal. Appl. Pyrolysis* **2014**, *106*, 63–70. [CrossRef]

60. De Miguel Mercader, F.; Groeneveld, M.J.; Kersten, S.R.A.; Venderbosch, R.H.; Hogendoorn, J.A. Pyrolysis oil upgrading by high pressure thermal treatment. *Fuel* **2010**, *89*, 2829–2837. [CrossRef]

61. Li, X.; Gunawan, R.; Wang, Y.; Chaiwat, W.; Hu, X.; Gholizadeh, M.; Mourant, D.; Bromly, J.; Li, C.-Z. Upgrading of bio-oil into advanced biofuels and chemicals. Part III. Changes in aromatic structure and coke forming propensity during the catalytic hydrotreatment of a fast pyrolysis bio-oil with Pd/C catalyst. *Fuel* **2014**, *116*, 642–649. [CrossRef]

62. Garcia-Perez, M.; Chaala, A.; Pakdel, H.; Kretschmer, D.; Roy, C. Characterization of bio-oils in chemical families. *Biomass Bioenergy* **2007**, *31*, 222–242. [CrossRef]

63. Nguyen, M.T.; Sengupta, D.; Raspoet, G.; Vanquickenborne, L.G. Theoretical Study of the Thermal Decomposition of Acetic Acid: Decarboxylation Versus Dehydration. *J. Phys. Chem.* **1995**, *99*, 11883–11888. [CrossRef]

64. Guo, Y.; Song, W.; Lu, J.; Ma, Q.; Xu, D.; Wang, S. Hydrothermal liquefaction of Cyanophyta: Evaluation of potential bio-crude oil production and component analysis. *Algal Res.* **2015**, *11*, 242–247. [CrossRef]

65. Bordoloi, N.; Narzari, R.; Chutia, R.S.; Bhaskar, T.; Kataki, R. Pyrolysis of Mesua ferrea and Pongamia glabra seed cover: Characterization of bio-oil and its sub-fractions. *Bioresour. Technol.* **2015**, *178*, 83–89. [CrossRef] [PubMed]

66. Lazdovica, K.; Liepina, L.; Kampars, V. Comparative wheat straw catalytic pyrolysis in the presence of zeolites, Pt/C, and Pd/C by using TGA-FTIR method. *Fuel Process. Technol.* **2015**, *138*, 645–653. [CrossRef]

67. Yorgun, S.; Yildiz, D. Slow pyrolysis of paulownia wood: Effects of pyrolysis parameters on product yields and bio-oil characterization. *J. Anal. Appl. Pyrolysis* **2015**, *114*, 68–78. [CrossRef]

68. Ben, H.; Ragauskas, A.J. Torrefaction of Loblolly Pine. *Green Chem.* **2012**, *14*, 72–76. [CrossRef]

69. Gudka, B.; Jones, J.M.; Lea-Langton, A.R.; Williams, A.; Saddawi, A. A review of the mitigation of deposition and emission problems during biomass combustion through washing pre-treatment. *J. Energy Inst.* **2015**. in press. [CrossRef]

70. Zhang, H.; Xiao, R.; Jin, B.; Xiao, G.; Chen, R. Biomass catalytic pyrolysis to produce olefins and aromatics with a physically mixed catalyst. *Bioresour. Technol.* **2013**, *140*, 256–262. [CrossRef] [PubMed]

71. Rezaei, P.S.; Shafaghat, H.; Daud, W.M.A.W. Production of green aromatics and olefins by catalytic cracking of oxygenate compounds derived from biomass pyrolysis: A review. *Appl. Catal. A Gen.* **2014**, *469*, 490–511. [CrossRef]

72. Alaba, P.A.; Sani, Y.M.; Mohammed, I.Y.; Daud, W.M.A.W. Insight into catalyst deactivation mechanism and suppression techniques in thermocatalytic deoxygenation of bio-oil over zeolites. *Rev. Chem. Eng.* **2015**, *32*, 71–91. [CrossRef]

73. Puertolas, B.; Veses, A.; Callen, M.S.; Mitchell, S.; Garcia, T.; Perez-Ramirez, J. Porosity-Acidity Interplay in Hierarchical ZSM-5 Zeolites for Pyrolysis Oil Valorization to Aromatics. *Chem. Sus. Chem.* **2015**, *8*, 3283–3293. [CrossRef] [PubMed]

74. Fischer, A.; Du, S.; Valla, J.A.; Bollas, G.M. The effect of temperature, heating rate, and ZSM-5 catalyst on the product selectivity of the fast pyrolysis of spent coffee grounds. *RSC Adv.* **2015**, *5*, 29252–29261. [CrossRef]

energies

Article

Ethanol Production from Sweet Sorghum Juice at High Temperatures Using a Newly Isolated Thermotolerant Yeast *Saccharomyces cerevisiae* DBKKU Y-53

Sunan Nuanpeng [1,2], Sudarat Thanonkeo [3], Mamoru Yamada [4] and Pornthap Thanonkeo [2,5,*]

[1] Graduate School, Khon Kaen University, Khon Kaen 40002, Thailand; nuanpeng2523@gmail.com
[2] Department of Biotechnology, Faculty of Technology, Khon Kaen University, Khon Kaen 40002, Thailand
[3] Walai Rukhavej Botanical Research Institute, Mahasarakham University, Maha Sarakham 44150, Thailand; sthanonkeo@gmail.com
[4] Department of Biological Chemistry, Faculty of Agriculture, Yamaguchi University, Yamaguchi 753-8515, Japan; m-yamada@yamaguchi-u.ac.jp
[5] Fermantation Research Center for Value Added Agricultural Products, Khon Kaen University, Khon Kaen 40002, Thailand
[*] Correspondence: portha@kku.ac.th; Tel.: +66-81-9743340; Fax: +66-43-362-121

Academic Editor: Tariq Al-Shemmeri
Received: 21 February 2016; Accepted: 15 March 2016; Published: 31 March 2016

Abstract: Ethanol production at elevated temperatures requires high potential thermotolerant ethanol-producing yeast. In this study, nine isolates of thermotolerant yeasts capable of growth and ethanol production at high temperatures were successfully isolated. Among these isolates, the newly isolated thermotolerant yeast strain, which was designated as *Saccharomyces cerevisiae* DBKKU Y-53, exhibited great potential for ethanol production from sweet sorghum juice (SSJ) at high temperatures. The maximum ethanol concentrations produced by this newly isolated thermotolerant yeast at 37 °C and 40 °C under the optimum cultural condition were 106.82 g·L^{-1} and 85.01 g·L^{-1}, respectively, which are greater than values reported in the literatures. It should be noted from this study with SSJ at a sugar concentration of 250 g·L^{-1} and an initial pH of 5.5 without nitrogen supplementation can be used directly as substrate for ethanol production at high temperatures by thermotolerant yeast *S. cerevisiae* DBKKU Y-53. Gene expression analysis using real-time RT-PCR clearly indicated that growth and ethanol fermentation activities of the thermotolerant yeast *S. cerevisiae* DBKKU Y-53 at a high temperature (40 °C) were not only restricted to the expression of genes involved in the heat-shock response, but also to those genes involved in ATP production, trehalose and glycogen metabolism, and protein degradation processes were also involved.

Keywords: high temperature fermentation; real-time RT-PCR; *Saccharomyces cerevisiae*; sweet sorghum; thermotolerant yeast; gene expression

1. Introduction

Bioethanol is a clean, renewable, environmental friendly source of fuel energy that can be produced from different feedstocks and conversion technologies. It is one of the most promising substitutes for fossil energy and has high potential to replace petroleum-based fossil fuels [1–3]. Bioethanol can be used directly or blended with gasoline (known as "gasohol") to power engines without modification. Development of bioethanol is not only addressing climate change due to the burning of petroleum-based fuels but is also of great significance in protecting national energy security and promoting rural economic growth [3,4].

Approximately 60% of global bioethanol is produced from sugar crops, the remaining 40% is produced from starchy grains [5]. Sweet sorghum (*Sorghum bicolor* L. Moench) is one of the most promising sugar crops for industrial bioethanol production. It is a C4 plant, is similar to sugarcane, and belongs to the grass family. Due to a high photosynthetic efficiency, it is known for high carbon assimilation and the ability to store high levels of extractable sugars in its stalks [6]. Sweet sorghum is one of the most drought-resistant agricultural crops and can be cultivated in nearly all temperate and tropical climate areas in both irrigated and non-irrigated lands [7,8]. Unlike sugarcane, sweet sorghum has a short growing period (3–4 months). Therefore, it can be planted two or three times a year. Its stalks contain both soluble carbohydrates (such as sucrose, glucose, fructose) and insoluble carbohydrates (such as cellulose and hemicellulose) that can be converted into fuel ethanol using a biological fermentation process [9,10]. With respect to the production cost, sweet sorghum has lower production cost than that of sugarcane and sugar beets because it requires less fertilizer. Therefore, considering the potential of sweet sorghum for industry, it is an ideal crop for commercial ethanol production [11]. Although sweet sorghum is considered as an important food resource in some countries, such as India, China, which uses sweet sorghum juice (SSJ) to produce syrup, the Thai government promotes sweet sorghum to be used as an energy crop for large-scale ethanol production together with sugarcane and cassava.

Fermentation at high temperature is a key requirement for effective bioethanol production in tropical countries where average daytime temperatures are usually high throughout the year. The advantages of fermentation at high temperatures are not only an increased the rate of fermentation but also a decreased risk of contamination by mesophilic microorganisms, such as *Williopsis* sp., *Candida* sp., *Zygosaccharomyces* sp., a reduced cost of the cooling system, and the possible use of simultaneous saccharification and fermentation (SSF) when coupled with a continuous stripping system for ethanol recovery. Utilization of a high potential thermotolerant yeast strain is a key to success in ethanol production at high temperatures [12,13]. There are several reports in the literature on the ethanol production at high temperatures using the thermotolerant yeast *Kluyveromyces marxianus* [13–18]; however, very few reports have considered the thermotolerant yeast *Saccharomyces cerevisiae* [12,19].

Under stressful conditions, such as heat, ethanol, or osmotic stress, several stress-responsive genes including those encoding for the heat shock proteins (HSPs), enzymes involved in protein degradation, such as ubiquitin ligase, and proteins involved in trehalose and glycogen metabolism in yeast have been reported to be stimulated [20,21]. HSPs play a key role as molecular chaperones by either stabilizing new proteins to ensure correct folding or refolding of proteins to the proper conformation, or degrading misfolded proteins that are damaged by stress conditions. HSPs also help transport proteins across membranes within the cell [22,23]. Trehalose, which is one of the compatible solutes synthesized during adverse environmental conditions, has been reported to protect the cell by replacing water at the surfaces of macromolecules, which holds proteins and membranes in their native conformation [24,25]. Glycogen, which is a reserve carbohydrate in *S. cerevisiae*, has also been reported to be involved with tolerance towards several stresses [26,27]. Although a number of genes responsible for the prevention of protein denaturation in yeast cells have been reported, the molecular mechanism conferring thermotolerance during ethanol fermentation at high temperatures is not fully understood.

In this study, the isolation and screening of highly efficient thermotolerant yeast strains capable of producing high levels of ethanol at high temperatures from SSJ were carried out. The optimum condition for ethanol production for the selected thermotolerant yeast strain was also investigated. Furthermore, to gain a better understanding of the molecular mechanism by which yeast cells adapt to adverse environmental conditions and acquire thermotolerance during high temperature ethanol fermentation, the expression of genes encoding the HSP26, HSP70, HSP90, HSP104, pyruvate kinase, trehalose-6-phosphate synthase, neutral trehalase, glycogen synthase, and ubiquitin ligase was evaluated. This work is the first to demonstrate the physiological changes related to the expression of genes involved in heat-shock response, ATP production, trehalose and glycogen metabolism, and protein degradation in the newly isolated thermotolerant yeast *S. cerevisiae* DBKKU Y-53 during ethanol fermentation at high temperatures.

2. Experimental Section

2.1. Isolation of Thermotolerant Yeast Strains

Samples including sugarcane juice, SSJ, rotten fruits, soils from sugarcane and sweet sorghum plantations collected from the Khon Kaen, Udon Thani, Nakhon Ratchasima, Maha Sarakham, Kalasin, Chaiyaphum, and Roi Et provinces of Thailand were used for the isolation of thermotolerant yeasts using the enrichment method described by Limtong *et al.* [13]. YM broth (0.3% yeast extract, 0.3% malt extract, 0.5% peptone, and 1% glucose) supplemented with 4% (*v*/*v*) ethanol was used as a selective medium in this study. After incubation at 35 °C for 3 days on a rotary incubator shaker at 100 rpm, the enriched cultures were then spread on YM agar supplemented with 4% (*v*/*v*) ethanol and subsequently incubated at 35 °C. Pure cultures were collected and maintained on YM agar at 4 °C for short-term storage and in 50% (*v*/*v*) glycerol at −20 °C for long-term storage.

2.2. Screening of Thermotolerant Yeast for Ethanol Fermentation at High Temperatures

SSJ containing 100 g·L^{-1} total sugars was used as substrate for screening thermotolerant ethanol-producing yeast strains. The ethanol production capability was tested by culturing the isolated thermotolerant yeast strains in 16 × 160 mm test tube containing 10 mL SSJ, which were then incubated at high temperatures (37 °C to 50 °C) on a rotary incubator shaker at 150 rpm. The Durham tube was placed inside the test tube, and strains that produced high levels of CO_2 in the Durham tube were selected for further study.

2.3. Identification of The Selected Thermotolerant Yeast Strains

Identification of the yeast strains was carried out using morphological and the D1/D2 domain of the 26S rDNA gene sequencing analysis [28]. Genomic DNA was isolated from the yeast cells using the method described by Harju *et al.* [29]. Amplification of the D1/D2 domain of the 26S rDNA gene was carried out using the specific primers NL-1 (5′-GCA TAT CAA TAA GCG GAG GAA AAG) and NL-4 (5′-GGT CCG TGT TTC AAG ACG G) [30] with the genomic DNA isolated from yeast cells as the template. After the PCR reaction, the amplified product was separated on a 1.0% agarose gel and purified using Invisorb® Fragment CleanUp Kit (Invitek GmbH, Berlin, Germany). All procedures for DNA amplification and purification were carried out according to the manufacturer's instructions. DNA sequencing was performed in the First BASE Laboratories Sdn Bhd (Seri Kembangan, Selangor Darul Ehsan, Malaysia). The sequences of the D1/D2 domain of the 26S rDNA gene were analyzed using GENETYX (Software Development, Tokyo, Japan), whereas homology searching was performed using the FASTA and BLAST programs in the GenBank and DDBJ databases. Phylogenetic analysis was performed using MEGA5 [31], and the tree topologies were analyzed using bootstrap analysis based on the neighbor-joining method [32].

2.4. Comparative Study of Ethanol Production by the Selected Thermotolerant Yeast Strains

The ethanol fermentation efficiencies of the selected thermotolerant yeast strains were compared using SSJ with an initial sugar concentration of 220 g·L^{-1} as substrate. Batch ethanol fermentation was performed in a 500-mL air-locked flask with a final working volume of 400 mL, and the initial cell concentration was 10^6 cells·mL^{-1}. Fermentation was carried out at 30 °C, 37 °C, 40 °C, 42 °C and 45 °C under static conditions. During ethanol fermentation, samples were withdrawn at certain time intervals and analyzed. All experiments were performed in duplicated and repeated twice, and the average values ± SD are presented.

2.5. Factors Influencing Ethanol Production by the Selected Thermotolerant Yeast

The SSJ was sterilized at 110 °C for 28 min and used as raw material for ethanol production using the selected thermotolerant yeast. In this study, *S. cerevisiae* SC90, which is commonly used in industrial ethanol production in Thailand, was used as a reference strain. The batch ethanol fermentation was carried out in 500-mL Erlenmeyer flasks equipped with an air-lock with a final working volume of 400 mL. The effects of pH (4.0, 4.5, 5.0, 5.5, 6.0), initial cell concentration $(1 \times 10^6, 1 \times 10^7, 1 \times 10^8$ cells· $mL^{-1})$, initial sugar concentration (200, 250, 300 g· L^{-1}), and nitrogen source (yeast extract, urea, $(NH_4)_2SO_4$) at various concentrations on ethanol production at high temperatures (37 °C and 40 °C) by the selected thermotolerant yeast were investigated. During ethanol fermentation, samples were withdrawn at certain time intervals and analyzed. All experiments were performed in duplicated and were repeated twice. The average values ± SD are presented.

2.6. Ethanol Production in a 2-L Bioreactor

The ethanol production potential of the selected thermotolerant yeast strain was evaluated in a 2-L bioreactor (Biostat®B, B. Braun Biotech, Melsungen, Germany) with a working volume of 1.5 L. The fermentation conditions at the bioreactor scale were selected from the results obtained from the flask-scale trials. The fermentation was carried out at high temperatures (37 °C and 40 °C) with an agitation speed of 100 rpm, and samples were periodically collected and analyzed during ethanol fermentation.

2.7. Real-Time RT-PCR Analysis of Gene Expression during Ethanol Fermentation

The expressions of genes encoding HSP26 (*hsp26*), HSP70 (*hsp70*), HSP90 (*hsp90*), HSP104 (*hsp104*), pyruvate kinase (*cdc*), trehalose-6-phosphate synthase (*tps*), neutral trehalase (*nth*), glycogen synthase (*gsy*), ubiquitin ligase (*rsp*) during ethanol fermentation were evaluated using real-time RT-PCR. Cells were grown in SSJ containing 220 g· L^{-1} total sugars at 40 °C until growth reached the exponential phase, and then the cells were harvested by centrifugation at 4 °C and 5000 rpm for 5 min. Total RNA was extracted from the yeast cells using Trizol reagent. The concentration of RNA was measured and adjusted by Nanodrop (Nanodrop Technologies, Wilmington, DE, USA). The real-time RT-PCR amplifications were performed using the Biorad-I-Cycler with the qPCRBIO SyGreen One-Step Lo-ROX (PCRBIOSYSTEMS, London, UK). The reactions (final volume 20 µL) were composed of 1 µL RNA template (100 ng RNA), 0.8 µL of each forward and reverse primer, 1 µL 20X RTase, 10 µL 2X qPCRBIO Sygreen One-Step mix, and 6.4 µL RNase-free water. The thermal cycling conditions for gene amplification were initial denaturation at 95 °C for 2 min, followed by 40 cycles each of denaturation at 95 °C for 5 s and annealing at 55 °C for 30 s. The primers used in this study are listed in Table 1. The actin gene (*act*) was used as an internal control. As a negative control, RNase-free water was used instead of the RNA template. All experiments were independently carried out in triplicate, and data from real-time RT-PCR analysis were determined using the CFX Manager Software (Bio-Rad, Hercules, CA, USA). Calculation of relative gene expression was performed using the comparative critical threshold ($\Delta\Delta C_T$) method in which the amount of the target genes was adjusted to the reference gene (*act* gene) [33].

Table 1. Primer used in this study.

Gene Name	Primer Name	Sequence (5′→3′)
hsp26	HSP26-F	AAGGCGGCTTAAGAGGCTAC
	HSP26-R	TGTTGTCTGCATCCACACCT
hsp70	SSA4-F	CGGTTCCAGCCTATTTCAAC
	SSA4-R	TGTCTGAGCAGACGAAGACAG
hsp90	HSP82-F	TTCAAACGACTGGGAAGACC
	HSP82-R	AGCAGCCCTGTTTTGGGTAT
hsp104	HSP104-F	CATATGGAACGTGACTTATCATCTGA
	HSP104-R	ACGGCATTGGAAACAGCTTT
cdc	CDC19-F	TGCTTTGAGAAAGGCTGGTT
	CDC19-R	AAAGCTGGCAAATCGACATC
tps	TPS1-F	TGTCTTCCGTGCAAAGAGTG
	TPS1-R	CTTGTGCATGAAATGGATGG
nth	NTH1-F	CCGTACGAGGACTATGAGTGTTT
	NTH1-R	GCAATTTCGCCTACGTTGTT
gsy	GSY1-F	ACGACTGTGTCGCAAATCACT
	GSY1-R	TGCGGTGACCTCATTAACAG
rsp	RSP5-F	CCTTCTGGCCATACTGCATC
	RSP5-R	CCACCTCCCACTTGAACTGT

2.8. Analytical Methods

The yeast cell numbers and total soluble solids of the fermentation broth were determined using a direct counting method using hemacytometer and a hand-held refractometer, respectively [34]. The fermentation broth was centrifuged at 13,000 rpm for 10 min. The supernatant was then determined for total residual sugars using the phenol sulfuric acid method [35]. The ethanol concentration (P, g·L^{-1}) was analyzed using gas chromatography (Shimadzu GC-14B, Kyoto, Japan) using a polyethylene glycol (PEG-20M) packed column with a flame ionization detector. N$_2$ was used as a carrier gas, and 2-propanol was used as an internal standard. The ethanol yield (Yp/s) was calculated as the actual ethanol produced and was expressed as g ethanol per g glucose utilized (g·g^{-1}). The volumetric ethanol productivity (Qp, g·L^{-1}·h^{-1}) was calculated using the following equations: $Qp = P/t$, where P is the ethanol concentration (g·L^{-1}), and t is the fermentation time (h) giving the greatest ethanol concentration.

3. Results and Discussion

3.1. Isolation and Selection of Thermotolerant Yeast Strains

Natural habitats are a major source of useful microorganisms for biorefinery production [36]. In this study, soil and plant samples from different locations in Northeastern Thailand were collected, and the isolation of thermotolerant yeasts was performed using the enrichment culture technique as described by Limtong *et al.* [13]. The utilization of this technique under selection pressures has been widely used to isolate several thermotolerant yeast strains [13,17]. It has been reported that the effects of high temperatures on microbial growth have often been aggravated by ethanol concentrations greater than 3% [37]. Therefore, ethanol at 4% (*v*/*v*) and an incubation temperature of 35 °C served as selection pressures for the isolation of thermotolerant yeast strains in this work. As a result, sixty-two yeast isolates were obtained, and most of these yeasts were from soil samples. Their ability to grow and produce ethanol at high temperatures were tested by culturing the yeast strains in SSJ incubated at temperatures between 30 and 50 °C. The results showed that nine isolates of yeast, which were designated as DBKKU Y-53, DBKKU Y-55, DBKKU Y-58, DBKKU Y-102, DBKKU Y-103, DBKKU Y-104,

DBKKU Y-105, DBKKU Y-106, and DBKKU Y-107, were able to grow and produce a relatively high level of ethanol at 45 °C. Of these, six isolates (*i.e.*, DBKKUY-102, DBKKUY-103, DBKKUY-104, DBKKUY-105, DBKKUY-106, DBKKUY-107) were capable of growth and ethanol production at 47 °C. Their ability to grow and produce ethanol at high temperatures were comparable to that of *K. marxianus* ATCC 8554, which is known as a thermotolerant yeast. Based on growth performance of the isolated yeast strains at high temperatures (above 40 °C), we speculated that these nine isolates of yeast were thermotolerant yeasts [38]. Therefore, all nine yeast isolates were selected for further study.

Successful isolation and selection of thermotolerant yeasts for ethanol production at high temperatures has been reported by several investigators. According to our literature review, SSJ has never been used directly as the substrate for yeast selection. Almost all previous research studies used glucose as a substrate, e.g., Ballesteros *et al.* [39] reported the screening and selection of thermotolerant yeast strains using glucose as the sole carbon source and found that *K. marxianus* L.G. was the most effective strain for ethanol production at high temperatures. This strain produced an ethanol concentration of approximately 3.76% (*w/v*) when the fermentation was carried out at 42 °C. Banat *et al.* [40] isolated and selected the thermotolerant yeast *K. marxianus* using an enrichment method with glucose as the carbon source and demonstrated that the selected *K. marxianus* strain produced an ethanol concentration of approximately 5.6%–6.0% (*w/v*) at 45–50 °C when using glucose as the substrate and ethanol concentration of approximately 7.5%–8.0% (*w/v*) and 6.5%–7.0% (*w/v*) at 37 °C and 40 °C, respectively, when using molasses as a substrate. Kiran Sree *et al.* [12] screened thermotolerant yeast *S. cerevisiae* VS3 using glucose as the sole carbon source, and they reported that the selected yeast strain produced an ethanol concentration of approximately 75 g·L^{-1} from a glucose concentration of 150 g·L^{-1} at 40 °C. Edgardo *et al.* [19] used glucose for the selection of thermotolerant yeast strains and obtained a good potential thermotolerant yeast *S. cerevisiae* that was able to produce approximately 75% of the theoretical ethanol yield at 40 °C. In addition to using glucose as a substrate, other carbon sources, such as sugarcane juice [13], xylose [41], sugarcane blackstrap molasses [42], and inulin [43] have also been used for screening and selecting thermotolerant yeasts.

3.2. Identification of the Selected Thermotolerant Yeasts

Based on the morphological and physiological characteristics, three isolated yeasts (*i.e.*, DBKKU Y-53, DBKKU Y-55, DBKKU Y-58) were found to be *Saccharomyces*, whereas six isolated yeasts (*i.e.*, DBKKU Y-102, DBKKU Y-103, DBKKU Y-104, DBKKU Y-105, DBKKU Y-106, DBKKU Y-107) were revealed to be *Kluyveromyces*. To confirm this finding, molecular taxonomic analyses based on the nucleotide sequences of the D1/D2 domain of the 26S rDNA gene were carried out. As a result, the nucleotide sequences from the yeast strains DBKKU Y-53, DBKKU Y-55, DBKKU Y-58 and from *S. cerevisiae* (NL45 and NL51) were 99% identical, whereas strains DBKKU Y-102, DBKKU Y-103, DBKKU Y-104, DBKKU Y-105, DBKKU Y-106, DBKKU Y-107, and *K. marxianus* (FJ627963) were 99% identical. Phylogenetic analysis confirmed that strains DBKKU Y-53, DBKKU Y-55, and DBKKU Y-58 were clustered in the same group as *S. cerevisiae* whereas strains DBKKU Y-102, DBKKU Y-103, DBKKU Y-104, DBKKU Y-105, DBKKU Y-106, and DBKKU Y-107 were closely related to *K. marxianus* (Figure 1). Therefore, strains DBKKU Y-53, DBKKU Y-55, and DBKKU Y-58 were identified as *S. cerevisiae*, whereas the other strains were *K. marxianus*.

Kluyveromyces marxianus (FJ627963)

Kluyveromyces marxianus DBKKU Y-102

Kluyveromyces marxianus DBKKU Y-103

Kluyveromyces marxianus DBKKU Y-104

Kluyveromyces marxianus DBKKU Y-105

Kluyveromyces marxianus DBKKU Y-106

Kluyveromyces marxianus DBKKU Y-107

Saccharomyces cerevisiae strain NL45

Saccharomyces cerevisiae strain NL51

Saccharomyces cerevisiae DBKKU Y-53

Saccharomyces cerevisiae DBKKU Y-55

Saccharomyces cerevisiae DBKKU Y-58

Hanseniaspora uvarum strain NL73 (HM191672)

Pichia manshurica strain NL72 (HM191679)

0.02

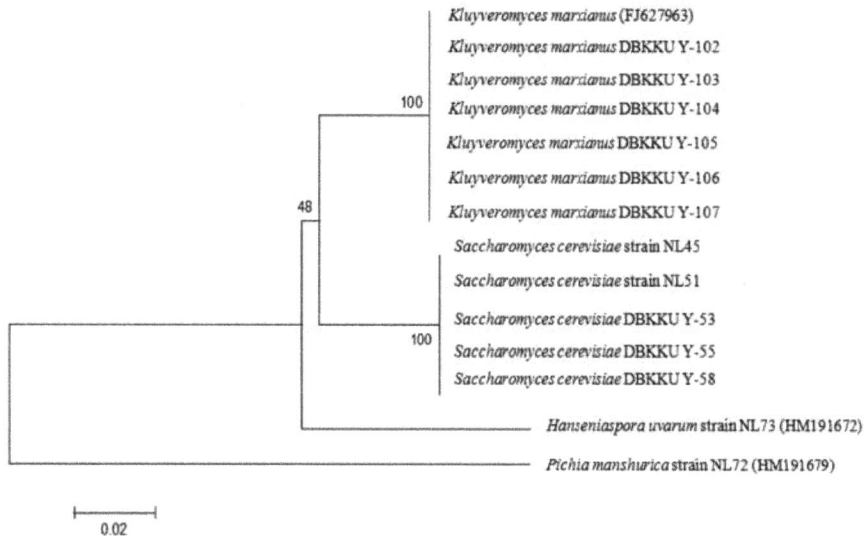

Figure 1. An unrooted phylogenetic tree showing the relationship between the isolated yeasts and the type strain *K. marxianus* and *S. cerevisiae* based on the D1/D2 domain of the 26s rDNA sequence analysis. The tree was built using the neighbor-joining method. Numbers given at the nodes indicate the percentage bootstrap values based on 1000 replications.

3.3. Comparative Study on the Ethanol Production by the Selected Thermotolerant Yeasts

A preliminary study of the ethanol production using SSJ by the selected thermotolerant yeasts revealed that only four strains, *i.e.*, DBKKU Y-53, DBKKU Y-102, DBKKU Y-103, and DBKKU Y-104, produced relatively high levels of ethanol at high temperatures after 24 h of fermentation. Therefore, these four strains were selected for further study of the ethanol production from SSJ at various incubation temperatures in a 500-mL flask scale. As shown in Table 2, strain DBKKU Y-53 produced the highest ethanol concentrations and volumetric ethanol productivities at 30 °C, 37 °C, 40 °C and 42 °C compared with the type strain, *K. marxianus* ATCC8554, and the other selected strains tested. At 45 °C, however, the ethanol concentration and volumetric ethanol productivity produced by this strain were lower than of the other strains tested. The ethanol concentrations and volumetric ethanol productivities produced by strains DBKKU Y-102, DBKKU Y-103, and DBKKU Y-104 were not significantly different compared with *K. marxianus* ATCC8554 at 30 °C, 37 °C, and 40 °C. At higher temperatures (42 °C and 45 °C), the ethanol concentrations and volumetric ethanol productivities from these strains were relatively greater than *K. marxianus* ATCC8554. It is evident from these results that increasing the fermentation temperature from 40 °C to 45 °C resulted in a reduction in the ethanol concentrations and productivities. Based on the maximum ethanol concentrations at 37 °C to 42 °C, which is a temperature commonly attained during fermentation in tropical regions, such as Thailand, *S. cerevisiae* DBKKU Y-53 was selected for further analysis.

Table 2. Ethanol production from sweet sorghum juice by isolated yeasts and the type strain, *K. marxianus* ATCC8554, at various temperatures.

Strain	30 °C		37 °C		40 °C		42 °C		45 °C	
	P	Q_p	P	Q_p	P	Q_p	P	Q_p	P	Q_p
ATCC 8554	19.55 ± 0.49 [a]	0.27 ± 0.01 [a]	25.06 ± 1.65 [a]	0.35 ± 0.02 [a]	25.83 ± 0.17 [a]	0.36 ± 0.00 [a]	12.45 ± 0.00 [a]	0.17 ± 0.00 [a]	12.45 ± 0.00 [b]	0.17 ± 0.00 [a]
DBKKU Y-53	63.67 ± 4.12 [b]	1.77 ± 0.11 [b]	61.99 ± 4.43 [b]	1.72 ± 0.12 [b]	58.20 ± 0.54 [d]	1.62 ± 0.02 [d]	38.77 ± 1.92 [c]	0.81 ± 0.04 [c]	7.67b ± 1.77 [a]	0.16 ± 0.04 [a]
DBKKU Y-102	19.39 ± 0.63 [a]	0.27 ± 0.01 [a]	34.30 ± 8.99 [a]	0.48 ± 0.12 [a]	29.51 ± 0.34 [b]	0.41 ± 0.00 [b]	29.46 ± 1.10 [b]	0.41 ± 0.02 [b]	21.78 ± 1.27 [c]	0.30 ± 0.02 [b]
DBKKU Y-103	21.89 ± 0.80 [a]	0.30 ± 0.01 [a]	30.78 ± 0.07 [a]	0.43 ± 0.00 [a]	31.63 ± 0.05 [c]	0.44 ± 0.00 [c]	30.70 ± 1.09 [b]	0.43 ± 0.02 [b]	25.32 ± 0.82 [cd]	0.35 ± 0.00 [bc]
DBKKU Y-104	19.24 ± 1.00 [a]	0.27 ± 0.01 [a]	31.14 ± 1.94 [a]	0.43 ± 0.03 [a]	27.38 ± 1.23 [a]	0.38 ± 0.02 [a]	29.98 ± 1.34 [b]	0.42 ± 0.02 [b]	27.65 ± 2.87 [d]	0.38 ± 0.04 [c]

P: ethanol concentration produced (g·L^{-1}); Q_p: volumetric ethanol productivity (g·L^{-1}·h^{-1}); Mean values ± SD with different letters in the same column are significant different at $p < 0.05$ based on DMRT analysis.

3.4. Factors Influencing Ethanol Production When Using Thermotolerant Yeast S. cerevisiae DBKKU Y-53

There are several factors influencing the growth and ethanol production of yeast, such as the incubation temperature, the pH of the fermentation medium, the cell concentration, the sugar concentration, the nitrogen sources, and the aeration [13,44]. Therefore, the effects of some of the major factors on ethanol production using SSJ by thermotolerant yeast *S. cerevisiae* DBKKU Y-53 were investigated. In this work, the ethanol production efficiency of *S. cerevisiae* DBKKU Y-53 was compared with that of *S. cerevisiae* SC90, which is one of the industrial yeast strains widely used to produce ethanol on a commercial scale in Thailand.

The effect of temperature on ethanol production from SSJ using *S. cerevisiae* DBKKU Y-53 and *S. cerevisiae* SC90 was analyzed, and the results are summarized in Table 3. There was no significantly difference in ethanol concentration produced by both strains at 30 °C. However, at 37 °C and 40 °C, *S. cerevisiae* DBKKU Y-53 produced greater ethanol concentrations than *S. cerevisiae* SC09. The maximum ethanol concentrations produced by *S. cerevisiae* DBKKU Y-53 at 37 °C and 40 °C were 71.73 ± 2.62 g·L^{-1} and 58.14 ± 7.71 g·L^{-1}, respectively. When the incubation temperature was shifted from 40 °C to 42 °C and 45 °C, ethanol concentrations and volumetric ethanol productivities produced by both strains were remarkably decreased. This might be due to the negative effect of high temperature on growth and metabolic processes in yeast cells. It has been reported that high temperature causes a modification of plasma membrane fluidity and a reduction in the effectiveness of the plasma membrane as a semipermeable barrier allowing leakage of essential cofactors and coenzymes required for the activity of enzymes involved in glucose metabolism and ethanol production [44]. Roukas [45] reported that high temperatures caused denaturation of cellular proteins, which resulted in the reduction of cell growth and fermentation activity. Moreover, the reduction in yeast growth at high temperatures was also due to the accumulation of intracellular ethanol, which modifies the cell membrane structure of yeast cell [46]. Our results were in good agreement with Banat *et al.* [40], who observed a reduction in ethanol concentration produced by the thermotolerant yeast *K. marxianus* at 40 °C. Kiran Sree *et al.* [12] reported the ethanol production from glucose (150 g·L^{-1}) using thermotolerant yeast *S. cerevisiae* VS3 and observed a reduction in ethanol concentration when the incubation temperature shifted from 30 °C to 40 °C. Limtong *et al.* [13] demonstrated that the ethanol concentration produced by the newly isolated thermotolerant yeast *K. marxianus* DMKU3-1042 decreased significantly when the incubation temperature was increased from 37 °C to 40 °C. Tofighi *et al.* [47] reported the reduction in cell mass productivity and ethanol fermentation ability of the thermotolerant yeast *S. cerevisiae* when the incubation temperature increased to greater than 40 °C. Recently, Charoensopharat *et al.* [43] found that the ethanol concentration produced by the newly isolated thermotolerant yeast *K. marxianus* decreased dramatically when ethanol fermentation was carried out at temperatures greater than 40 °C.

The pH of the medium is an important factor influencing ethanol yield. Generally, the optimum pH for yeast growth and ethanol production is in the range of 4.0 to 6.0 depending on growth conditions, such as the temperature, the presence of oxygen, the yeast species, and the type of raw material. For instance, the optimum pH for ethanol production from sugarcane juice at high temperatures using the thermotolerant yeast *K. marxianus* DMKU 3-1042 was 5.0 [13]. During ethanol fermentation, the pH of the fermentation medium is almost in the range of 4.0 to 5.5. This pH level typically prevents bacterial contamination during the fermentation process [48]. In this study, the effect of pH on ethanol production by the thermotolerant yeast *S. cerevisiae* DBKKU Y-53 and a reference strain was investigated, and the results are summarized in Tables 4 and 5. The optimum pH for ethanol production from SSJ by both strains at 37 °C and 40 °C was 5.5, which is in agreement with reports by Ercan *et al.* [49] and Sign and Bishnoi [50]. At a pH less than or greater than 5.5, ethanol concentrations and volumetric ethanol productivities tended to decrease. This might be related to the activity of enzymes involved in the ethanol production pathway. It has been reported that enzymes may be inactivated at a pH level that is less than or greater than the optimum value causing a reduction in cell growth and ethanol fermentation ability [48]. The initial pH of the SSJ was in the range of 5.2–5.5. Therefore, a pH of 5.5, which gave the highest ethanol concentration, was selected for further study.

Table 3. Kinetic parameters of ethanol production from sweet sorghum juice at various temperatures by *S. cerevisiae* DBKKUY-53 and *S. cerevisiae* SC90.

Temperature (°C)	*S. cerevisiae* DBKKU Y-53			*S. cerevisiae* SC90		
	P	Q_p	E_y	P	Q_p	E_y
30	82.77 ± 1.99 [a]	2.30 ± 0.06 [a]	88.40 ± 0.59	83.30 ± 2.26 [a]	1.74 ± 0.05 [a]	96.47 ± 2.30
37	71.73 ± 2.62 [b]	1.49 ± 0.05 [b]	87.76 ± 3.22	63.22 ± 2.42 [b]	1.32 ± 0.05 [b]	98.30 ± 0.46
40	58.14 ± 7.71 [c]	1.61 ± 0.20 [b]	88.25 ± 2.73	53.68 ± 1.17 [c]	1.12 ± 0.02 [c]	94.04 ± 0.70
42	32.30 ± 0.49 [d]	0.67 ± 0.01 [c]	88.20 ± 6.37	32.54 ± 1.56 [d]	0.68 ± 0.03 [d]	96.59 ± 0.97
45	9.08 ± 4.02 [e]	0.19 ± 0.08 [d]	62.64 ± 5.60	17.69 ± 2.54 [e]	0.37 ± 0.05 [e]	98.07 ± 0.58

P: ethanol concentration (g·L^{-1}); Q_p: volumetric ethanol productivity (g·L^{-1}·h^{-1}); E_y: percentage of ethanol production efficiency (%); Mean values ± SD with different letters in the same column are significant different at $p < 0.05$ based on DMRT analysis.

Table 4. Kinetic parameters of ethanol production from sweet sorghum juice under various pHs, cell concentrations, and sugar concentrations by *S. cerevisiae* DBKKUY-53 at 37 °C and 40 °C.

Conditions	37 °C			40 °C		
	P	Q_p	E_y	P	Q_p	E_y
pH						
4.0	68.21 ± 3.74 [a]	1.42 ± 0.08 [a]	87.47 ± 3.61	48.79 ± 2.09 [a]	1.02 ± 0.04 [a]	86.48 ± 0.94
4.5	67.80 ± 1.46 [a]	1.41 ± 0.03 [a]	91.27 ± 3.80	49.40 ± 0.13 [a]	1.03 ± 0.00 [a]	89.89 ± 4.43
5.0	71.73 ± 2.62 [a]	1.49 ± 0.05 [a]	87.76 ± 3.22	59.17 ± 5.33 [b]	1.61 ± 0.20 [c]	88.25 ± 2.73
5.5	74.23 ± 1.76 [a]	1.55 ± 0.08 [a]	92.06 ± 5.70	62.78 ± 0.67 [b]	1.31 ± 0.01 [b]	88.43 ± 0.53
6.0	72.14 ± 1.23 [ab]	1.50 ± 0.03 [a]	91.81 ± 0.66	61.98 ± 0.43 [b]	1.72 ± 0.01 [c]	92.35 ± 0.66
Cell concentration (cells·mL^{-1})						
1×10^6	74.23 ± 3.62 [a]	1.55 ± 0.08 [a]	92.06 ± 5.70	62.78 ± 0.67 [a]	1.31 ± 0.01 [a]	88.43 ± 0.53
1×10^7	84.06 ± 1.11 [b]	1.75 ± 0.02 [b]	87.72 ± 0.53	77.14 ± 1.58 [b]	2.14 ± 0.04 [b]	87.78 ± 1.37
1×10^8	88.17 ± 0.17 [b]	2.94 ± 0.01 [c]	84.73± 0.28	83.35 ± 2.37 [c]	3.47 ± 0.10 [c]	89.07 ± 0.04
Sugar concentration (g·L^{-1})						
200	88.17 ± 0.17 [a]	2.94 ± 0.01 [a]	90.42 ± 0.28	83.35 ± 2.37 [a]	3.47 ± 0.10 [a]	89.07 ± 0.04
250	99.56 ± 0.88 [c]	4.15 ± 0.04 [c]	98.44 ± 0.53	88.97 ± 1.23 [b]	3.71 ± 0.05 [b]	96.53 ± 2.55
300	92.69 ± 0.25 [b]	3.86 ± 0.01 [b]	98.89 ± 0.08	83.25 ± 0.43 [a]	3.47 ± 0.02 [a]	97.40 ± 1.11

P: ethanol concentration (g·L^{-1}); Q_p: volumetric ethanol productivity (g·L^{-1}·h^{-1}); E_y: percentage of ethanol production efficiency (%); Mean values ± SD with different letters in the same column are significant different at $p < 0.05$ based on DMRT analysis.

The initial cell concentration affects not only the ethanol yield but also the substrate consumption rate and ethanol fermentation rate. Generally, high initial cell concentrations reduce the lag phase of growth and increase the sugar consumption and ethanol fermentation rate, leading to a high ethanol yield and productivity. In this study, the effect of initial cell concentrations (1×10^6, 1×10^7, 1×10^8 cells·mL^{-1}) on ethanol fermentation from SSJ containing 220 g·L^{-1} of total sugars was investigated, and the results are summarized in Tables 4 and 5. As a result, increasing the cell concentration resulted in an increase in the ethanol concentration and the volumetric ethanol productivity. The maximum ethanol concentrations and volumetric ethanol productivities produced by *S. cerevisiae* DBKKU Y-53 and SC09 at 37 °C and 40 °C were achieved at an initial cell concentration of 1×10^8 cells·mL^{-1}, which is in good agreement with Charoensopharat *et al.* [43] and Laopaiboon *et al.* [51].

High sugar concentrations (more than 20% *w/v*) are not often used in industrial ethanol production because they may reduce the yeast cell viability, the substrate conversion rate, and the ethanol yield [52,53]. However, ethanol production with high sugar concentrations have also been reported, and the fermentation efficiencies vary depending on the yeast species and fermentation conditions. For example, Laopaiboon *et al.* [54] reported a maximum ethanol concentration of 120.68 g· L^{-1} and volumetric ethanol productivity of 2.01 g· L^{-1}· h^{-1} when *S. cerevisiae* NP01 was used to produce ethanol from SSJ under a very high gravity fermentation. Charoensopharat *et al.* [43] reported a maximum ethanol concentration of 104.83 g· L^{-1} and a volumetric ethanol productivity of 4.37 g· L^{-1}· h^{-1} when *K. marxianus* DBKKU Y-102 was used to produce ethanol from Jerusalem artichoke tubers at 37 °C during consolidated bioprocessing. To verify the effect of sugar concentration on ethanol production efficiency of the thermotolerant yeast *S. cerevisiae* DBKKU Y-53, SSJ containing various sugar concentrations (200, 250, 300 g· L^{-1}) was tested. As shown in Tables 4 and 5 increasing in the sugar concentration from 200 g· L^{-1} to 250 g· L^{-1} resulted in an increase in the ethanol concentration. However, at a sugar concentration of 300 g· L^{-1} the ethanol concentration was remarkably decreased, and a large amount of sugar remained in the fermentation broth at both 37 °C and 40 °C. High sugar concentrations have been reported to cause negative effects on cell viability and morphology due to an increase in the osmotic pressure, which leads to a reduction in the cell biomass and ethanol yield [53,55]. The maximum ethanol concentrations produced by *S. cerevisiae* DBKKU Y-53 and SC90 were achieved at a sugar concentration of 250 g· L^{-1}. Therefore, this sugar concentration was used for subsequent experiments.

Table 5. Kinetic parameters of ethanol production from sweet sorghum juice under various pHs, cell concentrations, and sugar concentrations by *S. cerevisiae* SC90 at 37 °C and 40 °C.

Conditions	37 °C			40 °C		
	P	*Q*$_p$	*E*$_y$	*P*	*Q*$_p$	*E*$_y$
			pH			
4.0	72.90 ± 2.31 [a]	1.22 ± 0.04 [bc]	96.21 ± 2.71	55.08 ± 2.67 [ab]	0.92 ± 0.04 [ab]	97.58 ± 0.28
4.5	70.70 ± 2.55 [a]	1.18 ± 0.04 [c]	97.15 ± 0.52	53.01 ± 1.35 [a]	0.88 ± 0.02 [a]	97.55 ± 1.77
5.0	63.22 ± 2.42 [b]	1.32 ± 0.05 [b]	98.30 ± 0.46	53.68 ± 1.17 [ab]	1.12 ± 0.02 [c]	94.04 ± 0.70
5.5	75.19 ± 2.40 [a]	1.25 ± 0.04 [bc]	95.08 ± 2.92	57.62 ± 0.64 [b]	0.96 ± 0.01 [b]	98.26 ± 1.90
6.0	72.41 ± 1.75 [a]	1.51 ± 0.04 [a]	75.99 ± 0.05	57.09 ± 0.42 [b]	0.96 ± 0.01 [b]	80.29 ± 1.89
			Cell concentration (cells· mL^{-1})			
1 × 10^6	75.19 ± 2.40 [a]	1.25 ± 0.04 [a]	95.08 ± 2.92	62.78± 0.67 [a]	1.31 ± 0.01 [a]	88.43 ± 0.53
1 × 10^7	75.16 ± 0.06 [a]	1.57 ± 0.00 [b]	90.50 ± 1.06	77.14 ± 1.58 [b]	2.14 ± 0.04 [b]	87.78 ± 1.37
1 × 10^8	76.41 ± 1.99 [a]	2.55 ± 0.07 [c]	79.98 ± 4.31	83.35 ± 2.37 [c]	3.47 ± 0.10 [c]	89.07 ± 0.04
			Sugar concentration (g· L^{-1})			
200	76.41 ± 1.99 [a]	2.55 ± 0.07 [c]	79.98 ± 4.31	70.91 ± 0.96 [a]	2.96 ± 0.04 [a]	88.45 ± 1.81
250	83.05 ± 0.02 [b]	1.73 ± 0.00 [a]	93.79 ± 2.20	73.00 ± 0.28 [a]	2.03 ± 0.01 [b]	94.34 ± 0.21
300	78.98 ± 0.78 [a]	2.19 ± 0.02 [b]	94.79 ± 0.65	71.66 ± 0.96 [a]	1.99 ± 0.03 [b]	90.93 ± 1.15

P: ethanol concentration (g· L^{-1}); *Q*$_p$: volumetric ethanol productivity (g· L^{-1}· h^{-1}); *E*$_y$: percentage of ethanol production efficiency (%); Mean values ± SD with different letters in the same column are significant different at *p* < 0.05 based on DMRT analysis.

Approximately 10% of yeast's dry weight is nitrogen [44]. Therefore, nitrogen is one of the important constituents for cell growth and synthesis of structural and functional proteins involved in metabolic processes. Various types of nitrogen sources have been used to supplement ethanol fermentation medium both in organic (yeast extract, corn steep liquor) and inorganic forms (ammonium sulfate, ammonium nitrate, urea, diammonium phosphate). In this study, the effects on ethanol production of using yeast extract, urea, and ammonium sulfate at different concentrations from SSJ by *S. cerevisiae* DBKKU Y-53 and SC90 were determined, and the results are summarized in Tables 6 and 7. Supplementation of urea in the SSJ was shown to significantly enhance the ethanol production by both strains of yeast at 37 °C and 40 °C, which is in good agreement with reports from Yue *et al.* [56]. There was no significant difference in ethanol production at 37 °C when comparing yeast extract supplementation and supplementation-free fermentations. However, at 40 °C, supplementation of yeast extract tended to result in a lower ethanol concentration compared with the control condition. Yeast extract has been reported to be a good organic nitrogen source for ethanol production using *S. cerevisiae*. However, its availability to be utilized by yeast cells is depended on the fermentation conditions. For example, Laopaiboon *et al.* [54] reported that supplementation of yeast extract in SSJ improved the ethanol production by *S. cerevisiae* NP01 for very high gravity fermentation. With respect to ammonium sulfate, supplementation of this nitrogen compound resulted in a reduction of the ethanol concentration in all conditions tested compared with the control without nitrogen supplementation. These results clearly indicate that ammonium sulfate was not a good nitrogen source for ethanol production at high temperatures using *S. cerevisiae* DBKKU Y-53 and SC90 when SSJ was used as a substrate. One possibility is that this nitrogen compound was not taken up by the yeast cells during high temperature fermentation. Ter Schure *et al.* [57] and Magasanik and Kaiser [58] reported that the uptake of nitrogen by yeast cells is regulated by the mechanism known as nitrogen catabolite repression (NCR). NCR enables yeast cells to select the best nitrogen sources by repressing the transcription of genes involved in the utilization of the poorer nitrogen [59]. However, the effectiveness of NCR mechanism is influenced by many factors including fermentation temperature and the presence of ethanol. Normally, high temperatures and high ethanol concentrations cause the modification of the plasma membrane. Therefore, the nitrogen sources sensing system, which is mainly located in the plasma membrane of yeast cells, may be affected by these stress conditions, leading to the impairment of ammonium sulfate uptake [60,61]. In this study, urea proved to be the best nitrogen source for ethanol production from SSJ during high temperature fermentation using *S. cerevisiae* DBKKU Y-53 and SC90; therefore, it was chosen for further analysis.

Table 6. Kinetic parameters of ethanol production from sweet sorghum juice with various nitrogen sources by *S. cerevisiae* DBKKUY-53 at 37 °C and 40 °C.

Conditions	37 °C			40 °C		
	P	*Q$_p$*	*E$_y$*	*P*	*Q$_p$*	*E$_y$*
Control	99.56 ± 0.88 [bcd]	4.15 ± 0.04 [a]	98.44 ± 0.53	88.97 ± 1.23 [a]	3.71 ± 0.05 [a]	95.56 ± 1.83
			Yeast extract			
6	101.30 ± 0.84 [bc]	3.38 ± 0.03 [b]	93.43 ± 1.70	79.19 ± 5.53 [bcd]	3.30 ± 0.23 [bc]	91.78 ± 3.06
9	100.35 ± 0.43 [bcd]	3.35 ± 0.01 [b]	92.39 ± 2.90	76.81 ± 8.84 [cde]	3.20 ± 0.37 [bcd]	94.98 ± 1.66
12	97.05 ± 0.98 [d]	3.23 ± 0.03 [c]	88.20 ± 2.84	83.76 ± 0.26 [abc]	3.49 ± 0.01 [ab]	96.01 ± 1.27
			Urea			
0.25	97.97 ± 3.29 [cd]	2.72 ± 0.09 [d]	86.47 ± 1.70	91.18 ± 2.70 [a]	3.04 ± 0.09 [cde]	95.42 ± 3.93
0.50	102.72 ± 1.94 [ab]	2.14 ± 0.04 [e]	96.01 ± 1.16	88.27 ± 0.36 [a]	2.94 ± 0.01 [de]	95.30 ± 3.66
0.75	105.17 ± 1.37 [a]	2.19 ± 0.03 [e]	94.31 ± 4.16	88.66 ± 0.59 [a]	2.96 ± 0.02 [de]	87.82 ± 0.26
1.00	99.90 ± 1.41 [bcd]	2.78 ± 0.04 [d]	85.65 ± 7.36	86.41 ± 2.43 [ab]	2.88 ± 0.08 [e]	83.82 ± 2.15
			(NH$_4$)$_2$SO$_4$			
0.25	97.38 ± 0.23 [d]	3.25 ± 0.01 [c]	94.37 ± 2.64	67.25 ± 1.60 [f]	1.40 ± 0.02 [g]	94.51 ± 2.69
0.50	96.77 ± 2.41 [d]	3.23 ± 0.08 [c]	90.37 ± 5.24	69.90 ± 1.02 [ef]	1.94 ± 0.01 [f]	94.50 ± 1.04
0.75	97.70 ± 0.06 [cd]	2.71 ± 0.00 [d]	86.76 ± 6.82	70.29 ± 0.59 [ef]	1.95 ± 0.01 [f]	96.18 ± 1.51
1.00	99.81 ± 1.44 [bcd]	2.77 ± 0.04 [d]	87.76 ± 6.82	74.27 ± 1.14 [def]	2.06 ± 0.02 [f]	96.25 ± 0.48

P: ethanol concentration (g· L^{-1}); *Q$_p$*: volumetric ethanol productivity (g· L^{-1}· h^{-1}); *E$_y$*: percentage of ethanol production efficiency (%); Mean values ± SD with different letters in the same column are significant different at $p < 0.05$ based on DMRT analysis.

Table 7. Kinetic parameters of ethanol production from sweet sorghum juice with various nitrogen sources by *S. cerevisiae* SC90 at 37 °C and 40 °C.

Conditions	37 °C			40 °C		
	P	Q_p	E_y	*P*	Q_p	E_y
Control	83.05 ± 0.02 [c]	1.73 ± 0.00 [d]	93.79 ± 2.20	73.00 ± 0.28 [a]	2.03 ± 0.01 [c]	94.34 ± 0.21
		Yeast extract				
6.0	83.53 ± 0.55 [c]	2.78 ± 0.02 [b]	97.83 ± 0.52	67.17 ± 2.78 [b]	2.24 ± 0.09 [b]	92.54 ± 4.06
9.0	88.23 ± 1.71 [b]	2.94 ± 0.06 [a]	94.07 ± 0.58	67.47 ± 1.81 [b]	2.25 ± 0.06 [b]	97.98 ± 0.50
12.0	87.38 ± 0.96 [b]	2.91 ± 0.03 [a]	94.68 ± 2.30	69.25 ± 1.16 [ab]	2.31 ± 0.04 [b]	98.18 ± 1.50

Table 7. *Cont.*

Conditions	37 °C			40 °C		
	P	Q_p	E_y	*P*	Q_p	E_y
		Urea				
0.25	92.48 ± 0.87 [a]	2.57 ± 0.02 [c]	97.48 ± 0.95	72.87 ± 0.72 [a]	3.04 ± 0.03 [a]	87.17 ± 1.23
0.50	92.68 ± 0.29 [a]	2.57 ± 0.01 [c]	98.03 ± 0.08	72.20 ± 2.15 [ab]	3.01 ± 0.09 [a]	91.91 ± 0.60
0.75	86.76 ± 3.35 [b]	2.89 ± 0.11 [a]	97.86 ± 0.11	69.90 ± 1.31 [ab]	2.91 ± 0.05 [a]	94.36 ± 0.37
1.00	87.39 ± 0.39 [b]	2.91 ± 0.01 [a]	98.36 ± 0.84	73.78 ± 1.17 [a]	2.05 ± 0.03 [c]	91.65 ± 2.09
		$(NH_4)_2 SO_4$				
0.25	57.57 ± 1.67 [f]	1.20 ± 0.02 [g]	91.90 ± 0.53	54.03 ± 0.56 [d]	1.13 ± 0.01 [d]	92.66 ± 2.92
0.50	71.85 ± 0.36 [d]	1.00 ± 0.00 [h]	91.77 ± 0.42	57.82 ± 0.93 [cd]	1.20 ± 0.02 [d]	92.23 ± 4.00
0.75	66.54 ± 2.11 [e]	1.39 ± 0.04 [f]	97.53 ± 0.01	56.65 ± 5.91 [cd]	1.18 ± 0.12 [d]	86.90 ± 5.20
1.00	72.10 ± 0.78 [d]	1.50 ± 0.02 [e]	94.84 ± 1.20	59.81 ± 1.34 [c]	1.25 ± 0.03 [d]	97.68 ± 1.26

P: ethanol concentration (g·L^{-1}); Q_p: volumetric ethanol productivity (g·L^{-1}·h^{-1}); E_y: percentage of ethanol production efficiency (%); Mean values ± SD with different letters in the same column are significant different at $p < 0.05$ based on DMRT analysis.

3.5. Ethanol Production in 2-L Bioreactor

The time profiles of ethanol production from SSJ using *S. cerevisiae* DBKKU Y-53 and SC90 at 37 °C and 40 °C in a 2-L bioreactor are illustrated in Figure 2. During the first 12 h after fermentation, ethanol concentrations produced by both strains sharply increased and reached their maximum values at 24 h for *S. cerevisiae* SC90 and 48 h for *S. cerevisiae* DBKKU Y-53. Table 8 summarizes the kinetic parameters of ethanol production from the SSJ at high temperatures using *S. cerevisiae* DBKKU Y-53 and SC90. It can be seen from this finding that the newly isolated thermotolerant yeast *S. cerevisiae* DBKKU Y-53 resulted in a greater ethanol concentration as well as sugar utilization capability compared with SC90.

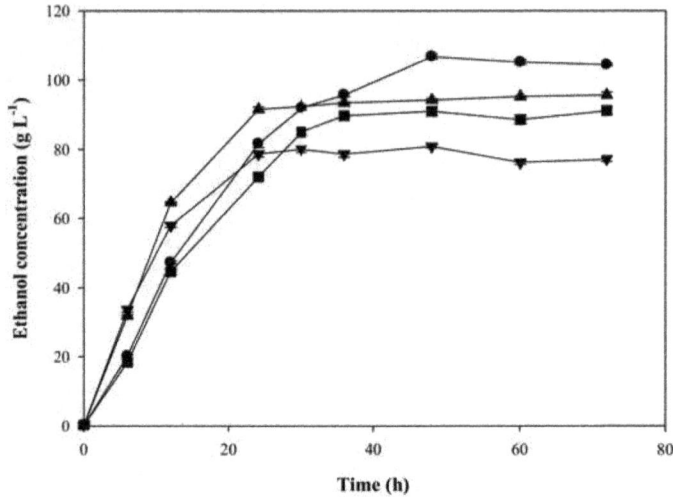

Figure 2. Ethanol production from sweet sorghum juice by *S. cerevisiae* DBKKUY-53 and *S. cerevisiae* SC90 in a 2-L bioreactor at high temperatures. *S. cerevisiae* DBKKUY-53 at 37 °C (●); *S. cerevisiae* DBKKUY-53 at 40 °C (■); *S. cerevisiae* SC90 at 37 °C (▲); *S. cerevisiae* SC90 at 40 °C (▼).

Table 8. Kinetic parameters of ethanol production from sweet sorghum juice by *S. cerevisiae* DBKKU Y-53 and *S. cerevisiae* SC90 in a 2-L bioreactor at 37 °C and 40 °C.

Strains	T (°C)	Parameters (mean ± SD)					
		P	Q_p	$Y_{p/s}$	E_y	t	Sugar Utilized (%)
DBKKU	37	106.82 ± 0.01 [a]	2.23 ± 0.01 [a]	0.45 ± 0.02	88.42 ± 0.11	48	91.84 ± 0.20
Y-53	40	85.01 ± 0.03 [a]	2.83 ± 0.02 [a]	0.42 ± 0.01	82.06 ± 0.09	30	79.35 ± 0.05
SC90	37	91.59 ± 0.01 [b]	3.82 ± 0.02 [b]	0.47 ± 0.01	91.52 ± 0.02	24	82.68 ± 0.02
	40	78.69 ± 0.02 [b]	3.28 ± 0.03 [b]	0.46 ± 0.02	89.08 ± 0.01	24	68.19 ± 0.02

T: incubation temperature; P: ethanol concentration (g·L^{-1}); Q_p: volumetric ethanol productivity (g·L^{-1}·h^{-1}); $Y_{p/s}$: ethanol yield (g·g^{-1}); E_y: percentage of ethanol production efficiency (%); *t*: fermentation time (h); Mean values ± SD with different letters in the same column are significant different at $p < 0.05$ based on DMRT analysis.

A comparative analysis of the ethanol production from SSJ using the thermotolerant yeast *S. cerevisiae* DBKKU Y-53 with values reported in the literatures using different raw materials and yeast strains was performed, and the results are summarized in Table 9. The ethanol concentration and volumetric ethanol productivity produced by *S. cerevisiae* DBKKU Y-53 were greater than values reported in other works, suggesting that the newly isolated thermotolerant yeast *S. cerevisiae* DBKKU Y-53 was a good candidate for ethanol production from SSJ at high temperatures. Although several isolates of the thermotolerant yeast *K. marxianus* have been reported to be capable of growth and ethanol production at temperatures greater than 45 °C, almost all of these isolates had relatively lower ethanol yields and were less tolerant to ethanol than *S. cerevisiae*. Furthermore, these isolates also required oxygen during ethanol fermentation resulting in an increase in the operating cost. Therefore, the thermotolerant yeast *S. cerevisiae* is more promising for ethanol production at high temperatures on a commercial scale compared with *K. marxianus*.

Table 9. Comparison of ethanol production by *S. cerevisiae* DBKKU Y-53 and other yeast strains reported in the literatures using different raw materials.

Strains	S_0 (g·L^{-1})	C-Source	T (°C)	P (g·L^{-1})	Q_p (g·L^{-1}·h^{-1})	References
S. cerevisiae UV-VS3 100	250	Glucose	30	98.0	2.04	Sridhar *et al.* [62]
			42	62.0	1.29	
S. cerevisiae VS3	150	Glucose	35	75.0	1.56	Kiran Sree *et al.* [12]
			40	60.0	1.25	
			42	58.0	1.21	
S. cerevisiae F111	180	Sugarcane molasses	43	84.0	2.33	Abdel-fattah *et al.* [63]
K. marxianus WR12	180	Sugarcane molasses	43	80.6	2.88	
Pichia kudriavzevii	170	Sugarcane juice	40	71.9	4.00	Dhaliwal *et al.* [64]
K. marxianus DMKU 3-1042	220	Sugarcane juice	37	87.0	1.45	Limtong *et al.* [13]
			40	67.8	1.13	
Issatchenkia orientalis IPE 100	150	Glucose	37	64.3	1.07	Kwon *et al.* [65]
			42	65.5	1.37	
S. cerevisiae C3751	100	Glucose	37	37.3	1.55	Auesukaree *et al.* [20]
			41	38.0	1.58	
S. cerevisiae DBKKU Y-53	250	SSJ	37	106.82	2.23	This study
			40	85.01	2.83	

S_0: initial sugar concentration; T: incubation temperature; P: ethanol concentration; Q_p: volumetric ethanol productivity.

This work will contribute a significant amount of information on ethanol production at an industrial scale. However, the ethanol production cost at a large-scale should be concerned. There are many techniques which can be employed to reduce the production cost; for example, cell recycling, various fermentation approaches, such as very high gravity fermentation, consolidated bioprocessing, and continuous ethanol fermentation using stirred tank bioreactor coupling with plug flow bioreactor [17,18,43,54].

3.6. Real-Time RT-PCR Analysis of Gene Expression

Stressful conditions during ethanol fermentation, such as high temperatures, and high concentrations of ethanol or sugar, have been reported to trigger the expression of several stress-responsive genes including those encoding HSPs, enzymes involved in protein degradation and in the glycolysis pathway, and other proteins involved in the synthesis of compatible solutes and reserve carbohydrates [22,23,25,66,67]. In the present study, the expression levels of *hsp26*, *hsp70*, *hsp90*, *hsp104*, *cdc*, *tps*, *nth*, *gsy*, and *rsp* in the thermotolerant yeast *S. cerevisiae* DBKKU Y-53 and SC90 were determined using real-time RT-PCR. As shown in Figure 3, the expression of all genes was activated in both strains of yeast during ethanol fermentation at 40 °C. Although the expression levels of *tps*, *nth*, *gsy*, and *rsp* in *S. cerevisiae* DBKKU Y-53 and SC90 were not dramatically different, the genes encoding HSPs (*hsp26*, *hsp70*, *hsp90*, *hsp104*) in DBKKU Y-53 were much greater than in SC90. This finding suggests that high growth and ethanol fermentation capabilities of DBKKU Y-53 at a high temperature might be related to increased expression levels of HSP genes. Conversely, the expression level of *cdc* encoding pyruvate kinase, which is involved in ATP production, in SC90 was greater than that in DBKKU Y-53, suggesting that SC90 required more ATP for growth and ethanol production at high temperature. The results in this study were in good agreement with Auesukaree *et al.* [20], who observed high expression levels of genes encoding the small HSP, HSP70, HSP90, and HSP100 family and those genes encoding trehalose-6-phosphate synthase, neutral trehalose, and glycogen synthase in *S. cerevisiae* after heat shock and long-term heat exposure at 37 °C. Piper *et al.* [68] reported that several HSPs are constitutively expressed at appropriate temperatures and play a crucial role in folding and assembling proteins. Many of the HSPs, such as HSP70, HSP90, and HSP104, play an important role as molecular chaperones. These molecular chaperones depend on the energy of

ATP hydrolysis for function [69]. Therefore, increasing the expression level of *cdc* during ethanol fermentation at a high temperature may provide sufficient energy not only for growth and ethanol fermentation activity but may also function as molecular chaperones to protect the structural integrity of the proteins in yeast cells. In addition to the HSPs, the accumulation of trehalose has also been reported to be associated with heat stress protection and thermotolerance [70]. Trehalose can stabilize the protein structure, reduce the aggregation of denatured proteins, and cooperate with HSPs to promote protein refolding [71–73].

In *S. cerevisiae*, ubiquitin ligase plays a key regulatory role in many cellular processes, such as trafficking, sorting, modifying gene expression, DNA repair, RNA transport as well as the degradation of a large number of proteins in multiple cellular compartments [74]. This protein is also involved in the pathways responsible for the regulation of chromatin function and ultimately controls gene expression under limited nutrient conditions [75]. Most recently, Shahsavarani *et al.* [76] also demonstrated that overexpression of *RSP5* encoding ubiquitin ligase improved the ability of *S. cerevisiae* to tolerate high temperatures. An increase in the ubiquitin ligase, which was observed in this study by the high expression level of *rsp* during ethanol fermentation at high temperature, might regulate the transcription of some genes and induce the heat stress response through the ubiquitination process. Therefore, the ability of *S. cerevisiae* DBKKU Y-53 and SC90 to grow and produce a relatively high level of ethanol at a high temperature might be explained by this mechanism.

In the thermotolerant yeast *K. marxianus*, the molecular mechanisms conferring thermotolerance are complicated and are controlled by multiple genes not only for HSPs biosynthesis but also for those genes encoding the proteins involved in DNA replication and repair, RNA processing, ribosome biogenesis, and carbohydrate metabolism process [21]. The expression of *hsp* genes and those functioning to prevent protein denaturation may be insufficient to allow growth and efficient ethanol fermentation of *S. cerevisiae* at high temperatures. To gain a better understanding and provide useful information for fundamental and applied research for innovative applications, further studies are required to clarify the precise mechanism conferring thermotolerance in *S. cerevisiae*.

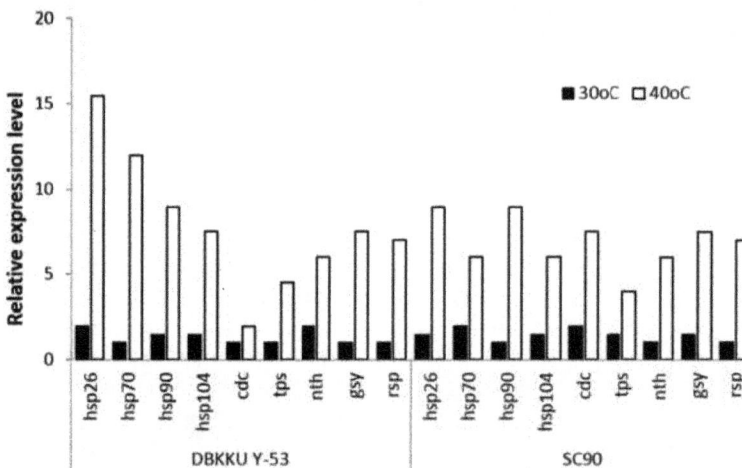

Figure 3. Expression levels of genes encoding HSP26 (*hsp26*), HSP70 (*hsp70*), HSP90 (*hsp90*), HSP104 (*hsp104*), pyruvate kinase (*cdc*), trehalose-6-phosphate synthase (*tps*), neutral trehalase (*nth*), glycogen synthase (*gsy*), and ubiquitin synthase (*rsp*) in the exponential-growth phase of the thermotolerant yeast *S. cerevisiae* DBKKU Y-53 and a reference strain *S. cerevisiae* SC90 during ethanol fermentation at 40 °C. Values presented are the means and relative expression levels of each gene as described in the Materials and Methods section.

4. Conclusions

Utilization of high-potential thermotolerant ethanol-producing yeast is a promising approach to reduce the energy used in cooling systems and to also reduce the operating cost of ethanol production at high temperatures. In this study, the newly isolated thermotolerant ethanol-producing yeast designated as *S. cerevisiae* DBKKU Y-53 exhibited high growth and ethanol production efficiencies at high temperatures (37 °C and 40 °C) compared with the other isolated strains and the industrial ethanol producer *S. cerevisiae* SC90. The optimum conditions for ethanol production by this thermotolerant yeast using SSJ as a raw material were the following: a pH of 5.5, a sugar concentration of 250 g· L^{-1}, a cell concentration of 1.0×10^8 cells· mL^{-1}. The SSJ without the addition of an exogenous nitrogen source can be used directly as substrate for ethanol production at high temperatures by the thermotolerant yeast *S. cerevisiae* DBKKU Y-53. During ethanol fermentation at 40 °C, genes encoding HSP26, HSP70, HSP90, HSP104, pyruvate kinase, trehalose-6-phosphate synthase, neutral trehalase, glycogen synthase, and ubiquitin ligase were highly expressed in *S. cerevisiae* DBKKU Y-53 and SC90 compared with the expression levels at 30 °C. This finding suggests that the growth and ethanol fermentation activity of yeast at high temperatures were not only correlated with the expression of genes involved in heat-stress response but were also correlated with genes involved in ATP production, trehalose and glycogen metabolism, and the protein degradation process.

Acknowledgments: This research was financially supported by the Energy Policy and Planning Office, Ministry of Energy, Thailand and the Research Fund for Supporting Lecturer to Admit High Potential Student to Study and Research on His Expert Program Year 2009, the Graduate School, Khon Kaen University, Thailand. Apart of this work was also supported by the Center for Alternative Energy Research and Development, the Fermentation Research Center for Value Added Agricultural Products (FerVAAP), The New Core to Core Program (CCP) A. Advanced Research Networks on "Establishment of an International Research Core for New Bio-research Fields with Microbes from Tropical Areas World Class Research Hub of Tropical Microbial Resources and Their Utilization", and a grant from Khon Kaen University.

Author Contributions: Sunan Nuanpeng performed the experiments; Sudarat Thanonkeo analyzed the data; Mamoru Yamada contributed reagents and analysis tools; Pornthap Thanonkeo conceived and designed the experiments, analyzed the data and wrote the paper.

Conflicts of Interest: The authors declare no conflict of interest.

References

1. Farrell, A.E.; Plevin, R.J.; Turner, B.T.; Jones, A.D.; O'Hare, M.; Kammen, D.M. Ethanol can contribute to energy and environmental goals. *Sciences* **2006**, *311*, 506–508. [CrossRef] [PubMed]
2. Hill, J.; Nelson, E.; Tilman, D.; Polasky, S.; Tiffany, D. Environmental, economic, and energetic costs and benefits of biodiesel and ethanol biofuels. *Proc. Natl. Acad. Sci. USA* **2006**, *103*, 11206–11210. [CrossRef] [PubMed]
3. Yan, J.; Lin, T. Biofuels in Asia. *Appl. Energy* **2009**, *86*, S1–S10. [CrossRef]
4. Zhou, A.; Thomson, E. The development of biofuels in Asia. *Appl. Energy* **2009**, *86*, S11–S20. [CrossRef]
5. Salassi, M.E. *The Economic Feasibility of Ethanol Production from Sugar Crops*; LSU AgCenter: Baton Rouge, LA, USA, 2007.
6. Daniel, P.H.; Lueschen, W.E.; Kanne, B.K.; Hoverstad, T.R. A comparison of sweet sorghum cultivars and maize for ethanol production. *J. Prod. Agric.* **1991**, *4*, 377–381.
7. Bvochora, J.M.; Read, J.S.; Zvauya, R. Application of very high gravity technology to the cofermentation of sweet stem sorghum juice and sorghum grain. *Ind. Crops Prod.* **2000**, *11*, 11–17. [CrossRef]
8. Woods, J. Integrating Sweet Sorghum and Sugarcane for Bioenergy: Modeling the Potential for Electricity and Ethanol Production in SE Zimbabwe. Ph.D. Thesis, University of London, London, UK, 2000.
9. Chan-u-tit, P.; Laopaiboon, L.; Jaisil, P.; Laopaiboon, P. High level ethanol production by nitrogen and osmoprotectant supplementation under very high gravity fermentation conditions. *Energies* **2013**, *6*, 884–899. [CrossRef]
10. Yu, M.; Li, J.; Chang, S.; Du, R.; Li, S.; Zhang, L.; Fan, G.; Yan, Z.; Cui, T.; Cong, G.; *et al.* Optimization of ethanol production from NaOH-pretreated solid state fermented sweet sorghum bagasse. *Energies* **2014**, *7*, 4054–4067. [CrossRef]

11. Gnansounou, E.; Dauriat, A.; Wyman, C.E. Refining sweet sorghum to ethanol and sugar: Economic trade-offs in the context of North China. *Bioresour. Technol.* **2005**, *96*, 985–1002. [CrossRef] [PubMed]

12. Kiran Sree, N.; Sridhar, M.; Suresh, K.; Banat, I.M.; Venkateswar, R.L. Isolation of thermotolerant, osmotolerant, flocculating *Saccharomyces cerevisiae* for ethanol production. *Bioresour. Technol.* **2000**, *72*, 43–46. [CrossRef]

13. Limtong, S.; Sringiew, C.; Yongmanitchai, W. Production of fuel ethanol at high temperature from sugar cane juice by a newly isolated *Kluyveromyces marxianus*. *Bioresour. Technol.* **2007**, *98*, 3367–3374. [CrossRef] [PubMed]

14. Kourkoutas, Y.; Bekatorou, A.; Banat, I.M.; Marchant, R.; Koutinas, A.A. Immobilization technologies and support materials suitable in alcohol beverages production: A review. *Food Microbiol.* **2004**, *21*, 377–397. [CrossRef]

15. Zafar, S.; Owais, M. Ethanol production from crude whey by *Kluyveromyces marxianus*. *Biochem. Eng. J.* **2006**, *27*, 295–298. [CrossRef]

16. Yuan, W.J.; Zhao, X.Q.; Ge, X.M.; Bai, F.W. Ethanol fermentation with *Kluyveromyces marxianus* from Jerusalem artichoke grown in salina and irrigated with a mixture of seawater and freshwater. *J. Appl. Microbiol.* **2008**, *105*, 2076–2083. [CrossRef] [PubMed]

17. Yuan, W.J.; Chang, B.L.; Ren, J.G.; Liu, J.P.; Bai, F.W.; Li, Y.Y. Consolidated bioprocessing strategy for ethanol production from Jerusalem artichoke tubers by *Kluyveromyces. marxianus* under high gravity conditions. *J. Appl. Microbiol.* **2011**, *112*, 38–44. [CrossRef] [PubMed]

18. Hu, N.; Yuan, B.; Sun, J.; Wang, S.A.; Li, F.L. Thermotolerant *Kluyveromyces marxianus* and *Saccharomyces cerevisiae* strains representing potentials for bioethanol production from Jerusalem artichoke by consolidated bioprocessing. *Appl. Microbiol. Biotechnol.* **2012**, *95*, 1359–1368. [CrossRef] [PubMed]

19. Edgardo, A.; Carolina, P.; Manuel, R.; Juanita, F.; Baeza, J. Selection of thermotolerant yeast strains *Saccharomyces cerevisiae* for bioethanol production. *Enzyme Microb. Technol.* **2008**, *43*, 120–123. [CrossRef]

20. Auesukaree, C.; Koedrith, P.; Saenpayavai, P.; Asvarak, T.; Benjaphokee, S.; Sugiyama, M.; Kaneko, Y.; Harashima, S.; Boonchird, C. Characterization and gene expression profiles of thermotolerant *Saccharomyces cerevisiae* isolates from Thai fruits. *J. Biosci. Bioeng.* **2012**, *114*, 144–149. [CrossRef] [PubMed]

21. Lertwattanasakul, N.; Kosaka, T.; Hosoyama, A.; Suzuki, Y.; Rodrussamee, N.; Matsutani, M.; Murata, M.; Fujimoto, N.; Suprayogi; Tsuchikane, K.; *et al.* Genetic basis of the highly efficient yeast *Kluyveromyces marxianus*: Complete genome sequence and transcriptome analyses. *Biotechnol. Biofuels* **2015**, *8*. [CrossRef] [PubMed]

22. Walter, S.; Buchner, J. Molecular chaperones-cellular machines for protein folding. *Angew. Chem. Int. Ed. Engl.* **2002**, *41*, 1098–1113. [CrossRef]

23. Borges, J.C.; Ramos, C.H. Protein folding assisted by chaperones. *Protein Pept. Lett.* **2005**, *12*, 257–261. [CrossRef] [PubMed]

24. Purvis, J.E.; Yomano, L.P.; Ingram, L.O. Enhanced trehalose production improved growth of *Escherichia coli* under osmotic stress. *Appl. Environ. Microbiol.* **2005**, *71*, 3761–3769. [CrossRef] [PubMed]

25. Li, L.L.; Ye, Y.R.; Pan, L.; Zhu, Y.; Zheng, S.P.; Lin, Y. The induction of trehalose and glycerol in *Saccharomyces cerevisiae* in response to various stresses. *Biochem. Biophy. Res. Commun.* **2009**, *387*, 778–783. [CrossRef] [PubMed]

26. Parrou, J.L.; Teste, M.A.; Francois, J. Effects of various types of stress on the metabolism of reserve carbohydrates in *Saccharomyces cerevisiae*: Genetic evidence for a stress-induced recycling of glycogen and trehalose. *Microbiology* **1997**, *143*, 1891–1900. [CrossRef] [PubMed]

27. Unnikrishnan, I.; Miller, S.; Meinke, M.; LaPorte, D.C. Multiple positive and negative elements involved in the regulation of expression of GSY1 in *Saccharomyces cerevisiae*. *J. Biol. Chem.* **2003**, *278*, 26450–26457. [CrossRef] [PubMed]

28. Kurtzman, C.; Robnett, C.J. Identification and phylogeny of ascomycetous yeasts from analysis of nuclear large subunit (26S) ribosomal DNA partial sequences. *Antonie van Leeuwenhoek* **1998**, *73*, 331–371. [CrossRef] [PubMed]

29. Harju, S.; Fedosyuk, H.; Peterson, K.R. Rapid isolation of yeast genomic DNA: Bust n' Grab. *BMC Biotechnol.* **2004**, *4*. [CrossRef] [PubMed]

30. O'Donnell, K. *Fusarium* and its near relatives. In *The Fungal Holomorph: Mitotic, Meiotic and Pleomorphic Speciation in Fungal Systematics*; Reynolds, D.R., Taylor, J.W., Eds.; CAB International: Wallingford, UK, 1993; pp. 225–233.

31. Tamura, K.; Peterson, D.; Peterson, N.; Stecher, G.; Nei, M.; Kumar, S. Mega 5: Molecular evolutionary genetics analysis using maximum likelihood evolutionary distance, and maximum parsimony methods. *Mol. Biol. Evol.* **2011**, *28*, 2731–2739. [CrossRef] [PubMed]

32. Saitou, N.; Nei, M. The neighbor-joining method. A new method for reconstructing phylogenetic trees. *Mol. Biol. Evol.* **1987**, *4*, 406–425. [PubMed]

33. Livak, K.J.; Schmittgen, T.D. Analysis of relative gene expression data using real-time quantitative PCR and the $2^{-\Delta\Delta C_T}$ method. *Methods* **2001**, *25*, 402–408. [CrossRef] [PubMed]

34. Zoecklein, B.W.; Fugelsang, K.C.; Gump, B.H.; Nury, F.S. *Wine Analysis and Production*; Chapman & Hall: New York, NY, USA, 1995.

35. Dubois, M.; Gilles, K.A.; Hamilton, J.R.; Robers, P.K.; Smith, F. Colourimetric method for determination of sugar and related substances. *Anal. Chem.* **1956**, *28*, 350–356. [CrossRef]

36. Banat, I.M.; Nigam, P.; Singh, D.; Marchant, R.; McHale, A.P. Review: Ethanol production at elevated temperatures and alcohol concentrations: Part I—Yeasts in general. *World J. Microbiol. Biotechnol.* **1998**, *14*, 809–821. [CrossRef]

37. Van Uden, N. Cardinal temperatures of yeast. In *Handbook of Microbiology*; Laskin, A.I., Lechevalier, H.H.A., Eds.; CRC Press: Boca Raton, FL, USA, 1984; Volume VI, pp. 1–8.

38. Lee, C.; Yamakawa, T.; Kodama, T. Rapid growth of a thermotolerant yeast on palm oil. *World J. Microbiol. Biotechnol.* **1993**, *9*, 187–190. [CrossRef] [PubMed]

39. Ballesteros, I.; Ballesteros, M.; Cabanas, A.; Carrasco, J.; Martin, C.; Negro, M.; Saez, F.; Saez, R. Selection of thermotolerant yeasts for simultaneous saccharification and fermentation (SSF) of cellulose to ethanol. *Appl. Biochem. Biotechnol.* **1991**, *28–29*, 307–315. [CrossRef] [PubMed]

40. Banat, I.M.; Nigam, P.; Marchant, R. Isolation of thermotolerant, fermentative yeasts growing at 52 °C and producing ethanol at 45 °C and 50 °C. *World J. Microbiol. Biotechnol.* **1992**, *8*, 259–263. [CrossRef] [PubMed]

41. Tanimura, A.; Nakamura, T.; Watanabe, I.; Ogawa, J.; Shima, J. Isolation of a novel strain of *Candida shehatae* for ethanol production at elevated temperature. *SpringerPlus* **2012**, *1*, 27. [CrossRef] [PubMed]

42. Kaewkrajay, C.; Dethoup, T.; Limtong, S. Ethanol production from cassava using a newly isolated thermotolerant yeast strain. *ScienceAsia* **2014**, *40*, 268–277. [CrossRef]

43. Charoensopharat, K.; Thanonkeo, P.; Thanonkeo, S.; Yamada, M. Ethanol production from Jerusalem artichoke tubers at high temperature by newly isolated thermotolerant inulin-utilizing yeast *Kluyveromyces marxianus* using consolidated bioprocessing. *Antonie Van Leeuwenhoek* **2015**, *108*, 173–190. [CrossRef] [PubMed]

44. Walker, G.M. The roles of magnesium in biotechnology. *Crit. Rev. Biotechnol.* **1994**, *14*, 311–354. [CrossRef] [PubMed]

45. Roukas, T. Solid-state fermentation of carob pods for ethanol production. *Appl. Microbiol. Biotechnol.* **1994**, *41*, 296–301. [CrossRef]

46. Torija, M.J.; Beltran, G.; Novo, M.; Poblet, M.; Guillamon, J.M.; Mas, A.; Rozes, N. Effects of fermentation temperature and *Saccharomyces* species on the cell fatty acid composition and presence of volatile compounds in wine. *Int. J. Food Microbiol.* **2003**, *85*, 127–136. [CrossRef]

47. Tofighi, A.; Assadi, M.M.; Asadirad, M.H.A.; Karizi, S.Z. Bio-ethanol production by a novel autochthonous thermo-tolerant yeast isolated from wastewater. *J. Environ. Health Sci. Eng.* **2014**, *12*, 107. [CrossRef] [PubMed]

48. Narendranath, N.V.; Power, R. Relationship between pH and medium dissolved solids in terms of growth and metabolism of *Lactobacilli* and *Saccharomyces cerevisiae* during ethanol production. *Appl. Environ. Microbiol.* **2005**, *71*, 2239–2243. [CrossRef] [PubMed]

49. Ercan, Y.; Irfan, T.; Mustafa, K. Optimization of ethanol production from carob pod extract using immobilized *Saccharomyces cerevisiae* cells in a stirred tank bioreactor. *Bioresour. Technol.* **2013**, *135*, 365–371. [CrossRef] [PubMed]

50. Singh, A.; Bishnoi, N.R. Ethanol production from pretreated wheat straw hydrolyzate by *Saccharomyces cerevisiae* via sequential statistical optimization. *Ind. Crops Prod.* **2013**, *41*, 221–226. [CrossRef]

51. Laopaiboon, L.; Thanonkeo, P.; Jaisil, P.; Laopaiboon, P. Ethanol production from sweet sorghum juice in batch and fed-batch fermentations by *Saccharomyces cerevisiae*. *World J. Microbiol. Biotechnol.* **2007**, *23*, 1497–1501. [CrossRef]

52. Ingledew, W.M. Alcohol production by *Saccharomyces cerevisiae*: A yeast primer. In *The Alcohol Textbook*, 3rd ed.; Jacques, K.A., Lyons, T.P., Kelsall, D.R., Eds.; Nottingham University Press: Nottingham, UK, 1999; pp. 49–87.

53. Ozmichi, S.; Kargi, F. Ethanol fermentation of cheese whey powder solution by repeated fed-batch operation. *Enzyme Microb. Technol.* **2007**, *41*, 169–174.

54. Laopaiboon, L.; Nuanpeng, S.; Srinophakun, P.; Klanrit, P.; Laopaiboon, P. Ethanol production from sweet sorghum juice using very high gravity technology: Effects of carbon and nitrogen supplementations. *Bioresour. Technol.* **2009**, *100*, 4176–4182. [CrossRef] [PubMed]

55. Bai, F.W.; Chen, L.J.; Zhang, Z.; Anderson, W.A.; Moo-Young, M. Continuous ethanol production and evaluation of yeast cell lysis and viability loss under very high gravity medium conditions. *J. Biotechnol.* **2004**, *110*, 287–293. [CrossRef] [PubMed]

56. Yue, G.; Yu, J.; Zhang, X.; Tan, T. The influence of nitrogen sources on ethanol production by yeast from concentrated sweet sorghum juice. *Biomass Bioenergy* **2012**, *39*, 48–52. [CrossRef]

57. Ter Schure, E.G.; van Riel, A.A.W.; Verrips, C.T. The role of ammonia metabolism in nitrogen catabolite repression in *Saccharomyces cerevisiae*. *FEMS Microbiol. Rev.* **2000**, *24*, 67–83. [CrossRef] [PubMed]

58. Magasanik, B.; Kaiser, C.A. Nitrogen regulation in *Saccharomyces cerevisiae*. *Gene* **2002**, *290*, 1–18. [CrossRef]

59. Magasanik, B. Regulation of nitrogen utilization. In *The Molecular and Cellular Biology of the Yeast Saccharomyces: Gene expression*; John, E.W., Pringle, J.R., Broach, J.R., Eds.; Cold Spring Harbor: New York, NY, USA, 1992; pp. 283–317.

60. Beltran, G.; Esteve-Zarzoso, B.; Rozès, N.; Mas, A.; Guillamón, J.M. Influence of the timing of nitrogen additions during synthetic grape must fermentations on fermentation kinetics and nitrogen consumption. *J. Agric. Food Chem.* **2005**, *53*, 996–1002. [CrossRef] [PubMed]

61. Beltran, G.; Rozès, N.; Mas, A.; Guillamón, J.M. Effect of low-temperature fermentation on yeast nitrogen metabolism. *World J. Microbiol. Biotechnol.* **2007**, *23*, 809–815. [CrossRef]

62. Sridhar, M.; Kiran Sree, N.; Venkateswar, R.L. Effect of UV radiation on thermotolerance, ethanol tolerance and osmotolerance of *Saccharomyces cerevisiae* VS1 and VS3 strains. *Bioresour. Technol.* **2002**, *83*, 199–202. [CrossRef]

63. Abdel-fattah, W.R.; Fadil, M.; Nigam, P.; Banat, I.M. Isolation of thermotolerant ethanologenic yeasts and use of selected strains in industrial scale fermentation in an Egyptian distillery. *Biotechnol. Bioeng.* **2000**, *68*, 531–535. [CrossRef]

64. Dhaliwal, S.S.; Oberoi, H.S.; Sandhu, S.K.; Nanda, D.; Kumar, D.; Uppal, S.K. Enhanced ethanol production from sugarcane juice by galactose adaptation of a newly isolated thermotolerant strain of *Pichia kudriavzevii*. *Bioresour. Technol.* **2011**, *102*, 5968–5975. [CrossRef] [PubMed]

65. Kwon, Y.J.; Ma, A.Z.; Li, Q.; Wang, F.; Zhuang, G.Q.; Liu, C.Z. Effect of lignocellulosic inhibitory compounds on growth and ethanol fermentation of newly-isolated thermotolerant *Issatchenkia. orientalis*. *Bioresour. Technol.* **2011**, *102*, 8099–8104. [CrossRef] [PubMed]

66. Ni, H.T.; LaPorte, D.C. Response of a yeast glycogen synthase gene to stress. *Mol. Microbiol.* **1995**, *16*, 1197–1205. [CrossRef] [PubMed]

67. Verghese, J.; Abrams, J.; Wang, Y.; Morano, K.A. Biology of the heat shock responses: Building yeast (*Saccharomyces cerevisiae*) as a model system. *Microbiol. Mol. Biol. Rev.* **2012**, *76*, 115–158. [CrossRef] [PubMed]

68. Piper, P.W. Molecular events associated with acquisition of heat tolerance by the yeast *Saccharomyces cerevisiae*. *FEMS Microbiol. Rev.* **1993**, *11*, 339–356. [CrossRef] [PubMed]

69. Lindquist, S. Heat-shock proteins and stress tolerance in microorganisms. *Curr. Opin. Genet. Dev.* **1992**, *2*, 748–755. [CrossRef]

70. Singer, M.A.; Lindquist, S. Thermotolerance in *Saccharomyces cerevisiae*: The Yin and Yang of trehalose. *Trends Biotechnol.* **1998**, *16*, 460–468. [CrossRef]

71. Hottiger, T.; De Virgilio, C.; Hall, M.N.; Boller, T.; Wiemken, A. The role of trehalose synthesis for the acquisition of thermotolerance in yeast II. Physiological concentrations of trehalose increase the thermal stability of proteins *in vitro*. *Eur. J. Biochem.* **1994**, *219*, 187–193. [CrossRef] [PubMed]

72. Zancan, P.; Sola-Penna, M. Trehalose and glycerol stabilize and renature yeast inorganic pyrophosphate inactivated by very high temperatures. *Arch. Biochem. Biophys.* **2005**, *444*, 52–60. [CrossRef] [PubMed]
73. Lindquist, S.; Kim, G. Heat-shock protein 104 expression is sufficient for thermotolerance in yeast. *Proc. Natl. Acad. Sci. USA* **1996**, *93*, 5301–5306. [CrossRef] [PubMed]
74. Krsmanović, T.; Kölling, R. The HECT E3 ubiquitin ligase Rsp5 is important for ubiquitin homeostasis in yeast. *FEBS Lett.* **2004**, *577*, 215–219. [CrossRef] [PubMed]
75. Cardona, F.; Aranda, A.; Del Olmo, L. Ubiquitin ligase Rsp5 is involved in the gene expression changes during nutrient limitation in *Saccharomyces cerevisiae*. *Yeast* **2009**, *26*, 1–15. [CrossRef] [PubMed]
76. Shahsavarani, H.; Sugiyama, M.; Kaneko, Y.; Chuenchit, B.; Harashima, S. Superior thermotolerance of *Saccharomyces cerevisiae* for efficient bioethanol fermentation can be achieved by overexpression of *RSP5* ubiquitin ligase. *Biotechnol. Adv.* **2012**, *30*, 1289–1300. [CrossRef] [PubMed]

![energies logo] *energies*

MDPI

Article

Simplification of a Mechanistic Model of Biomass Combustion for On-Line Computations

Alexandre Boriouchkine * and Sirkka-Liisa Jämsä-Jounela

Department of Biotechnology and Chemical Technology, School of Chemical Technology, Aalto University, Aalto 00076, Finland; sirkka-liisa.jamsa-jounela@aalto.fi
* Correspondence: alexandre.boriouchkine@aalto.fi; Tel.: +358-9-470-23846

Academic Editor: Tariq Al-Shemmeri
Received: 15 May 2016; Accepted: 19 August 2016; Published: 10 September 2016

Abstract: Increasing utilization of intermittent energy resources requires flexibility from energy boilers which can be achieved with advanced control methods employing dynamic process models. The performance of the model-based control methods depends on the ability of the underlying model to describe combustion phenomena under varying power demand. This paper presents an approach to the simplification of a mechanistic model developed for combustion phenomena investigation. The aim of the approach is to simplify the dynamic model of biomass combustion for applications requiring fast computational times while retaining the ability of the model to describe the underlying combustion phenomena. The approach for that comprises three phases. In the first phase, the main mechanisms of heat and mass transfer and limiting factors of the reactions are identified in each zone. In the second phase, each of the partial differential equations from the full scale model are reduced to a number of ordinary differential equations (ODEs) defining the overall balances of the zones. In the last phase, mathematical equations are formulated based on the mass and energy balances formed in the previous step. The simplified model for online computations was successfully built and validated against industrial data.

Keywords: biomass; combustion; mechanistic modeling; faster-than-real-time simulation; on-line computations

---

1. Introduction

Increasing flexibility demands force boilers to operate with wide ranges of load levels and fast transitions between these levels. Model-based control is an excellent technique to achieve optimal load transitions and wide operational ranges. The model-based control methods utilize the process model to obtain a reliable assessment of the key variables of the process state and to predict the future boiler behavior, which is required to ensure extensive control over the combustion process. Thus, further improvement in terms of combustion efficiency and flexibility of boilers demands that the model be able to provide accurate descriptions of combustion phenomena.

Gray-box modeling which comprises ordinary differential equations (ODEs) with parameters identified in order to fit a particular boiler system is a rather popular modeling technique for online applications. For instance, Hogg and El-Rabaie [1] utilized a gray box coal combustion model to develop a generalized predictive control strategy for a 200 MW boiler. Similarly, the most recent application of this technique discusses modelling of boilers of various scales. Bauer et al. [2] developed a gray box model whose structure was based on process knowledge obtained by analyzing mathematical correlations of char combustion. In order to fit the predictions of the model, to the process data empirical coefficients were obtained through model identification. Gölles et al. [3] modeled combustion in a 30 kW boiler using a black-box model to describe the key states present in a boiler. Similarly, Kortela and Jämsä-Jounela [4] used gray-box modeling to model biomass firing in a 16 MW

BioGrate boiler. Instead, Flynn and O'Malley [5], constructed a model based on air/mass flows and the adiabatic flame temperature for predicting the key process variables. Although, these models describe the amount of fuel on the grate, they lack important variables such as fuel and gas temperatures which are essential for prediction pollutant emissions.

In mechanistic modeling for real time and control applications two approaches have been presented in the literature. In the first approach, the burning fuel bed of a sloping-grate boiler is represented as a lumped model, developed by Paces and Kozek [6]. The fuel layer is divided in horizontal direction and each lump describes fuel drying, pyrolysis, char combustion and cooling zone. The model comprises gaseous and solid phases and the chemical reactions in these phases are governed by linearized Arrhenius equations. The model developed by Paces and Kozek [6] relied on parameter identification in order to achieve satisfactory accuracy in describing biomass combustion. In the second approach, developed by Belkhir et al. [7], the layer was also divided into gaseous and solid phase, in contrast, only the fuel oxidation reaction was described in the model assuming mass transfer controlled reactions.

Fuzzy logic—neural networks have also been proposed for the modeling of energy boilers [8]. For instance, Peng et al. [9] used a Gaussian radial basis function (RBF), autoregressive model with exogenous inputs (ARX) to describe combustion in a coal-fired boiler for the development of a nonlinear model predictive control strategy for efficiency improvement of the de-NO_x reaction. This approach proved to be effective in minimizing ammonia consumption in the de-NO_x reaction. The main approach of describing fuel combustion in ARX models is to relate fuel and air flows to boiler power output [10] or to steam production [11,12].

The models proposed for on-line applications can be divided into two categories based on the degree of involvement of identification methods: (1) fully data-based methods—ARX, ANN-ARX, ANN-fuzzy; and (2) models based on ODEs with identified parameter functions. Nevertheless, both these model categories suffer from limitations, such as data-overfitting and insufficient reproducibility of nonlinear process behavior. In addition, industrial processes tend to operate at a limited number of regimes and thus the identification of less frequently occurring operational conditions is inherently difficult. For non-linear processes, such as biomass combustion, mechanistic modeling is the only method which can guarantee acceptable accuracy and robustness, since non-linear system identification suffers from several shortcomings. Morari and Lee [13] highlighted numerous issues related to nonlinear model identification, such as model structure determination, test input signal design, multiple input-multiple output model fitting algorithm and uncertainty quantification for robust control. However, mechanistic models developed for real time applications neglect important phenomena such as moisture evaporation due to convective and radiative heat transfer which is particularly important in case of moist biomass fuel and in boiler operating under concurrent combustion conditions and overfed fuel beds where hot char comes into contact with moist cold fuel.

On the other hand, several mechanistic models were developed for process phenomena investigation that accurately reproduce the process behavior. Unfortunately, these models are excessively detailed and contain variables and phenomena which have insignificant impact on the descriptive ability of the model. Moreover, in these mechanistic models, all parts of the burning fuel layer are described equally in detail, even the ones which remain at constant conditions for the long duration of time and do not affect the overall burning behavior of a fuel layer. Thus, by only describing the relevant phenomena and the parts of the fuel layer, a significant reduction in the model complexity can be achieved which leads to simpler model implementation, calibration, short computational times and consequently considerably lower hardware requirements which is especial relevant for industrial control systems.

The aim of this study is to simplify mechanistic model to decrease the excessive complexity while retaining the ability of a model to describe combustion phenomena. With the method, we determine the factors, such as heat and mass transfer mechanisms having the largest impact on the model accuracy and disregard the factors with low contribution. In addition, based on the method, we divide the layer into a few zones which eliminates the need for the spatial discretization of the model and considerably shortens the computational time. Based on this, we obtain a novel dynamic combustion model which provides the insight into the burning fuel layer with accuracy comparable to that of the original model, but with a simpler mathematical formulation and significantly shorter computational times. The simplified model was evaluated against the original mechanistic model to ensure its ability to describe combustion phenomena. Furthermore, the model developed in this work has been successfully validated against industrial data from a BioGrate boiler suggesting the applicability of the proposed simplification approach. As a result, this work finally closes the gap between the detailed modeling of biomass combustion for phenomena investigation and the modeling of combustion processes for applications which include process control and monitoring.

The remainder of the paper is structured as follows. Section 2 describes a BioGrate process. Section 3 presents the simplification approach along with the description of the model upon which simplification is based and the simplified model validation. In Section 4, the simplified model is compared against the original and a linear one. Conclusions and future research are presented in Section 5.

2. Process Description of a BioGrate Boiler

The BioGrate boiler is used for electricity and steam production as well as to supply heat to a local municipal hot water network. BioGrate boilers are based on a conical grate technology that is designed to utilize various fuels with high moisture content. A BioGrate furnace consists of the following functional parts: grate rings, a water-filled ash space below the grate and heat-insulating refractory walls around the grate rings which form the combustion chamber.

To improve the spread of fuel within the combustion chamber, the grate consists of several ring zones, which are further divided into two types of alternating rings: rotating and fixed. Half the rotating rings move clockwise and the rest counter clockwise, distributing the fuel evenly on the grate. As the fuel is fed into the center of the grate from below, the surface of the fuel starts to dry in the center of the cone as a result of heat radiation, which is emitted by the combusting flue gas and reflected back to the grate by the grate walls. The dry fuel then proceeds to the outer rings of the grate, where pyrolysis char gasification and combustion occur. Finally, the resulting ash and carbon residues fall off the edge of the grate into the water-filled annular ash space and Figure 1 shows an outline of a BioGrate furnace.

Figure 1. BioGrate boiler furnace.

The air required for combustion is distributed via annular primary air registers below the grate to the nozzles in each grate plate (primary air) and through nozzles in the combustion chamber wall (secondary air). Burning produces heat that is absorbed in several steps: first, the evaporator and water-tube walls of the boiler absorb the energy from the flue gases, which is followed by energy transfer to steam through superheaters. In the third phase, heat is transferred from the flue gas to the feed water by the convective evaporator, before finally the economizers remove the remaining flue-gas energy.

3. Model Development for Process Control and Monitoring

Model development for process control and monitoring purposes proceeds through the simplification of a mechanistic model. Therefore, this section is divided into three main parts. Firstly, the mechanistic model developed by the authors [14,15] is presented, including proposed simplifications based on the subsequent simulation results. The simplified model is then presented and discussed prior to the last section which outlines the validation of the developed model.

3.1. Mechanistic Model

In a BioGrate boiler, reactions of biomass occur through three main reaction pathways that can take place either in parallel or sequentially with respect to each other: drying, pyrolysis and char conversion and the mechanistic modeling of the BioGrate boiler considers these three stages of the biomass combustion. As the model accounts for mass and energy conservation, it considers two phases, a solid and a gaseous one. The burning fuel bed in a BioGrate boiler is modeled in one dimension using the walking grate concept [14]. The walking grate approach divides the fuel layer into vertical, one-dimensional portions with the fuel movement on a grate assumed to resemble that observed on a traveling grate. As a result, the developed model describes traveling grate combustion and the structure of the model with the solving algorithm is presented in Figure 2.

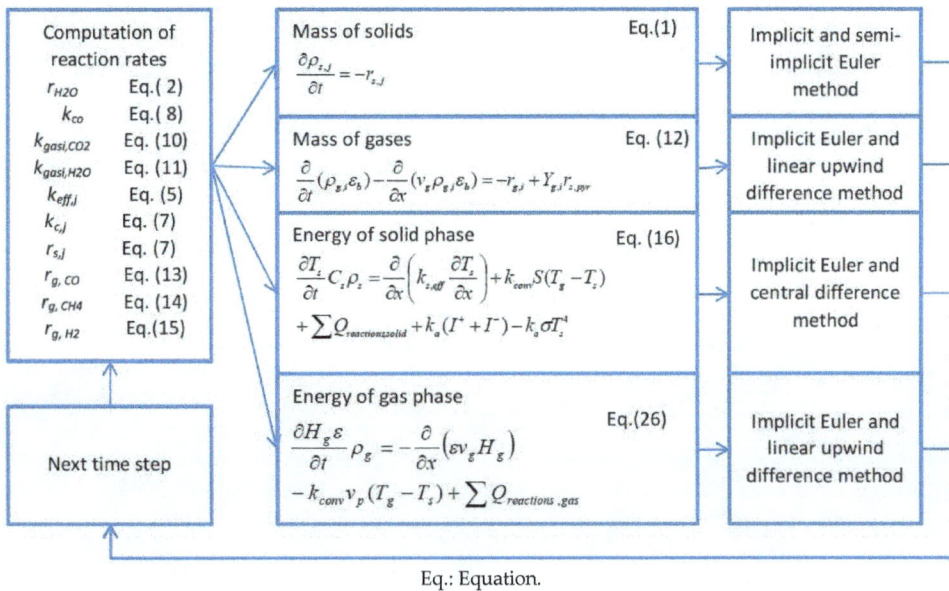

Eq.: Equation.

Figure 2. Outline of the model used for process phenomena investigation.

3.1.1. Modelling of Mass Conservation in the Solid Phase

The solid phase reactions considered by the model include drying, pyrolysis as well as char oxidation and gasification. In general, these can be described by Equation (1) and the reactions considered in the solid phase are presented in Table 1:

$$\frac{\partial \rho_{s,j}}{\partial t} = -r_{s,j} \tag{1}$$

The rate of drying, Equation (2), is defined by the energy available for evaporation:

$$r_{s,H_2O} = \max \left(0, \frac{\rho_{H_2O}}{\max \left(\rho_{H_2O} \right)} \frac{C_s (T_s - 378(K))}{\Delta t \Delta H_{evap}} \right) \tag{2}$$

Table 1. Solid phase reactions.

H_2O (l) $\rightarrow$ H_2O (g)
Wood $\rightarrow$ Gas, Tar, Char
C (s) $+ \alpha O_2$ (g) $\rightarrow 2(1-\alpha)$ CO (g) $+ (2\alpha - 1)$ CO_2 (g)
C (s) $+ CO_2$ (g) $\rightarrow$ 2CO (g)
C (s) $+ H_2O$ (g) $\rightarrow$ CO (g) $+ H_2$ (g)

The pyrolysis reaction rates are described by Equations (3) and (4). The volatile matter of the fuel was assumed to contain two pseudo-components with different combination of cellulose, lignin and hemicellulose. The first component mainly contains lignin and slowly decomposes over a large range of temperatures while the second contains mainly cellulose and hemicellulose. The mass fraction of the first volatile component in wood is 0.15.

$$r_{s,pyr,1} = 315.6 \exp(-80(kJ/mol) / (RT_s)) \rho_{pyr,1} \tag{3}$$

$$r_{s,pyr,2} = 2.205 \cdot 10^4 \exp\left(-78(kJ/mol)\right) / (RT_s)) \rho_{pyr,2} \tag{4}$$

Pomerantsev [16] has proposed to calculate the rate of char burning as a function of either char or oxygen concentration, depending on whichever is limiting. Thus the rates for the char reactions can be calculated by Equation (5):

$$r_{s,i} = k_{eff,i} \rho_{g,i} \tag{5}$$

For each heterogeneous char reaction, an effective reaction constant—given by Equation (6)—is then calculated as follows:

$$k_{eff,i} = \frac{S k_{m,i} k_{r,i}}{S k_{m,i} + k_{r,i}} \tag{6}$$

where S is the density number, the particle surface area per unit volume [17], given by Equation (7), $k_{m,i}$ is the mass transfer coefficient [17], calculated from Equation (8) and $k_{r,i}$ is the reaction constant:

$$S = \frac{6(1 - \varepsilon_b)}{d_p} \tag{7}$$

where ε_b is the fuel bed porosity and d_p is the particle diameter.

$$k_{m,i} = D_{g,i}/d_p \left(2 + Re^{1/2} Sc^{1/3}\right) \tag{8}$$

The reaction constants, $k_{r,i}$, are determined for three different chemical reactions: oxidation, Equation (9) [18], and reduction with H_2O, Equation (10) [19], and CO_2, Equation (11) [20]:

$$k_{r,C} = 1.1 \cdot 10^6 \cdot \exp\left(-114.5(kJ/mol)/R/T_s\right) \tag{9}$$

$$k_{r,H_2O} = 9.99 \cdot 10^4 \cdot \exp\left(-136(kJ/mol)/R/T_s\right)(1-X)\sqrt{1-10\ln(1-X)} \tag{10}$$

$$k_{r,CO_2} = 1.1 \cdot 10^9 \exp\left(-260(kJ/mol)/R/T_s\right)X \tag{11}$$

with the associated CO/CO_2 ratio given by Equation (12) [21]:

$$CO/CO_2 = 4.3\exp(-3390/T_s) \tag{12}$$

3.1.2. Modelling of Mass Conservation in the Gas Phase

The gas phase modelling, presented in Equation (13), describes gas convection, combustion of carbon monoxide (Equation (14)), methane (Equation (15)) [16] and hydrogen (Equation (16)) [16], as well as gas formation in the pyrolysis reaction, $Y_{g,i}r_{pyr}$:

$$\frac{\partial}{\partial t}(\rho_{g,i}\varepsilon_b) - \frac{\partial}{\partial x}(v_g\rho_{g,i}\varepsilon_b) = -r_{g,i} + Y_{g,i}r_{pyr} \tag{13}$$

where r_{pyr} is the pyrolysis reaction rate, while $Y_{g,i}$ is the mass fraction of the gaseous pyrolysis product formed:

$$r_{g,CO} = 1.3 \cdot 10^{14}\exp\left(-125.5(kJ/mol)\right)/R/T_g) C_{CO} C_{O_2}{}^{0.5} C_{H_2O}{}^{0.5} \tag{14}$$

$$r_{g,CH_4} = 5.6 \cdot 10^{12}\exp\left(-103.8(J/mol)\right)/R/T_g) C_{O_2} \tag{15}$$

$$r_{g,H_2} = 2.14 \cdot 10^{14}\exp\left(-129\,(J/mol)\right)/R/T_g) C_{O_2} \tag{16}$$

The following gaseous components are considered by the model: water vapor, tar, oxygen, carbon monoxide, carbon dioxide, nitrogen, methane, and hydrogen.

3.1.3. Modelling of Energy Conservation in the Solid Phase

The energy equation for the solid phase describes heat conduction, heat exchange between the phases, energy consumed in the drying and pyrolysis reactions and energy gained by char combustion (Equation (17)):

$$\frac{\partial T_s}{\partial t}C_s\rho_s = \frac{\partial}{\partial x}\left(\kappa_{s,eff}\frac{\partial T_s}{\partial x}\right) + k_{conv}S(T_g - T_s) + \sum Q_{s,i} + k_a(I^+ + I^-) - k_a\sigma T_s^4 \tag{17}$$

The heat conduction in solid particles is described by Equations (18)–(23) [22]:

$$\kappa_{s,eff} = \varepsilon_p k_{max} + (1-\varepsilon_p)k_{min} + k_{s,rad} \tag{18}$$

$$k_{s,rad} = \frac{4\varepsilon_p\sigma T_s{}^3 d_{cavity}}{1-\varepsilon_p} \tag{19}$$

$$d_{cavity} = 3.5 \cdot 10^{-5}\sqrt{\varepsilon_p} \tag{20}$$

$$\varepsilon_p = 1 - (\rho_v + \rho_c)/1500 - \rho_w/1000 \tag{21}$$

$$k_{min} = \frac{k_g k_{fiber}}{\varepsilon_p k_{fiber} + (1-\varepsilon_p)k_g} \tag{22}$$

$$k_{\text{max}} = \varepsilon_p k_g + (1 - \varepsilon_p) k_{\text{fiber}} \tag{23}$$

The radiative heat transfer inside the bed is described with a two-flux model given by Equations (24) and (25), where the absorption and scattering coefficients are given by Equation (26) [23]:

$$\frac{dI^+}{dx} = -(k_a + k_s)I^+ + k_s I^- + \frac{1}{2}k_a \sigma T_s^4 \tag{24}$$

$$-\frac{dI^-}{dx} = -(k_a + k_s)I^- + k_s I^+ + \frac{1}{2}k_a \sigma T_s^4 \tag{25}$$

$$k_a = -\frac{1}{d_p}\ln(\varepsilon_b), \ k_s = 0 \tag{26}$$

3.1.4. Modelling of the Gas Phase Energy Conservation

The energy continuity equation—Equation (27)—of the gas phase considers the heat exchange between the gas and solid phases, the energy received through gas convection, and the energy gained from carbon monoxide, methane and hydrogen oxidation:

$$\frac{\partial H_g \varepsilon_b}{\partial t}\rho_g = -\frac{\partial}{\partial x}\left(\varepsilon_b v_g H_g\right) - k_{\text{conv}} S(T_g - T_s) + \sum Q_{g,i} \tag{27}$$

3.2. Development of a Simplified Dynamic Model

The main idea of the simplification is to divide the fuel layer into zones, located between the reaction fronts, and to assume uniform conditions within each zone. This is based on the observation that large temperature and composition gradients exist at the reaction fronts while the degree of variation is limited in between the fronts. Such an approach allows a significant reduction in the number of variables by neglecting the ones which are not critical in a mathematical description of a combustion process.

Simplification of the model developed in this study involves the following three main phases. In the first phase, the main mechanisms of heat and mass transfer and limiting factors of the reactions, such as oxygen concentration and the energy available, are identified in each zone. In the seconds phase, each of the partial differential equations from the full scale model, representing conservation laws at each spatial point, are reduced to a number of ODEs defining the overall balances of the zones. In the last phase, mathematical equations are formulated based on the mass and energy balances formed in the previous step.

3.3. Approach for Simplification of the Mechanistic Model

The simplification of the model developed in this work involves three main phases. The purpose of the first phase is to divide the fuel layer into zones to reduce computational time. This is performed based on the simulation results from the mechanistic model by identifying largest gradients with respect to the char, volatile and moisture content which indicate the end of each respective zone.

In the second phase, the reaction-limiting factors are determined among oxygen concentration, convective, radiative and conductive heat transfer, such that non-restrictive factors can be omitted from the model. This is done by comparing the magnitude of each heat transfer mechanism with respect to each other and by removing the least significant ones. The char oxidation reaction is generally known to be limited by the primary air feed, since almost all oxygen is consumed close to the grate and hence the propagation of the front is limited by oxygen supply [15]. The area on top of the fuel layer is not included into the consideration, since it does not receive oxygen from below. With the absence of oxygen, the low heat conductivity of char can be assumed to prevent the heat flux downwards and to effectively stop the front from propagating. In case of pyrolysis and drying, the front propagation of these reactions is limited by heat transfer from hotter areas as these reactions require sufficient temperature to become active. Thus, the identification of reaction limiting factors can be done based on the gradients in the temperature profile as well as on the radiative heat flux profile.

These considerations result in a significant simplification of the temperature profile as only the temperature of the char layer is assumed to be a variable. Instead, concentration-wise the fuel layer is divided into three zones. This results in a great reduction in computational time by avoiding the need for spatial discretization which results in a significant decrease in the number of equations to be solved.

Finally, the detailed description of the third phase of the simplification approach, namely, the mathematical formulation of the model, is provided in the following.

3.4. Mathematical Formulation of the Simplified Model

This section presents the development of a simplified dynamic model for on-line computations. First, based on the simulation results from the enhanced model the reaction-limiting factors are identified. Then, the equations of the enhanced model are simplified by considering these factors and the model is validated. The simplified model was implemented in the MATLAB computation environment and solved with the implicit Euler method. An overview of the model structure is presented in Figure 3.

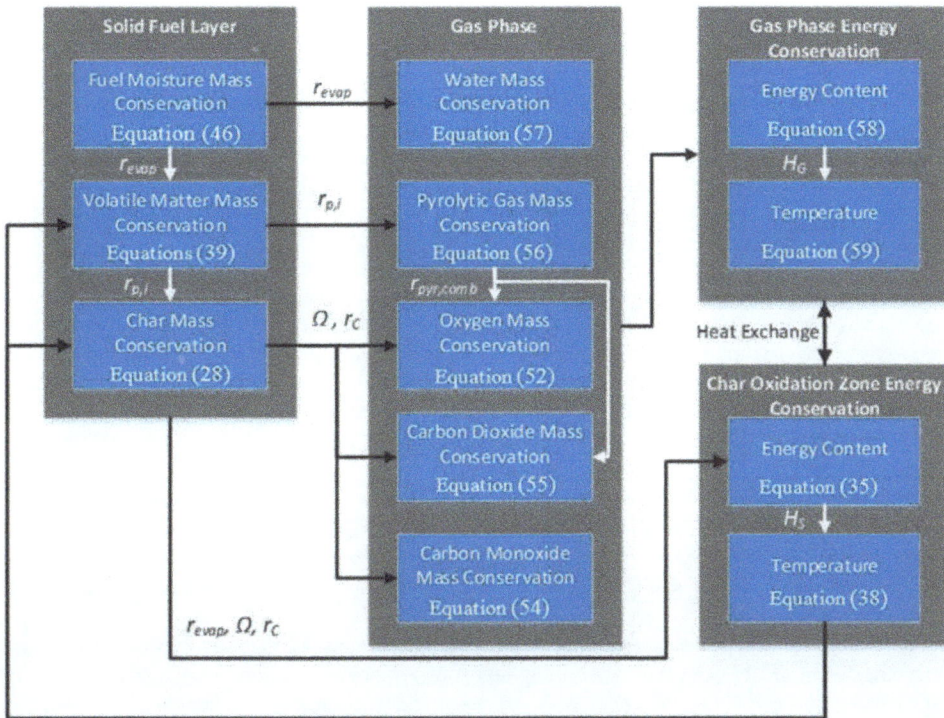

Figure 3. Overview of the model structure.

Mathematical Formulation of the Dynamic Model for On-Line Computations

In the first phase, the reaction-limiting factors have to be identified, which is done based on simulations of the enhanced mechanistic model. The results from the mechanistic model (Figure 4) suggest the presence of three reaction fronts corresponding to each major reaction: char combustion, pyrolysis, and evaporation zones. This means that the walking grate concept, which described fuel movement on the grate and the temperature and concentration gradients in the horizontal direction in the dynamic model, has to be omitted because of the assumption of uniform conditions inside each zone.

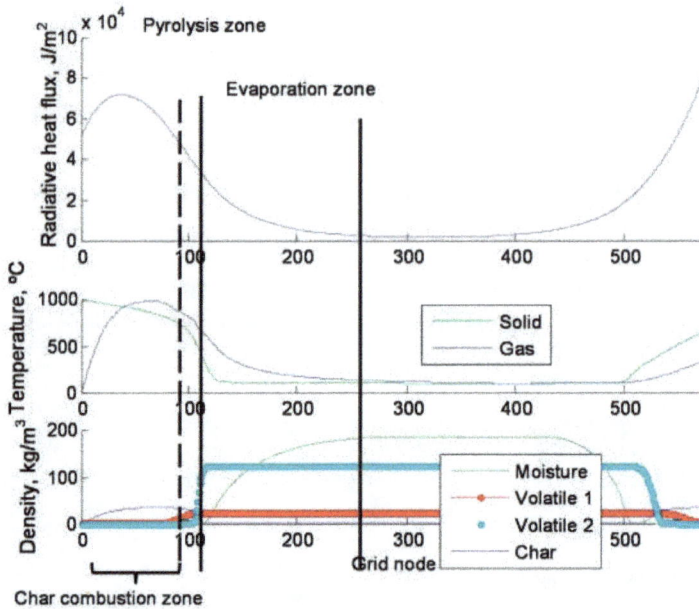

Figure 4. Simulation results from the enhanced model.

In the second phase, the reaction-limiting factors are identified from the simulation results presented in Figure 4 starting from char oxidation zone. In that zone, oxygen is consumed almost completely by the oxidation reaction. However, the primary air flow initially decreases the temperature of char as it enters the furnace. This effect is shown in Figure 4 as a decrease in temperature which becomes more significant towards the surface of the grate. Thus, in addition to the oxidation reaction also the heat exchange between char and the primary air is significant and can be up to 1000 °C close to the grate.

The pyrolysis zone, which lies above the char oxidation zone, is rather thin and, thus, the temperature of the reacting fuel is close to the char combustion temperature. The simulation results in Figure 4 indicate that the char oxidation and pyrolysis zones overlap since char forms during the devolatilization reaction. Thus, it is assumed that the temperature of the pyrolysis zone equals that of the char oxidation zone.

Moisture evaporation, being an endothermic reaction, is limited by the radiative as well as by the convective heat transfer from the char combustion zone. The heat conduction is neglected due to its low contribution since, conductive heat flux over one char particle and 100 °C temperature gradient is 670 W/m² while in case of radiative heat transfer it is nine times higher (calculated by differentiating the heat flux over the interface between the drying and pyrolysis zones over one particle diameter). However, conduction contributes only slightly to the evaporation due to low heat conductivity of char and thus its effect is neglected. The contribution of the radiative heat transfer mechanism is only significant at the interface of the pyrolysis and the drying zones where temperature exceeds 500 °C. Thus, both heat transfer mechanisms can be assumed to contribute to the moisture evaporation. Next, mathematical equations based on these considerations are presented for each zone, in accordance with the phase three of the approach.

- Char Combustion Zone

In char combustion zone, both the mass and energy are considered in accordance with the general simplification approach.

- Mass Conservation

Char forms during the pyrolysis phase in which cellulose, hemicellulose and lignin decompose to produce volatile gas and char which is subsequently consumed by the oxidation reaction:

$$\frac{dm_C}{dt} = X_C \sum_{i=1}^{3} r_{p,i} - r_C \tag{28}$$

where m_C is the amount of char on the grate, X_C is the char yield from wood pyrolysis and r_C is the char reaction rate, $r_{p,i}$ accounts for individual pyrolysis reactions of cellulose, lignin and hemicellulose. The effective char consumption rate is described by Equation (29) and, depending on the conditions, char oxidation is kinetically or mass transfer controlled. Owing to the omission of the spatial coordinates, the term $h_C SV$ was introduced to model the effect of the char layer thickness on the consumption of oxygen. Thus, Equation (29), which describes the effective rate constant of char combustion, assuming that the process can be both reaction and mass transfer controlled, in the mechanistic model, is modified as:

$$k_{C,eff} = h_C SV k_{O_2} k_C / (k_C + h_C S V k_{O_2}) \tag{29}$$

where $k_{C,eff}$ is the effective reaction rate of char, V is the fuel layer volume, k_{O2} is the mass transfer coefficient of oxygen onto the surface of char particle, h_C is the mass fraction of char on the grate where the mass transfer coefficient is calculated from the Sherwood number:

$$k_{O_2} = D_g / d_p \left(2 + Re^{1/2} Sc^{1/3}\right) \tag{30}$$

where D_g is the oxygen diffusivity in air, and d_p is the particle size

The volume of the fuel layer is defined by Equation (31), which assumes that both fuel and char contribute to the overall volume of the fuel bed:

$$V = \frac{m_{fuel}}{\rho_{fuel\ bed\ dry}} + \frac{m_C}{\rho_{char\ bed}} \tag{31}$$

where $\rho_{fuel\ bed\ dry}$ is the density of dry fuel, and $\rho_{char\ bed}$ is the density of char bed

Parameter S is the density number which describes surface to volume ratio of the fuel layer:

$$S = \frac{6(1 - \varepsilon)}{d_p} \tag{32}$$

where ε is the porosity of fuel bed and h_C is the mass fraction of char:

$$h_C = m_C / (m_{fuel} + m_{H_2O,l} + m_C) \tag{33}$$

where m_C, m_{fuel} and $m_{H2O,l}$ are the amounts of char, fuel and water, respectively. Thus the term $h_C SV$ in Equation (29) effectively describes the overall surface area of char available for the oxidation reaction.

The char combustion kinetics are defined by Equation (34) [24]:

$$k_C = 1.1 \cdot 10^6 \exp(-114.5(kJ/mol) / (RT_S)) \tag{34}$$

- Energy Conservation

The energy balance of the solid phase, as outlined in Equation (35), includes the energy content of char and of minute amounts of volatile components and depends on the energy lost due to emitted radiation and heat transferred to the incoming primary air flow. A fraction of the energy of the solid phase is transferred with the gases formed in pyrolysis and oxidation reactions. Moreover, the enthalpy of the solid phase is increased by oxidation reactions and heat from pyrolysis reactions. As the air is supplied from under the grate, it is heated by the burning char layer, such that:

$$\frac{dH_S}{dt} = Q_{in} - Q_{out} - \Delta H_{evap} r_{evap,rad} + \sum_{i=1}^{3} (\Delta H_{p,i} - C_{p,w} T) \, r_{p,i} + Q_C - Ah(T_S - T_{G,In}) \tag{35}$$

where Q_{in} is the energy of contained in primary air, Q_{out} is the energy of the gas out flow, ΔH_s is the enthalpy of the solid phase, $\Delta H_{p,i}$ is the pyrolysis heat of component i, $C_{p,w}$ is the heat capacity of wood, Q_C is the energy form char combustion, A is the total outer surface of the fuel, and h is the heat transfer coefficient.

It is evident from the Equation (35) that depending on the intensity of thermal decomposition, the effect of pyrolysis heat can have a significant effect on the energy balance and thus needs to be estimated. Moreover, the overall pyrolysis enthalpy is fuel specific and is largely dependent on chemical composition of fuel and can be endothermic, neutral or exothermic. The endothermicity or exothermicity is essentially defined by the pyrolytic secondary reactions. In case of the debarking residue used as fuel in BioGrate boiler the pyrolysis heats for cellulose, lignin and hemicellulose were determined to be -100, -600 and -200 kJ/mol, respectively [25]. These are the total reaction heats which include the heat from primary and secondary reactions for each component. Modeling of lignin decomposition is, especially, important since it decomposes slowly and the pyrolysis temperature can increase up to 900 °C before lignin has completely decomposed. In contrast, hemicellulose and cellulose complete decomposition at relatively moderate temperatures.

The char is assumed to react to carbon monoxide and carbon dioxide and the overall energy contribution is described by Equation (36):

$$Q_C = \left(\frac{\Delta H_{CO}}{M_{CO}} \frac{2(\Omega - 1)}{\Omega} r_C + \frac{\Delta H_{CO_2}}{M_{CO_2}} \frac{2 - \Omega}{\Omega} - C_{p,C} T_S \right) r_C \tag{36}$$

where ΔH_{CO} and ΔH_{CO2} is the reaction enthalpy of char oxidation to carbon monoxide and carbon dioxide, respectively, $C_{p,c}$ is the heat capacity of char, Ω is the stoichiometric coefficient.

The coefficient is given by Equation (37) [21]. This equation accounts for the fact that at high temperatures the breakage of oxygen-carbon complex on the surface of the char particle tends to produce mostly CO [16]:

$$\Omega = \frac{2 + 2 \cdot 4.3 \exp(-3390/T_S)}{2 + 4.3 \exp(-3390/T_S)} \tag{37}$$

The temperature of the solid phase can be calculated by dividing the energy content of the char by its enthalpy:

$$T = \frac{H_S}{m_C C_{p,c} + C_{p,w} \sum_{i}^{3} m_{p,i}} \tag{38}$$

- Fuel Pyrolysis Zone

For each volatile component of the fuel (cellulose, hemicellulose, and lignin), the following mass conservation equation is devised:

$$\frac{dm_{p,i}}{dt} = r_{evap} \frac{X_{p,i}(1 - X_m)}{X_m} - r_{p,i} \tag{39}$$

where $X_{p,i}$ is the fraction of cellulose, hemicellulose and lignin, X_m, the moisture content of the fuel in the furnace, is defined by Equation (40):

$$X_m = \frac{m_{H_2O,l}}{m_{fuel} + m_{H_2O,l}} \tag{40}$$

Wood pyrolysis, Equation (41), consists of individual devolatilization reactions of cellulose, lignin and hemicellulose the kinetic expressions of which are given by Equations (42)–(44), respectively [26]. These kinetic expressions were shown to reproduce the pyrolysis of debarking residue, the material composition ($X_{p,i}$) of which was as follows: cellulose 36 wt%, lignin 40 wt%, hemicellulose 24 wt% [25]:

$$r_{p,i} = k_{p,i} m_{p,i} \tag{41}$$

$$k_{p,1} = 8.75 \cdot 10^{18} \exp\left(-233(kJ/mol)/(RT_{pyr})\right) \tag{42}$$

$$k_{p,2} = 25 \exp\left(-30(kJ/mol)/(RT_{pyr})\right) \tag{43}$$

$$k_{p,3} = 5 \cdot 10^{8} \exp\left(-105(kJ/mol)/(RT_{pyr})\right) \tag{44}$$

where T_{pyr} is the temperature of pyrolysis, according to the simplification approach it is assumed to be equal to the char combustion temperature.

It was experimentally determined that pyrolysis of fuel produces 74 wt% pyrolytic gas and 26 wt% of char [25]. Due to partial overlapping between char and pyrolysis zones, char combustion and volatile gas combustion might compete if the temperature in char oxidation zone will become excessively low:

$$k_{pyr,comb} = T_g/T_s k_C \tag{45}$$

- Fuel Drying Zone

Evaporation of moisture is most active at temperatures exceeding 100 °C. However, according to the simplification approach, the temperature profile described by Equation (17) is omitted and the heat flux becomes the determinant factor of moisture evaporation rather than local temperature. Therefore, instead of temperature, the evaporation rate (Equation (46)) is now determined by radiative heat transfer from burning char (Equation (47)) and by convective heat transfer from the gas phase (Equation (48)) [27]:

$$\frac{dm_{H_2O,l}}{dt} = X_m \dot{m}_{In} - r_{evap,conv} - r_{evap,rad} \tag{46}$$

The walking grate concept which described fuel movement on the grate is replaced with a continuous fuel input, the term $X_m \dot{m}_{In}$, which describes the moisture content in the fuel feed. $r_{evap,conv}$ and $r_{evap,rad}$ are the drying rates due to convective (Equation (47)) and radiative heat transfer (Equation (48)), respectively:

$$r_{evap,rad} = \sigma A_{moist}\left(T_S^4 - 373^4\right)/\Delta H_{evap} \tag{47}$$

where σ is the Stefan-Boltzmann constant and ΔH_{evap} is the heat of evaporation:

$$r_{evap,conv} = h A_{moist}\left(T_G - 373\right)/\Delta H_{evap} \tag{48}$$

where h is the heat transfer coefficient:

$$h = \frac{2 + 1.1 Pr^{1/3} Re^{3/5}}{d_p} k_{air} \tag{49}$$

where D_p is the particle diameter.

A_{moist} is the area of moist fuel:

$$A_{moist} = A_{moist,0} \frac{m_{H_2O}}{m_{H_2O} + m_C + m_{fuel}} \tag{50}$$

The amount of wet fuel (Equation (51)) depends on the fuel drying and pyrolysis rates, as described by Equation (41). However, since the walking grate concept is omitted, a variable, m_{In}, describing fuel feeding rate has to be introduced:

$$\frac{dm_{fuel}}{dt} = (1 - X_m)\dot{m}_{In} - \sum_{i=1}^{3} r_{p,i} \tag{51}$$

where m_{fuel} is the amount of dry fuel on the grate, X_{H2O} is the moisture content of fuel, m_{In} is the feed of the dry fuel, and $r_{p,i}$ is the reaction rate of the individual wood component (cellulose, hemicellulose and lignin).

- Mass Conservation Equations in the Gas Phase

The amount of oxygen in the fuel layer comprises the oxygen supplied by the primary and secondary air flows, as well as the char reaction rate and the flow of oxygen out of the layer:

$$\frac{dm_{O_2}}{dt} = 0.23\rho_{Air}F_{In} - \left(k_{eff,c} + k_{pyr,comb} + F_{Out}/V\right)m_{O_2} \tag{52}$$

where ϱ_{Air} is the density of air at normal pressure at 20 °C, F_{In} and F_{Out} are flow into and from the furnace, respectively.

It is assumed that the gas density depends only on the temperature. Therefore, the outflow of gas from the fuel layer depends on the gas expansion, as outlined in Equation (53):

$$F_{Out} = F_{in}\frac{\rho_G(T_G)}{\rho_G(T_{In})} \tag{53}$$

where $\varrho_G(T_G)$ is the gas density at the temperature of gas, T_G and $\varrho_G(T_{In})$ is the gas density at 20 °C.

The amount of carbon monoxide, dioxide and pyrolytic gas are described by Equations (54)–(56) respectively:

$$\frac{dm_{CO}}{dt} = \frac{2(\Omega - 1)}{\Omega}r_C - (F_{Out}/V)m_{CO} \tag{54}$$

$$\frac{dm_{CO_2}}{dt} = \frac{2 - \Omega}{\Omega}r_C + \Omega_{pyr}r_{pyr,comb} - (F_{Out}/V)m_{CO_2} \tag{55}$$

The transport equations of the pyrolytic components (Equation (56)) were combined into one equation to describe the evolution of volatiles, including gas formation, combustion and flow out of the fuel layer—Equation (57):

$$\frac{dm_{p,G}}{dt} = X_{p,G}m_{p,G} - \Omega_{pyr}r_{pyr,comb} - (F_{Out}/V)m_{p,G} \tag{56}$$

where Ω_{pyr} is the stoichiometric coefficient of pyrolytic gas combustion:

$$\frac{dm_{H_2O}}{dt} = r_{evap} - (F_{Out}/V)m_{H_2O} \tag{57}$$

- Energy Conservation Equations in the Gas Phase

The enthalpy of the gas phase, defined by Equation (58), is affected by the flow of the primary air and the flow of the reacted gas out of the fuel layer. In addition, the energy is consumed to heat the inflowing primary air and in the heat transfer between the gas and the drying fuel. The enthalpy is increased by evaporated moisture, gases formed in the oxidation reaction, and energy gained by heat transfer between gas and char:

$$\frac{dH_G}{dt} = Q_{In} - Q_{Out} + T_s C_{p,c} r_C + \sum_{i=1}^{3} (C_W T_S) \, r_{p,i} + 373 C_{p,H_2O,g} r_{evap}$$
$$- Ah(T_s - T_{g,In}) - \Delta H_{evap} r_{evap,conv} \tag{58}$$

where Q_{In} is the enthalpy of the primary air, and Q_{Out} is the enthalpy of gas flowing out of the fuel layer.

The temperature of the gas phase is calculated as follows:

$$T_G = \frac{H_G}{m_{CO}C_{p,CO} + m_{CO_2}C_{p,CO_2} + m_{p,G}C_{p,G} + m_{H_2O,G}C_{p,H_2O,G} + m_{O_2}C_{p,O_2} + m_{N_2}C_{p,N_2}} \tag{59}$$

3.5. Simplified Model Validation with Industrial Data from a BioGrate Boiler

The objective of the model validation was to evaluate the model prediction of the combustion rate m_d and water evaporation rate m_w. Validation was performed against industrial data obtained during full-scale plant experiments at BioPower 5, 16 MW CHP, a BioGrate boiler located at Mänttä-Vilppula, Finland. The aim of the experiments was to study the effect of fuel moisture content variations on the boiler operations. For this purpose, three changes in the fuel moisture content were induced by switching the fuel feed to the auxiliary fuel bunker loaded with 5 m³, 10 m³ and 25 m³ wood chips with approximately 22 wt% moisture content. After the bunker contents were fully discharged, the fuel feed was switched back to the typical fuel, having 55 wt% moisture content. Thus, the experiment induced fast changes in the process operations, which are suitable to test the ability of the simplified model to describe combustion phenomena during rapid transitions in fuel quality.

During the experiments, the residual oxygen content, X_{O2}, was directly available from the database of the automation system. The values for the molar fraction of water vapor, X_{H2O}, were obtained with a Servomex 2500 FTIR Analyzer (Crowborough, UK) by probing the flue gas after the economizer. The data was collected with the one second sampling interval and then utilized for validation purposes along with the measured primary and secondary air flowrates and stoker speed.

The model prediction of the burning rates was calculated from the amount of consumed char and pyrolysis gas, while the evaporation rate was computed through the summation of evaporation rate due to convection and due to radiation, Equations (47) and (48), respectively. The combustion and evaporation rates predicted by the model were compared to the ones calculated from the experimental data which were obtained by solving Equations (60)–(65) for the evaporation rate, m_w, as well as the combustion rate, m_d:

$$X_{O_2} = \frac{\dot{n}_{O_2}}{\dot{n}_{N_2} + \dot{n}_{H_2O} + \dot{n}_{O_2} + \dot{n}_{CO_2}} \tag{60}$$

$$X_{H_2O} = \frac{\dot{n}_{H_2O}}{\dot{n}_{N_2} + \dot{n}_{H_2O} + \dot{n}_{O_2} + \dot{n}_{CO_2}} \tag{61}$$

where $\dot{n}_{N_2}$, $\dot{n}_{H_2O}$, $\dot{n}_{O_2}$ and $\dot{n}_{CO_2}$ are the molar contents of the components in the flue gas, and X_{O2} and X_{H2O} are the volumetric fractions of oxygen and water vapor in flue gas. The molar composition of the flue gas is defined as follows:

$$\dot{n}_{O_2} = 0.23 \frac{m_{Air}}{M_{O_2}} + \left(-\frac{0.51}{M_C} - \frac{0.06}{2M_{H_2}} + \frac{0.41}{M_{O_2}}\right) \cdot \dot{m}_d \tag{62}$$

$$\dot{n}_{H_2O} = 0.06\frac{\dot{m}_d}{M_{H_2}} + \frac{\dot{m}_W}{M_{H_2O}} \tag{63}$$

$$\dot{n}_{CO_2} = 0.51\frac{\dot{m}_d}{M_C} \tag{64}$$

$$\dot{n}_{N_2} = 0.77\frac{\dot{m}_{Air}}{M_{N_2}} \tag{65}$$

where $\dot{m}_{Air}$ is the mass flow of the combined primary and secondary air, $\dot{m}_d$ is the combustion rate of fuel, and $\dot{m}_W$ is the evaporation rate of fuel moisture. M_{H2}, M_{H2O}, M_C, and M_{N2} are the molar weights of hydrogen, water, carbon, and nitrogen, respectively.

The process inputs during the experiment, presented in Figure 5, were used as model inputs to obtain the model prediction of the process variables. The validation results presented in Figure 6 confirm a sufficient descriptive ability of the model. In particular, the variations in the air flows result in slightly lower fuel burning rate while dryer fuel is used, which is reproduced well by the simplified model. The amount of water in the fuel feed directly affects the evaporation rate, the dynamics of which follows closely the model prediction. Furthermore, only a minor difference can be observed between the model prediction for the oxygen content in the flue gas and its measured value, maintained at the constant level by the plant control system. In fact, the model prediction error for the flue gas oxygen is correlated with the prediction error of the burning rate. For example, the model slightly overestimates the burning rate between 30,000 s and 45,000 s of the simulation time, which results in overestimating the oxygen consumption and underestimating the oxygen content in the flue gas. This presents a possibility of utilizing the flue gas oxygen measurement for further improvement in the accuracy of the on-line predictions.

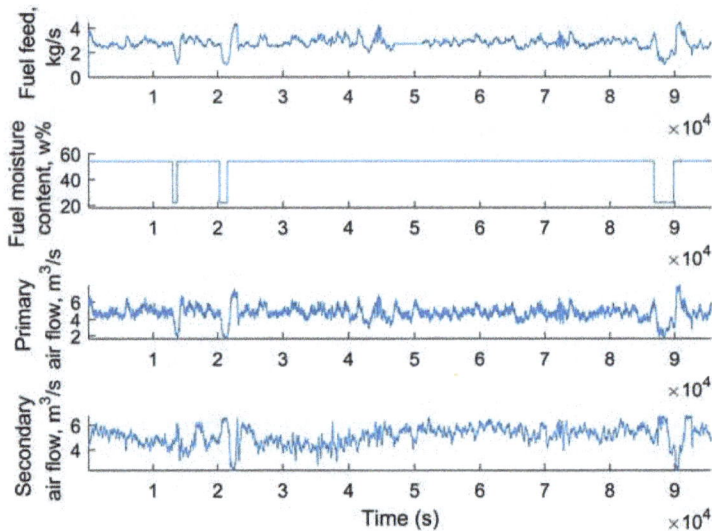

Figure 5. Inputs: fuel feed, fuel moisture, primary air, secondary air.

Energies **2016**, *9*, 735

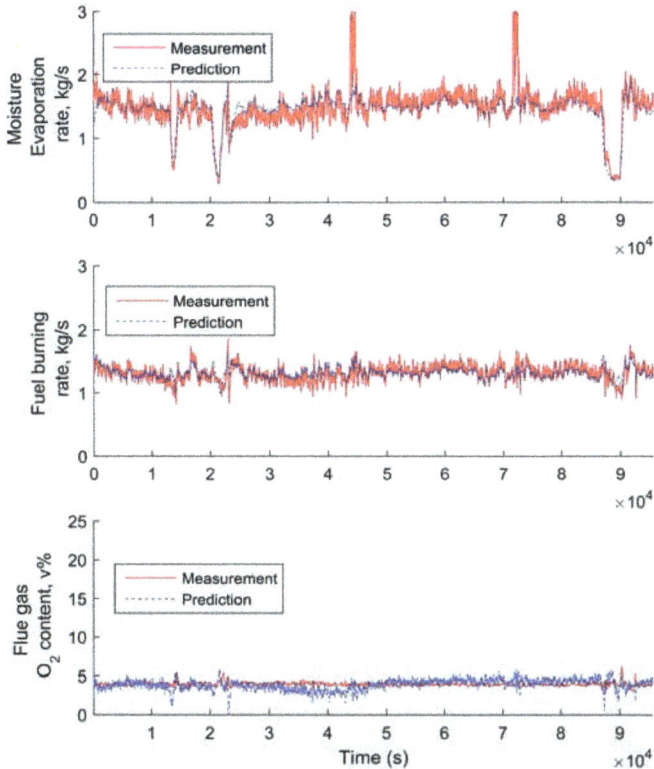

Figure 6. Measured and predicted outputs: moisture evaporation rate, fuel burning rate and flue gas content.

The dynamics of the model states, including the temperature of char and gas, as well as the amount of char, fuel and water on the grate, are presented in Figure 7. The predicted gas and burning char (solid) temperatures are in a good agreement with the values previously measured by Cooper and Hallett [28]. The variations in fuel moisture content result in large changes in char and gas temperature, as presented in Figure 7, which have a major impact on the rates of combustion reactions. Moreover, switching to a dryer fuel raises the amount of char on the grate and simultaneously decreases the quantity of fuel and moisture. Thus, the increased char to fuel ratio requires the adjustment of primary to secondary air ratio in order to achieve clean combustion.

Summarizing, the model demonstrated the ability to describe the fuel burning and evaporation rates that determine the power production of the boiler. Furthermore, the insight provided by the simplified model is especially attractive for the implementation of an efficient model-based process control the boiler. This efficiency can be achieved through the model-based estimation of the attainable power load transitions and load limits could be estimated to avoid the risk of combustion extinguishing or inefficient biomass firing. Thus, the validation results support the use of the simplified model for flexible control of power generation.

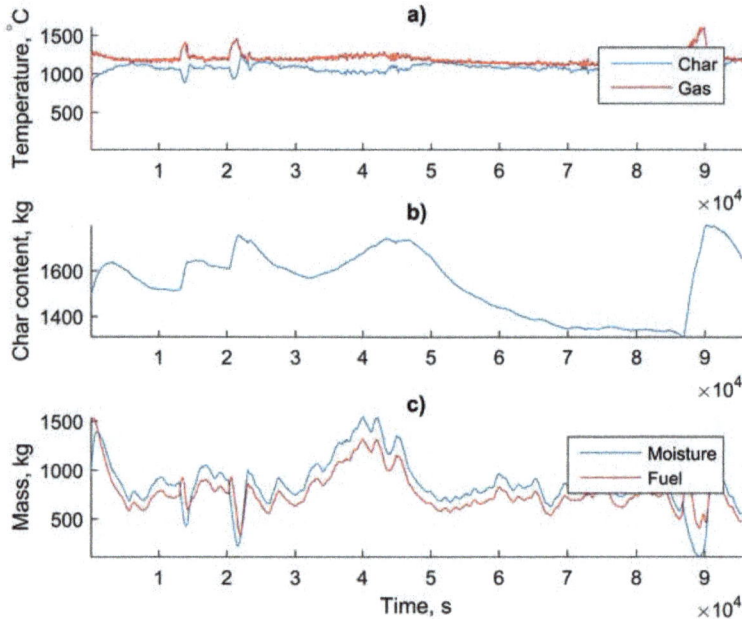

Figure 7. Model states during the validation: (**a**) char and gas temperatures, (**b**) amount of char, and (**c**) fuel and moisture content.

4. Performance Evaluation of the Simplified Model Predictions against the Predictions from the Mechanistic Model and Autoregressive Model with Exogenous Inputs

4.1. Comparison of the Simplified Dynamic Model against the Mechanistic Model

This section discusses the comparison of the results from the full-scale and the simplified models for a fuel moisture change scenario. In this scenario, the fuel moisture was switched from 55 wt% to 22 wt% at 160 s and back to 55 wt% at 840 s of the simulation time, as shown in Figure 8. The decreased primary air flow between 200 s and 800 s delivered less oxygen to the furnace for the combustion reactions. This reduced the burning rate, as predicted by both the full-scale and the simplified models, as demonstrated in Figure 8. In contrast, as the moisture content of the fuel is increased back to the initial level, the primary air flow and the burning rate rise to provide increased amount energy for the fuel drying. This is visible as an increase in the combustion rate in the period that occurred at the time instance of 800 s from the beginning of the simulation.

In general, both models predict similar burning rate trends, agreeing with the process knowledge. However, the prediction of the full-scale model is more sensitive to the high-frequency variations of the primary air flow, as confirmed by Figure 8. This is due to faster local temperature changes in the full model since in the simplified model the temperature is averaged over the whole char layer resulting in slower temperature variations. The temperatures of char layer from both models are presented in Figure 9, in case of the mechanistic model, the temperature was averaged over the char layer. The temperature of the char layer follows the primary air flow dynamics, namely, larger airflows result in larger temperatures while lower flow rate produces lower temperatures. The effect of convective cooling on char by the primary air has lower effect on the temperature compared to the effect of decreased combustion intensity, resulting in lower combustion temperatures at low flow rates.

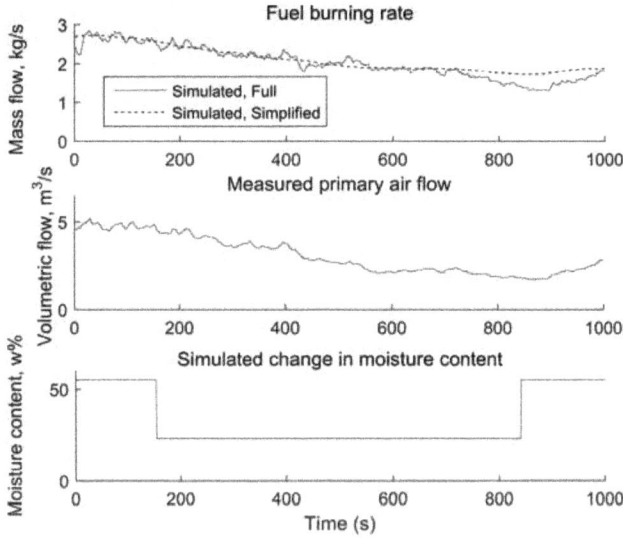

Figure 8. Comparison of the predictions from mechanistic to the ones from the simplified model.

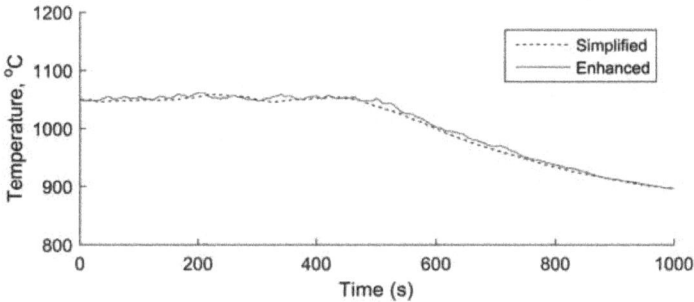

Figure 9. Comparison of the predictions of averaged char layer temperatures from the mechanistic to the ones from the simplified model.

4.2. Comparison of the Simplified Dynamic Model against a Linear Autoregressive Model with Exogenous Inputs Model

In this section, the predictions from the simplified model are compared to the predictions from linear data based-model. The ARX model structure was selected for the linear data based-model due to its satisfactory ability in capturing process dynamics, fast identification and straightforward model diagnostics. The comparison allows the evaluation of the degree of the process nonlinearity effect on the accuracy of a linear model. For the comparison, two ARX models were identified from process data for the moisture evaporation and the fuel burning rates, respectively. The autocorrelation of the model residuals and its cross-correlation with the model inputs were evaluated to ensure the assumptions of the identification methods. Different model orders were tried, and fourth order delivered the best results in both cases.

The step responses are presented in Figure 10 for the identified burning-rate-ARX model and in Figure 11 for the evaporation model. For the burning rate ARX model, the secondary air has twice the effect on the combustion rate compared to the primary air, which seems to be unreasonably high. In fact, the secondary air could be more sensitive to the burning rate variations compared other process inputs because the process control adjusts the secondary air according to the oxygen consumption to maintain the flue gas oxygen constant. Furthermore, the stoker speed has negligible effect on the burning rate, because the stoker speed is coupled with the primary air flow rate in the control strategy of the boiler and the data-based modeling is unable to separate the effect of these two inputs on the process dynamics. However, the stoker speed essentially controls the amount of fuel on the grate and subsequently the combustion rate, thus contradicting the degree of the effect of stoker speed identified by the model. The significance of the moisture content on the combustion rate appears reasonable, although according to the model, the increase in moisture content increases burning rate which is in conflict with the process knowledge. The moisture content affects the ignition delay of the fuel rather than having a direct effect on the magnitude of the combustion rate.

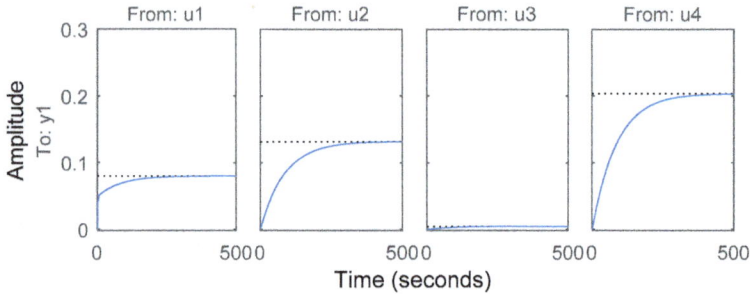

Figure 10. Step responses of the autoregressive model with exogenous inputs (ARX) model for burning rate prediction. u1: primary air; u2: secondary air; u3: stoker speed; and u4: fuel moisture content.

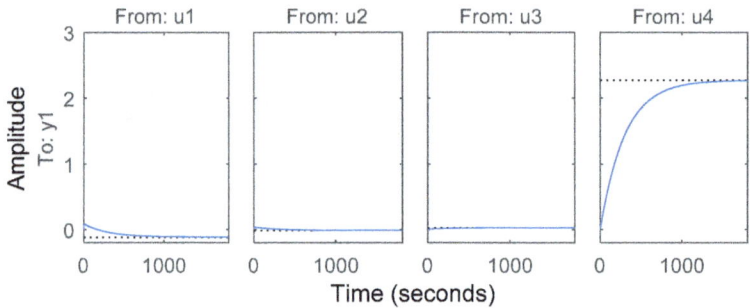

Figure 11. Step responses of the ARX model for evaporation rate prediction. u1: primary air; u2: secondary air; u3: stoker speed; and u4: fuel moisture content.

For the second model, the step responses shown in Figure 11 indicated that only fuel moisture content has a significant effect on the evaporation rate, while other inputs have virtually no influence on the model output. This contradicts to the process knowledge, as the fuel feed rate, determined by the stoker speed, directly influences the amount of water delivered to the furnace, and therefore, certainly affects the evaporation rate. Furthermore, as increasing the primary airflow intensifies the burning rate and the amount of heat released, it should thereby increase the evaporation rate as well.

As discussed in Section 3.5, a correlation exists between the simplified model prediction errors of the flue gas oxygen content and the burning rate. Thus, an adaptive modification of the simplified model was developed correcting its estimation of the combustion coefficient in Equation (45) by means of a PI controller. In more detail, the coefficient is adjusted on-line as follows:

$$k_{\text{pyr,comb}} = u(t) T_{\text{gas}} / T_{\text{char}} k_C \qquad (66)$$

where $u(t)$ is the output signal of the PI controller:

$$u(t) = \max\left(\frac{5}{1000} \cdot \left(e(t) + \frac{1}{10}\int e(\tau)\,dt\right), 0\right) \tag{67}$$

receiving the flue gas oxygen prediction error defined by Equation (68):

$$e(t) = X_{O_2,\text{meas}}(t) - X_{O_2,\text{pred}}(t) \tag{68}$$

The comparison indicates that all three models: the simplified, simplified adaptive and data based-model are able to predict the fuel combustion rate due to a linear relation of fuel consumption to primary and secondary air flow during the experiment (Figure 12). The adaptive modification of the simplified model, utilizing the flue gas oxygen measurement to correct the burning rate prediction, provides a little more accurate results compared to the original simplified model. The results shown in Figure 12 confirm that the ARX model can be accurate in certain operating conditions, when the burning rate, not restricted by the amount of combustible materials available, depends linearly on the air flows. However, the adequacy of the ARX model at largely varying operating conditions and rapid transitions is questionable, as it is unable to monitor the amount of fuel and char in the furnace. This conclusion is confirmed by the step responses shown in Figure 10, not fully agreeing with the process knowledge.

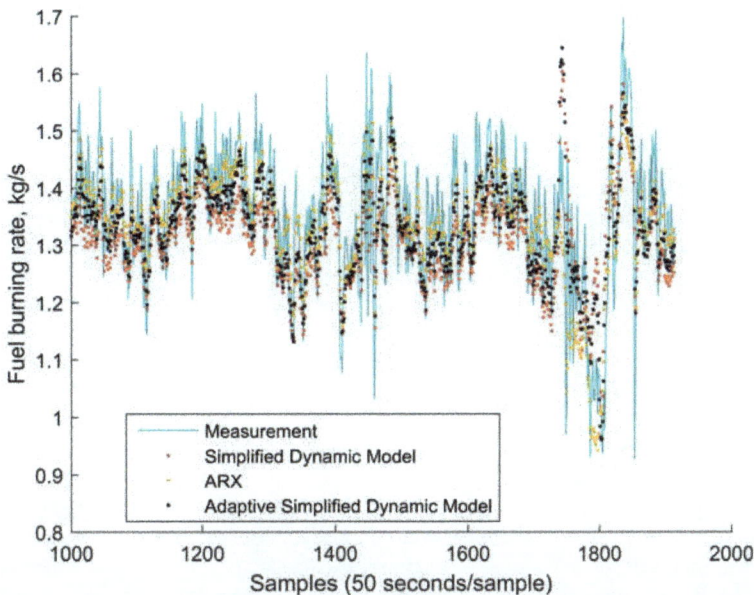

Figure 12. Comparison of burning rates predicted by the simplified dynamic model, ARX, and the adaptive simplified dynamic model.

The evaporation rate, presented in Figure 13, was predicted sufficiently well only by the simplified dynamic models. ARX model demonstrated only a minor ability to follow the variations observed in the evaporation rate. One explanation to the observed ARX model deficiency is that drying depends on nonlinear factors such as heat radiation from oxidation zone and to properly account for these factors, a nonlinear model is required for a proper description of combustion phenomena.

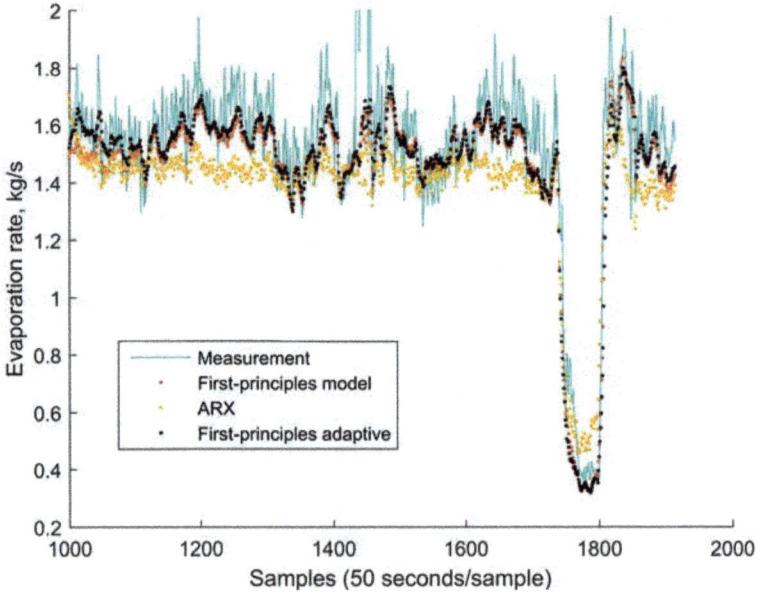

Figure 13. Comparison of evaporation rates predicted by the simplified dynamic model, ARX, and the adaptive simplified dynamic model.

4.3. Comparison of Computational Times of Different Models

The models—including the mechanistic model, simplified model and ARX models—were compared in terms of their run times and the results are presented in Table 2 while the associated computer configurations are shown in Table 3. The computational times of the mechanistic model largely benefit from the parallel computations in Matlab versions below R2015b and both clock frequency and the number of CPU cores. A larger number of processor cores and higher clock frequencies allowed Desktop-1 to perform almost three times faster compared to the laptop computer. On these two computers, computational times of ARX and the simplified model benefitted less from a larger core number and higher clock frequencies. Matlab version R2015b demonstrated significantly longer computational times for the mechanistic model and were almost six times longer when compared to the simulation time obtained on computer Desktop-1 running R2014b, possibly due to some deficiencies in the parallel computing toolbox. Nevertheless, the computational times of the simplified model on the newer Matlab version on Desktop-1 were almost two times faster than on Desktop-1 running an older version. Furthermore, the Desktop-2 demonstrated even further improvements in computational times due to the upgrades in the CPU architecture and these advancements resulted in faster computational times of the ARX model.

Table 2. Run times of different models on different computer configurations.

Computer	Mechanistic (Real-Time Seconds/Simulated Second)	Simplified	ARX
Laptop	14.173497	6.23×10^{-4}	7.21×10^{-7}
Desktop-1 (R2014b)	5.403695	5.34×10^{-4}	6.477×10^{-7}
Desktop-1 (R2015b)	31.443311	2.81×10^{-4}	7.066×10^{-7}
Desktop-2	28.240536	2.11×10^{-4}	5.927×10^{-7}

Table 3. Computer configurations.

Computer	Matlab Version	RAM (GB)	CPU Model	Frequency (GHz)	# of Cores	# of Threads
Laptop	R2015a	8	Intel i5-3340M	2.7	2	4
Desktop-1	R2014b	12	Intel i7 920	3.6	4	8
Desktop-2	R2015b	8	Intel Xeon E3-1230	3.2	4	8

5. Conclusions

In this work, a mechanistic biomass combustion model developed for process phenomena investigation was simplified to obtain a model describing the combustion phenomena and suitable for applications requiring fast computational time. The simplified model was validated with industrial data from the process and demonstrated sufficient accuracy in describing the combustion of biomass fuel. The adaptive modification introduced to the simplified model provided further improvement of the burning and evaporation rates prediction.

Increasing demand for power production flexibility and handling the unmeasured variations of fuel properties require advanced process control techniques, such as model predictive control. Based on an underlying model able to describe the combustion phenomena, this control technique would allow one to avoid fuel burnout and to prevent incomplete fuel combustion and formation of pollutants in the furnace. As was demonstrated by the results, the deep insight into the process conditions provided by the model allows to ensure that prerequisites for a fast load transition, such as sufficient amount of fuel in the furnace. Furthermore, the power production flexibility can be supported through on-line monitoring of the furnace state, including fuel amount and composition, temperature as well as the composition of the forming flue gas. In contrast, as was demonstrated in this article, the models identified from process data fail to capture the key relations between the variables which cause performance deficiencies if utilized for process control. In this work, the simplified model of the biomass combustion is obtained for control and monitoring algorithms implementation, whereas its low computational load allows its utilization in on-line calculations.

Acknowledgments: The first author would like to thank the Finnish Automation Foundation, Emil Aaltonen foundation and Finnish Foundation for Technology Promotion for the financial support.

Author Contributions: Alexandre Boriouchkine developed the simplified mechanistic model and validated it. Alexandre Boriouchkine also identified and validated the ARX models, compared them with the mechanistic model and analyzed the results. Sirkka-Liisa Jämsä-Jounela supervised the research and together with Alexandre Boriouchkine wrote the manuscript.

Conflicts of Interest: The authors declare no conflict of interest.

Abbreviations

ΔH_{CO}	Enthalpy of CO formation (kJ/mol)
ΔH_{CO_2}	Enthalpy of CO_2 formation (kJ/mol)
ΔH_{evap}	Enthalpy of vaporization (kJ/kg)
$\Delta H_{p,i}$	Pyrolysis enthalpy of component i (kJ/kg)
A	Pre-exponential factor (s^{-1})
C_{CH_4}	Concentration of methane (mol/cm^3)
C_{CO}	Concentration of carbon monoxide (mol/cm^3)
C_{H_2O}	Concentration of steam (mol/cm^3)
C_{O_2}	Concentration of oxygen (mol/cm^3)
C_{p,O_2}	Heat capacity of oxygen (J/(kg K))
$C_{p,H_2O,g}$	Heat capacity of water vapor (J/(kg K))
C_{p,N_2}	Heat capacity of nitrogen (J/(kg K))
$C_{p,c}$	Heat capacity of char (J/(kg·K))
$C_{p,G}$	Heat capacity of pyrolytic gas (J/(kg K))
C_{p,H_2O}	Heat capacity of liquid water (J/(kg·K))

$C_{p,w}$	Heat capacity of wood (J/(kg·K))
C_s	Heat capacity of the solid phase (J/(kg·K))
d_{cavity}	Average cavity diameter (m)
$D_{g,i}$	Gas-phase diffusivity of component i in air (m^2/s)
d_p	Particle diameter (m)
E	Activation energy (J/mol)
F_{In}	Total volumetric air flow to the furnace (m^3/s)
F_{out}	Total volumetric flow out of the furnace (m^3/s)
h	Heat transfer coefficient (W m^{-2})
h_C	Fraction of char in the total amount of material on the grate
H_g	Enthalpy of the gas phase (J/kg)
H_G	Energy content of the gas phase (J)
$h_{i,eff}$	Effective mass transfer coefficient
I^-	Energy flux in a negative direction (W/m^2)
I^+	Energy flux in a positive direction (W/m^2)
k_a	Absorption coefficient (m^{-1})
k_{air}	Thermal conductivity of air
k_{bed}	Effective heat conduction coefficient of the packed bed (W/(m·K))
k_C	Reaction rate of char combustion reaction (1/s)
$k_{eff,i}$	Effective reaction constant of a heterogeneous reaction i (kg/(m^3·s))
k_{fiber}	Heat conductivity of wood fiber (W/(m·K))
k_g	Heat conductivity of the gas (W/(m·K))
k_{max}	Maximum heat transfer coefficient (W/(m·K))
k_{min}	Minimal heat conduction coefficient (W/(m·K))
k_{O_2}	Mass transfer coefficient of oxygen to the char particle (m/s)
$k_{p,i}$	Rate constant for pyrolysis of component i
$k_{pyr,comb}$	Rate constant of combustion of pyrolytic gas (1/s)
k_s	Scattering coefficient (m^{-1})
$k_{r,C}$	Rate constant for the char reaction with oxygen (kg/(m^3·s))
k_{R,CO_2}	Rate constant for the char reaction with carbon dioxide (kg/(m^3·s))
$k_{s,eff}$	Heat conduction coefficient of the solid matter (W/(m·K))
k_{r,H_2O}	Rate constant for the char reaction with water (kg/(m^3·s))
$k_{r,i}$	Reaction rate constant for the component i (kg/(m^3·s))
$k_{s,rad}$	Heat radiation coefficient of the solid matter (W/(m·K))
M_{CO}	Molar mass of CO (kg/mol)
M_{CO_2}	Molar mass of CO$_2$ (kg/mol)
m_C	Amount of char in the furnace (kg)
m_{CO}	Amount of carbon monoxide in the furnace (kg)
m_{fuel}	Amount of fuel in the furnace (kg)
$m_{H_2O,i}$	Amount of moisture in the furnace (kg)
m_{In}	Fuel feed to the furnace (kg/s)
m_m	Predicted weight of component m (kg)
m_{O_2}	Amount of air in the furnace (kg)
$m_{p,G}$	Total amount of pyrolytic gas in the furnace (kg)
$m_{p,i}$	Amount of volatile component i in the furnace (kg)
$m_{H_2O,1}$	Amount of moisture on the grate (kg)
m_{H_2O}	Amount of water vapor in the fuel layer (kg)
Pr	Prandatl number
$Q_{g,i}$	Energy produced or consumed by a gas phase reaction i (J/(m^3·s))
Q_s	Energy released through radiation and convection (W)
$Q_{s,i}$	Energy produced or consumed by a solid phase reaction i (J/(m^3·s))
Q_C	Energy from char combustion (J/s)
Q_{in}	Energy contained in primary air (J/s)
Q_{out}	Energy of outflowing gas (J/s)
r_C	Reaction rate of char (kg/s)
Re	Reynolds number

r_{evap}	Drying rate (kg/(m^3·s))
$r_{evap,conv}$	Rate of moisture evaporation due to the heat transfer between gas and solid
$r_{evap,rad}$	Rate of moisture evaporation due to the radiative heat transfer from char layer
r_{g,CH_4}	Oxidation rate of methane (kg/(m^3·s))
$r_{g,CO}$	Oxidation rate of the carbon monoxide (kg/(m^3·s))
r_{g,H_2}	Oxidation rate of hydrogen (kg/(m^3·s))
$r_{g,i}$	Reaction rate of the gaseous component i (kg/(m^3·s))
$r_{p,i}$	Rate of pyrolysis of component i (kg/s)
r_{pyr}	Rate of pyrolysis reaction (kg/s)
$r_{pyr,comb}$	Rate of combustion of pyrolytic gas (kg/s)
r_{s,H_2O}	Drying rate of fuel (kg/(m^3·s))
$r_{s,j}$	Rate of reaction of the solid component j (kg/(m^3·s))
$r_{s,pyr}$	Reaction rate of pyrolysis (kg/(m^3·s))
S	Density number (m^{-1})
Sc	Schmidt number
T_g	Temperature of the gas phase (K)
T_{In}	Temperature of the fed air flow (K)
T_s	Temperature of the solid (K)
t	Time variable (s)
V	Volume of the material on the grate (m^3)
v_g	Gas flow velocity (m/s)
X	Degree of conversion of char
X_C	Char fraction in the pyrolysis products
X_{H_2O}	Moisture content of the fuel
X_m	Moisture content of the fuel in the furnace
$X_{p,G}$	Fraction of pyrolysis gas in the wood pyrolysis
x	Vertical coordinate (m)
$Y_{g,i}$	Mass fraction of the gaseous component i
ε_b	Bed porosity
ε_p	Particle porosity
κ_{conv}	Heat convection coefficient (W/(m^2·K))
$\kappa_{s,eff}$	Effective heat conduction coefficient of the solid matter (W/(m·K))
ϱ	Density of the fluid (kg/m^3)
ϱ_{Air}	Density of air (kg/m^3)
ϱ_c	Mass concentration of char (kg/m^3)
ϱ_{CO}	Mass concentration of carbon monoxide (kg/m^3)
ϱ_g	Mass concentration of the gas phase (kg/m^3)
ϱ_{H_2}	Mass concentration of hydrogen (kg/m^3)
ϱ_m	Mass concentration of component m (kg/m^3)
ϱ_{O_2}	Mass concentration of oxygen (kg/m^3)
ϱ_s	Total mass concentration of the solid phase (kg/m^3)
$\varrho_{s,j}$	Mass concentration of the solid component j (kg/m^3)
ϱ_v	Mass concentration of volatiles (kg/m^3)
ϱ_w	Mass concentration of water (kg/m^3)
σ	Stefan-Boltzman constant (W/(m^2·K^4))
Ω	Ratio of carbon monoxide to carbon dioxide
Ω_{pyr}	Stoichiometric coefficient of pyrolytic gas combustion

References

1. Hogg, B.; El-Rabaie, N. Multivariable generalized predictive control of a boiler system. *IEEE Trans. Energy Convers.* **1991**, *6*, 282–288. [CrossRef]
2. Bauer, R.; Gölles, M.; Brunner, T.; Dourdoumas, N.; Obernberger, I. Modelling of grate combustion in a medium scale biomass furnace for control purposes. *Biomass Bioenergy* **2010**, *34*, 417–427. [CrossRef]

3.   Gölles, M.; Reiter, S.; Brunner, T.; Dourdoumas, N.; Obernberger, I. Model based control of a small-scale biomass boiler. *Control Eng. Pract.* **2014**, *22*, 94–102. [CrossRef]
4.   Kortela, J.; Jämsä-Jounela, S. Model predictive control utilizing fuel and moisture soft-sensors for the BioPower 5 combined heat and power (CHP) plant. *Appl. Energy* **2014**, *131*, 189–200. [CrossRef]
5.   Flynn, M.; O'Malley, M. A drum boiler model for long term power system dynamic simulation. *IEEE Trans. Power Syst.* **1999**, *14*, 209–217. [CrossRef]
6.   Paces, N.; Kozek, M. Modeling of a Grate-Firing Biomass Furnace for Real-Time Application. In Proceedings of the International Symposium on Qualitative, Quatitative and Hybrid Models and Modeling Methodologies in Science and Engineering (MMMse 2011), Orlando, FL, USA, 27–30 March 2011.
7.   Belkhir, F.; Meiers, J.; Felgner, F.; Frey, G. A Biomass Combustion Plant Model for Optimal Control Applications—The Effect of Key Variables on Combustion Dynamics. In Proceedings of the 6th International Renewable Energy Congress (IREC), Sousse, Tunisia, 24–26 March 2015; pp. 1–6.
8.   Liu, X.; Kong, X.; Hou, G.; Wang, J. Modeling of a 1000 MW power plant ultra super-critical boiler system using fuzzy-neural network methods. *Energy Convers. Manag.* **2013**, *65*, 518–527. [CrossRef]
9.   Peng, H.; Nakano, K.; Shioya, H. Nonlinear predictive control using neural nets-based local linearization ARX model—Stability and industrial application. *IEEE Trans. Control Syst. Technol.* **2007**, *15*, 130–143. [CrossRef]
10. Havlena, V.; Findejs, J. Application of model predictive control to advanced combustion control. *Control Eng. Pract.* **2005**, *13*, 671–680. [CrossRef]
11. Leskens, M.; Van Kessel, L.; Van den Hof, P. MIMO closed-loop identification of an MSW incinerator. *Control Eng. Pract.* **2002**, *10*, 315–326. [CrossRef]
12. Leskens, M.; Van Kessel, L.; Bosgra, O. Model predictive control as a tool for improving the process operation of MSW combustion plants. *Waste Manag.* **2005**, *25*, 788–798. [CrossRef] [PubMed]
13. Morari, M.; Lee, J.H. Model predictive control: Past, present and future. *Comput. Chem. Eng.* **1999**, *23*, 667–682. [CrossRef]
14. Boriouchkine, A.; Zakharov, A.; Jämsä-Jounela, S. Dynamic modeling of combustion in a BioGrate furnace: The effect of operation parameters on biomass firing. *Chem. Eng. Sci.* **2012**, *69*, 669–678. [CrossRef]
15. Boriouchkine, A.; Sharifi, V.; Swithenbank, J.; Jämsä-Jounela, S. A study on the dynamic combustion behavior of a biomass fuel bed. *Fuel* **2014**, *135*, 468–481. [CrossRef]
16. Pomerantsev, V. *Fundamentals of Applied Combustion Theory*; Energoatomizdat: Moscow, Russia, 1986.
17. Fogler, S. *Elements of Chemical Reaction Engineering*, 4th ed.; Pearson Educations: Upper Saddle River, NJ, USA, 2006.
18. Branca, C.; Di Blasi, C. Devolatilization and combustion kinetics of wood chars. *Energy Fuels* **2003**, *17*, 1609–1615. [CrossRef]
19. Senneca, O. Kinetics of pyrolysis, combustion and gasification of three biomass fuels. *Fuel Process. Technol.* **2007**, *88*, 87–97. [CrossRef]
20. Matsumoto, K.; Takeno, K.; Ichinose, T.; Ogi, T.; Nakanishi, M. Gasification reaction kinetics on biomass char obtained as a by-product of gasification in an entrained-flow gasifier with steam and oxygen at 900–1000 °C. *Fuel* **2009**, *88*, 519–527. [CrossRef]
21. Evans, D.D.; Emmons, H. Combustion of wood charcoal. *Fire Saf. J.* **1977**, *1*, 57–66. [CrossRef]
22. Janssens, M.; Douglas, B. Wood and Wood Products. In *Handbook of Building Materials for Fire Protection*; Harper, C.A., Ed.; McGraw-Hill: New York, NY, USA, 2004.
23. Shin, D.; Choi, S. The combustion of simulated waste particles in a fixed bed. *Combust. Flame* **2000**, *121*, 167–180. [CrossRef]
24. Branca, C.; Blasi, C.D.; Elefante, R. Devolatilization and heterogeneous combustion of wood fast pyrolysis oils. *Ind. Eng. Chem. Res.* **2005**, *44*, 799–810. [CrossRef]
25. Boriouchkine, A.; Sharifi, V.; Swithenbank, J.; Jämsä-Jounela, S. Experiments and modeling of fixed-bed debarking residue pyrolysis: The effect of fuel bed properties on product yields. *Chem. Eng. Sci.* **2015**, *138*, 581–591. [CrossRef]
26. Garcìa-Pérez, M.; Chaala, A.; Pakdel, H.; Kretschmer, D.; Roy, C. Vacuum pyrolysis of softwood and hardwood biomass: Comparison between product yields and bio-oil properties. *J. Anal. Appl. Pyrolysis* **2007**, *78*, 104–116. [CrossRef]

27. Wakao, N.; Kaguei, S.; Funazkri, T. Effect of fluid dispersion coefficients on particle-to-fluid heat transfer coefficients in packed beds: Correlation of nusselt numbers. *Chem. Eng. Sci.* **1979**, *34*, 325–336. [CrossRef]
28. Cooper, J.; Hallett, W.L.H. A numerical model for packed-bed combustion of char particles. *Chem. Eng. Sci.* **2000**, *55*, 4451–4460. [CrossRef]

Article

An Economic and Policy Analysis of a District Heating System Using Corn Straw Densified Fuel: A Case Study in Nong'an County in Jilin Province, China

Shizhong Song [1], Pei Liu [1], Jing Xu [2], Linwei Ma [1,]*, Chinhao Chong [1], Min He [1], Xianzheng Huang [1], Zheng Li [1] and Weidou Ni [1]

[1]   State Key Laboratory of Power Systems, Department of Thermal Engineering, Tsinghua-BP Clean Energy Centre, Tsinghua University, Beijing 100084, China; ssz777777@163.com (S.S.); liu_pei@tsinghua.edu.cn (P.L.); chongchinhao@hotmail.com (C.C.); mina23805@163.com (M.H.); huangxz425@sina.com (X.H.); lz-dte@tsinghua.edu.cn (Z.L.); niwd@tsinghua.edu.cn (W.N.)

[2]   Department of Mechanical Engineering, Jinan Vocational College, Jinan 250104, China; 9915142@163.com

*   Correspondence: malinwei@tsinghua.edu.cn; Tel.: +86-10-6279-5734; Fax: +86-10-6279-5736

Academic Editor: Tariq Al-Shemmeri
Received: 23 November 2016; Accepted: 16 December 2016; Published: 23 December 2016

Abstract: The development of district heating systems of corn straw densified fuel (CSDF-DHS) is an important option to promote the use of bioenergy on a large scale for sustainable development, especially in China. At present, China's biomass densified solid fuel (BSDF) development lags behind previously planned target, main barriers of which are economic and policy support problems. Accurate case studies are key to analyze these problems. This manuscript takes Nong'an County in Jilin Province of China as an example to establish a techno-economic model to evaluate the economic performance of a CSDF-DHS under two policy scenarios. It calculates the economic performance under a benchmark market scenario (BMS) and the current policy scenario (CPS) and analyzes the influence of various policy instruments, including subsidies, carbon trading, and preferential taxation. The results indicate that: (1) The CSDF-DHS option is not competitive under the BMS or CPS compared to the traditional energy system based mainly on coal and liquefied petroleum gas; (2) Comparatively, the economic performance of corn straw briquette fuel (CSBF) is better than that of corn straw pellet fuel (CSPF); and (3) further policy support can make CSDF-DHSs competitive in the market, especially with subsidies for concentrated heating services and CSDF, carbon trading, and economic compensation to reduce the profit margin of enterprises, which can make both CSPF-DHSs and CSBF-DHSs competitive. The research results could provide scientific basis for relevant policy making and project decision.

Keywords: corn straw; solid densified fuel; district heating system; economic model; policy influence

1. Introduction

The use of biomass for the production of low carbon energy is recognized as an important goal of sustainable development [1]. Among the various ways to use biomass for energy production, one of the most widely used and commonly available in the market is BSDF [2,3]. BSDF refers to fuel of a certain shape that is of high density and is obtained from loose biomass pressed at a certain temperature and under a certain pressure. The general shape of BSDF can be a pellet, briquette or rod. Its volume is 1/8–1/6 of the biomass raw materials, and the density is 1.0–1.4 t/m^3. With an energy density equal to that of intermediate soft coal, BSDF has attracted widespread attention in recent years [4], including in China. In China, the raw materials of BSDF are mainly forestry and agricultural residues and BSDF is utilized mainly as a clean fuel for heating boilers in communities, industrial parks and other public or industrial facilities. China's total production of BSDF in 2015 was 8 million tonnes [5], which was less than the planned 2015 target of 10 million tonnes [6]. For the year 2020, China's planned target of BSDF production is 30 million tonnes [5], which requires an annual 30% increase of BSDF production, so it is necessary to take measures to promote the healthy and rapid development of China's BSDF industry [3]. Moreover, being a large corn producer with abundant crop straw resources, especially in Liaoning Province, Jilin Province and Heilongjiang Province in northeastern China, China has seen serious air pollution problems arise due to the large scale field incineration of crop straw waste [6–9]. As estimated, China's agricultural residues available for energy use are about 0.2 billion tce [6], which accounts for 4.7% of China's total energy consumption of 4.3 billion tce [10] in 2015. Therefore, to promote the energy use of biomass and also reduce the field incineration of corn straw, the development of corn straw solid densified fuel (CSDF) for heating should be prioritized in the development of China's BSDF industry.

Referring to a literature review, economic analysis is a popular area of international research on the development of BSDF [11], and many studies have applied economic analyses of BSDF heating systems on the district level or building level. For example, Thomson [12] reviewed the suitability of wood pellet heating for domestic households and discussed the advantages, issues, and barriers. Vallios [13] designed biomass district heating systems considering the optimum design of building structures and urban settlements around the plant and carried out an environmental and economic evaluation. Chau [14,15], Michopoulos [16], and Stolarski [17] analyzed the economic performance and other performance of BSDF utilization in buildings heating systems. Hendricks [18] evaluated the cost-effectiveness of biomass district heating in rural communities. Stephen [19] analyzed the economics and influence factors of biomass use for residential and commercial heating in a remote Canadian aboriginal community. Tabata [20] discussed the effectiveness of a woody biomass utilization system with wood pellet production and energy recovery processes for household energy demand, taking the case of Gifu Japan as an example. Ren et al. [21] analyzed and compared the logistics cost of corn stover feedstock supply systems based on China's case. Zhao et al. [22] researched the techno-economic performance of bioethanol production from corn straw in China.

In addition, it is also well recognized that BSDF development policies have a major impact on its economic performance. For example, Toka [23] researched how to manage the diffusion of biomass in the residential energy sector and discussed an illustrative real-world case study. Moiseyev et al. [24] indicated that subsidies were likely to be the major driving force to increase the energy use of woody biomass and found that subsidies and carbon prices can effectively promote the energy use of woody biomass in the E.U. and reduce carbon dioxide emissions. Madlener [25] investigated the innovation diffusion, public policy, and local initiative of using biomass for energy production, taking the case of wood-fueled district heating systems in Austria as an example. Gan [26] researched policy options and co-benefits of bioenergy transition in rural China, pointing out that there is great potential for developing and disseminating household-based biomass technologies in rural areas, especially with energy-efficient modern biomass stoves, which can produce far more economic, social and environmental benefits. Shan [27] proposed a novel and viable village-level BSDF utilization mode based on the field survey of China's BSDF industry and the results from demonstration projects, which is helpful to boost the utilization of BSDF mainly in rural China. Wang et al. [28] assessed densified biomass solid fuel utilization policies and strategies in China based on the supply chain framework. Therefore, it is concluded that economic and policy analysis for district heating system is a key area of international research on BSDF development. Scientific economic analysis system need to be established with field survey and literature review results to analyze the influence of relevant support policies. However, currently, few studies have been published on the economic and policy analysis methods and case studies of CSDF for district heating systems, especially in China.

The aim of this manuscript is to develop a techno-economic model to evaluate the economic performance of a corn straw densified fuel-district heating system (CSDF-DHS) and analyze the influence of polices using a case study of Nong'an County in Jilin Province, China. First, we investigate the system description of a CSDF-DHS and basic information about the case, which are introduced in Section 2. Then, we develop a technical model of a CSDF-DHS, construct a two-stage economic model to evaluate its economic performance, and set two scenarios to analyze the influence of policies. Those methods and data are introduced in Section 3. Finally, we provide the results and discussion in Section 4 and main conclusions and policy implications in Section 5.

2. System Description and Case Information

2.1. System Description of a CSDF-DHS

The basic function of a CSDF-DHS is to turn corn straw into solid densified fuel to provide fuel for heating and cooking in rural areas or for concentrated heating services in urban area. In a CSDF-DHS, corn straw is collected, packaged, transported, and then produced as CSDF. Then, a part of CSDFs is sold to rural residents, and the rest is sold to urban users. The boundaries and correlations of a CSDF-DHS are illustrated in Figure 1. To obtain fuel and heating services in a CSDF-DHS, various inputs with economic costs by various operators are needed, as listed in the left of Figure 1, and these economic costs can be changed by various types of policy support, such as subsidies, tax preferences, and carbon trading.

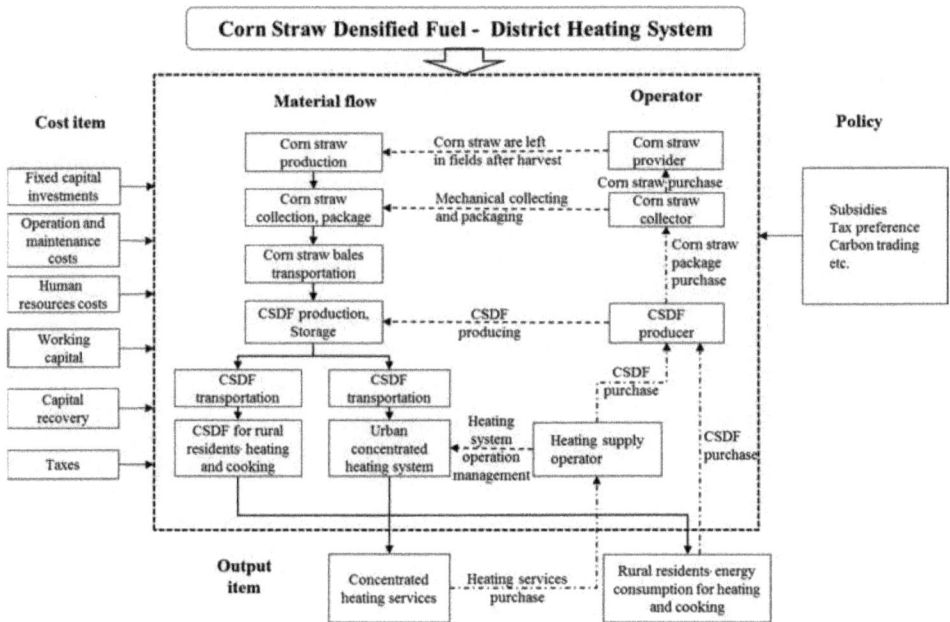

Figure 1. System boundary and correlations of a CSDF-DHS.

2.2. Case Information

Nong'an County in Jilin Province is taken as the example in this study. Its geological location is illustrated in Figure 2. In 2015, the per capita GDP of Nong'an County was $5068, substantially lower than national average level of $7095. The county has a population of 1,150,000, covering an area of 5400 km², and is located in a severe cold area [29] with 167 days of heating per year. The existing concentrated heating systems for the urban area and fuel systems for rural residents' heating and cooking mainly rely on coal and liquefied petroleum gas (LPG). Statistically [30], one household in the rural area consumes 1.53 tonnes of coal, 0.02 tonnes of LPG and 2.19 tonnes of firewood and corn straw annually on average in Jilin Province. In the downtown area of Nong'an County, the concentrated heating system covers an area of 8,180,000 m², with 20 kgce of coal consumption per unit area and approximately 164,000 tce of total coal consumption annually.

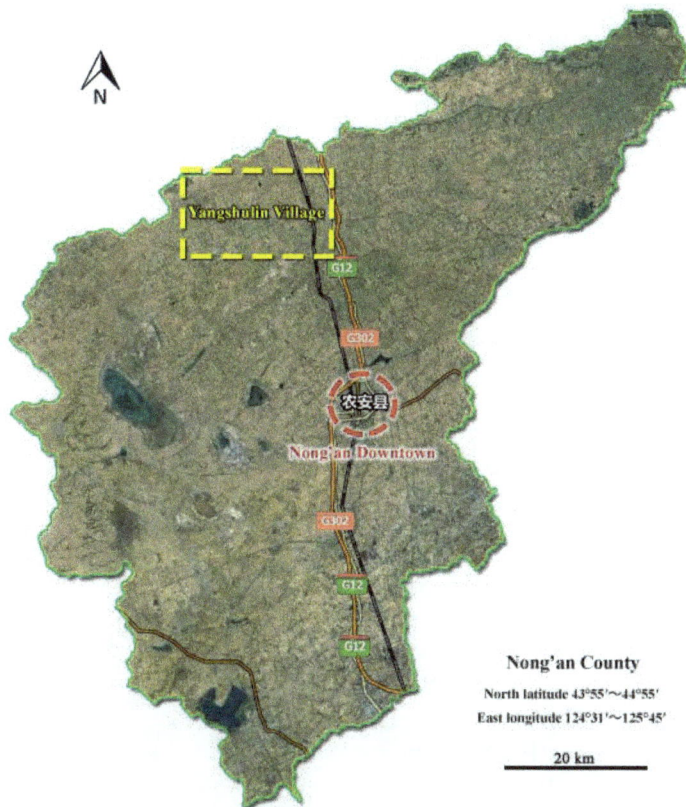

Figure 2. Geological location of Nong'an Country, Jilin Province.

Whereas coal burning causes high CO_2 emissions and air pollution, Nong'an Country, located in the World Gold Corn Belt, has abundant resources of corn straw, amounting to approximately 3 million t/a in total. However, currently, these resources are mostly abandoned and incinerated in the field, which is both a serious waste of resources and a cause of seasonal air pollution. To promote the energy use of corn straw resources, Yangshulin Village of Nong'an Country has established a CSDF-DHS demonstration project, with a scale of 10,000 t/a CSDF. It is urgent for Nong'an Country to further build a larger scale CSDF-DHS and use more corn straw resources to optimize the energy structure, reduce air pollution, create new jobs and increase the incomes of local residents. In this case study, we designed a 55,000 t/a CSDF production in Yangsulin Village, where a part of the CSDF will be sold to the downtown area for urban concentrated heating services, and the rest will be used for the heating and cooking of rural residents in Yangsulin Village.

3. Methods and data

In the following, we first introduce a technical model of a CSDF-DHS and then an economic model in two stages, the first stage from corn straw to CSDF and the second stage from CSDF to heating service. Finally, we present the basic settings of two scenarios for policy analysis.

3.1. Technical Model of the CSDF-DHS

According to Figure 1, the technical processes of a CSDF-DHS include many details, not only the parameters of various stages but also the type and numbers of main machines collecting corn straw and producing CSDF. The technical model is introduced below by stages, including corn straw resources, corn straw collection, CSDF production, and the final use of CSDF in the rural area of Yangshulin Village and the urban area of downtown Nong'an's.

3.1.1. Corn Straw Energy Utilization Resources Assessment

Corn straw resources can be used as energy in Yangshulin Village, as calculated by Equation (1), in which P is the corn straw energy utilization resources, P_{yield} is the output of corn, α is the ratio of straw to grain, β is the actual collection ratio of corn straw, and β' is the energy utilization ratio of corn straw.

$$P = P_{yield} \cdot \alpha \cdot \beta \cdot \beta'$$
(1)

According to a field survey and literature review, the basic data for calculating available resources for energy utilization are listed in Table 1.

Table 1. Data of corn straw energy utilization resources assessment in the case district.

Output of Corn (10,000 t)	Unit Yield of Corn (t/ha)	Ratio of Straw to Grain	Total Resources of Straw (10,000 t)	Actual Collectable Ratio	Total Collectable Resources (10,000 t)	Energy Utilization Ratio of Corn Straw
18.0 [31]	7.80 * [32]	1.6 [33]	28.98	0.6 [33]	17.39	1/3

* The unit yield of corn in the literature [32] is 8.37 t/ha, but according to the field survey, the collectable corn straw is 7.50 t/ha. To make the data consistent, we adjusted the unit yield of corn to 7.80 t/ha.

3.1.2. Corn Straw Collection

Referring to the literature [34,35], the mechanical collection of corn straw, including mechanical packaging, collecting, and transporting, is more economically feasible compared with manual collection. In this case, mechanical collection includes a rake machine used to put corn straw together and the machine used for packaging bales. Based on the field survey and interviews with machine producers, mechanical harvesting of corn puts corn and straw into separated places. Then, straw is spread on fields exposed to the sun to dry out. In Nong'an Country, Jilin Province, there are 7.50 tonnes of collectible corn straw per ha. The collection of corn straw is influenced by the season. Each year, operation begins with collecting corn straw already dried out and ends when the fields are covered by heavy snow. Given weather conditions, generally every year, a rake machine can collect corn straw of 1333 ha and a packaging machine can collect corn straw of 667 ha. The corn straw bales are stored in the corn straw collection stations and the storage yards of CSDF production plants. As existing commercial transportation can be used for corn straw bale transportation, a separate design of transportation facilities is not required.

3.1.3. CSDF Production

According to field survey and interviews with machine producers, two main products are determined for our research on CSDF, including CSPF, which is small cylinders with diameters of 5–12 mm and lengths of 10–30 mm, and CSBF, which has square sections of 30 × 30 mm² and lengths of 30–80 mm. Moreover, we suppose that CSDF is produced by a 3.5–4 t/h CSPF production line or a 7–8 t/h CSBF production line. The total production capacities of CSPF and CSBF for each line are estimated as 10,800 t/a and 21,600 t/a, respectively, by taking the average production rate and operation time as 16 h/day and 180 days/a, respectively. Figure 3 illustrates a typical schematic diagram of CSDF, which can both be applied to CSPF and CSBF, and the main process includes raw material pretreating and drying, processing, and storing. We only consider the case that the production line is used to produce CSPF only or CSBF only in this case study.

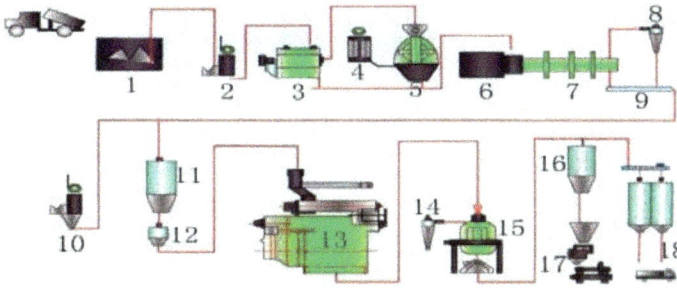

Figure 3. Typical schematic diagram of CSDF processing [36]. 1: Storage yard; 2: Conveyor; 3: Sieving machine; 4: Cyclone dust removal; 5: Intermediate bunker; 6: Hot-blast stove; 7: Dryer; 8: Separator; 9: Spiral conveyor; 10: Conveyor; 11: Processed material bin; 12: Magnetic material cleaners; 13: Forming machine; 14: Cooling fan; 15: Cooler; 16: CSDF storage tank; 17: Wrapper; and 18: Vehicle for transporting CSDF.

The produced CSDFs are stored in the storehouse of CSDF plants. Like the corn straw bale transportation, it is assumed the transportation of CSDF is based on available commercial transportation services.

3.1.4. Final Use for Heating and Cooking in Rural Area

CSDF should firstly meet the needs of rural residents for heating and cooking, as illustrated in Figure 1. CSDF consumption by rural residents for heating and cooking is determined by Equation (2):

$$M_{rural} = m_{rural} \cdot n_{household} \cdot \eta_{rural} \tag{2}$$

In Equation (2), M_{rural} is the rural consumption of CSDF, m_{rural} is the CSDF consumption for heating and cooking per household, $n_{household}$ is the number of total households in a rural area, and η_{rural} is the penetration rate of household heating and cooking by CSDF in the rural area.

Using a field survey, it is estimated that 3.5 t/a of CSDF consumption per household can meet residents' basic needs for heating and cooking in Yangshulin Village, and there are a total of 9428 households. Referring to the document issued by Jilin Province [37], the target penetration rate of CSDF in rural areas is set as 50% in 2016, which means 4714 households should use CSDF.

3.1.5. Final Use for Concentrated Heating Service in Urban Area

The amount of CSDF utilized for concentrated heating services in urban areas is determined by Equation (3):

$$M_{city} = P \cdot (1 - n_{loss}) - M_{rural} \tag{3}$$

In Equation (3), M_{city} is the amount of CSDF available for urban concentrated heating, P and M_{rural} have the same meaning as in Equations (1) and (2), and n_{loss} is the rate of loss of corn straw during CSDF processing, which is assumed to be 5%.

(a) Heating load calculation [38]

$$Q_h = q_h A_c \cdot 10^{-3} \tag{4}$$

In Equation (4), Q_h is the heating load of the building in kW, A_c is the area of the building in m², and q_h is the thermal index of the heating area of the building, which equals the heating load per square meter (W/m²).

According to the field survey on concentrated heating enterprises in Nong'an County, q_h is 45 W/m², and A_c is 1.08 million m² in this CSDF-DHS.

(b) Annual heat consumption [38]

$$Q_h^a = 0.0864NQ_h \left(\frac{t_i - t_a}{t_i - t_{o,h}} \right)$$ (5)

In Equation (5), Q_h^a is the annual heating consumption on space-heating in GJ/a, N is the days of the heating period in d/a, Q_h is the heating load of the building in kW, t_i is the heating room calculated temperature in °C, t_a is the outdoor average temperature during the heating period in °C, and $t_{o,h}$ is the outdoor average calculated temperature during the heating period in °C.

The heating period is 167 days in this case. According to Nong'an County meteorological data and the survey on heating enterprises, the outdoor average temperature during the heating period is −7.6 °C, the calculated heating temperature is −21.1 °C, and the heating room calculated temperature is 18 °C.

(c) The heat demand for concentrated heating of CSDF

The heat demand for concentrated heating of CSDF in urban areas is determined by Equation (6):

$$m_{district} = \frac{Q_h^a}{\eta \cdot F_{LHV}}$$ (6)

In Equation (6), $m_{district}$ is the CSDF consumption per unit area in t/a, η is the thermal efficiency of the boiler, and F_{LHV} is the low heat value of CSDF in GJ/t. According to the field survey on heating enterprises, the thermal efficiency of a biomass boiler is 75%–85%. This study chooses 80% as the thermal efficiency. Based on industrial analysis, the low heat value of CSDF is 14.93 GJ/t.

(d) Concentrated heating area and the heating station

We use the amount of available CSDF for concentrated heating and the annual heat demand per unit area during heating period as parameters to measure the CSDF consumption per unit heating area and total heating area (see Equation (7)). The total concentrated heating area by a CSDF-DHS is the sum of all heating areas contributed by various heating stations:

$$A_{total} = \frac{M_{city}}{m_{district}}$$ (7)

In Equation (7), A_{total} is the heating area supported by the concentrated heating system in m^2, and the other two symbols refer to Equations (3) and (6).

According to the documents of the National Energy Administration of China [39], this case takes 20 t vapor/h CSDF boiler as the basic heating station unit, calculates the maximum heating area, determines the number of basic heating station units, and then equally distributes the total heating area to each basic heating station unit. M_{city} and $m_{district}$ are 38,600 t/a and 35.59 kg/a, respectively, in this CSDF-DHS.

3.2. Economic Model from Corn Straw to CSDF

The first stage model includes 4 sub-stages from corn straw to CSDF, and the cost is calculated by Equation (8), where C_{fuel} is the total cost from corn straw to CSDF, $C_{collect}$ is the cost of corn straw collection, C_{trans} is the cost of corn straw bale transportation, C_{prod} is the cost of CSDF production, and C'_{trans} is the cost of CSDF transportation:

$$C_{fuel} = C_{collect} + C_{trans} + C_{prod} + C'_{trans}$$ (8)

3.2.1. Corn Straw Collection

The cost of corn straw collection mainly includes the investments in agricultural machinery, the corn straw purchase fee, operation and maintenance costs, human resource fees, profits of collection, and taxes, referring to Equation (9):

$$C_{collect} = \frac{C_{equi} \cdot CRF}{M_{agv}} + C_{purchase} + C_{O\&M} + C_{labor} + C_{management} + C_{profit} + C_{tax} \tag{9}$$

In Equation (9), C_{equi} is the investment of agricultural machinery, M_{agv} is the annual collection corns straw weight, $C_{purchase}$ is the corn straw purchase fee, $C_{O\&M}$ is the operation and maintenance cost, C_{labor} is the human resource fee, $C_{management}$ is the management fee, C_{profit} is the profits of collection, and C_{tax} is the taxes.

The capital recovery factor (CRF) [40,41] is the ratio of constant annuity to the present value of receiving and is calculated by $CRF = \frac{i(1+i)^n}{(1+i)^n - 1}$, where i is the discount rate and n is the payback period.

According to the field survey, the summarized calculation of this stage are listed in Table 2.

Table 2. The summarized cost table of corn straw collection ($i = 0.08$ and $n = 10$).

Category	Agricultural Machinery Capital Recovery	Corn Straw Price	Operation and Maintenance Costs	Human Resources Costs	Management Fees
Unit price ($/t)	3.0	0.0	6.9	2.9	1.1

3.2.2. Corn Straw Bale Transportation

The cost of corn straw bales transportation includes the tonne-kilometer price of the transportation and handling charge, referring to Equation (10):

$$C_{trans} = W_{straw} \cdot l \cdot Trans_{price} + C_{handling} \tag{10}$$

In Equation (10), W_{straw} is the corn straw weight, l is the transportation distance, $Trans_{price}$ is the tonne-kilometer price of corn straw transportation, and $C_{handling}$ is the handling charge.

According to the field survey, the corn straw bale transportation distance is 10 km around, the price of transportation is $5.3/t, and the handling charge is $3.3/t.

3.2.3. CSDF Production

The cost of CSDF production consists of the investments in production machines and infrastructure, the corn straw bale purchase fee, operation and maintenance costs, human resource fees, profits from collection, and taxes, referring to Equation (11):

$$C_{prod} = \frac{C'_{equi} \cdot CRF}{M_{prod}} + C'_{purchase} + C'_{O\&M} + C'_{labor} + C'_{management} + C'_{profit} + C'_{tax} \tag{11}$$

In Equation (11), C'_{equi} is the investment of production machine and infrastructure, M_{prod} is the annual production CSDF weight, $C'_{purchase}$ is the purchase fee of corn straw bales, $C'_{O\&M}$ is the operation and maintenance cost, C'_{labor} is the human resource fee, $C'_{management}$ is the management fee, C'_{profit} is the profits of production, and C'_{tax} is the taxes.

According to enterprise interviews and field survey, the calculated costs of CSDF production are listed in Table 3.

Table 3. Calculated cost of CSDF production ($i = 0.08$ and $n = 12$).

CSDF Type	Capital Recovery	Operation and Maintenance Costs				Human Resources Costs	Management Fees
		Maintenance	Electricity Fees	Wearing Parts	Packing Fees		
CSPF ($/t)	5.6	1.1	9.6	3.6	2.9	2.3	3.0
CSBF ($/t)	1.6	0.3	5.6	2.2	2.9	1.7	2.5

3.2.4. CSDF Transportation

The cost of CSDF transportation includes the tonne-kilometer price of the transportation and handling charge, referring to Equation (12):

$$C'_{trans} = W'_{straw} \cdot l' \cdot Trans'_{price} + C'_{handing} \tag{12}$$

In Equation (12), W'_{straw} is the CSDF weight, l' is the CSDF transportation distance, $Trans'_{price}$ is the tonne-kilometer price of the transportation, and $C'_{handing}$ is the handling charge.

According to the field survey, the transportation distance is 10 km and the price of transportation is \$0.08/t·km in rural areas. When the CSDF produced by Yangshulin Village is transported to Nong'an County, according to geological information, the transportation distance is 48 km, the price of the transportation is \$0.04/t·km, and the handling charge is \$3.62/t.

3.3. Economic Model from CSDF to Heating Service

3.3.1. Rural Residents Cost for Heating and Cooking

The costs from CSDF for heating and cooking rural residents per household consist of the investment recovery of stoves and CSDF cost per household, referring to Equation (13).

$$C_{rural} = C_{furnace} \cdot CRF + m_{household} \cdot C_{fuel} \tag{13}$$

In Equation (13), C_{rural} is the cost from CSDF for heating and cooking of rural residents per household, $C_{furnace}$ is the stove investment per household, $m_{household}$ is the CSDF consumption per household, and C_{fuel} is the CSDF price.

Based on the per capita housing area of rural residents, 24.71 m^2 [42], every household has 4 people in a housing area of 100 m^2, so the heating load of a stove is estimated to be 100 m^2, and the price of one stove is \$194 [43]. Here, i is 0.08 and n is 12 years.

3.3.2. Urban Concentrated Heating Cost

The costs from CSDF for urban concentrated heating per unit area consist of the investment recovery of the concentrated heating system and infrastructure, the CSDF costs, operation and maintenance costs, human resource fees, enterprises profits, and taxes, referring to Equation (14).

$$C_{district} = \frac{C''_{equi} \cdot CRF + M_{district} \cdot C''_{fuel} + C''_{O\&M} + C''_{labor} + C''_{profit} + C''_{tax}}{heat_{area}} \tag{14}$$

In Equation (14), $C_{district}$ is the cost from CSDF to urban concentrated heating per unit area, C''_{equi} is the concentrated heating system and infrastructure investment, $M_{district}$ is the CSDF consumption of the heating stations, C''_{fuel} is the CSDF price for heating stations, $C''_{O\&M}$ is the operation and maintenance costs, C''_{labor} is the human resource fees, C''_{profit} is the profits of concentrated heating stations, C''_{tax} is the taxes, and $heat_{area}$ is the service heating area of heating stations.

According to the field survey, the concentrated heating system and infrastructure investment is shown in Table 4. In the research, the heating system investment is estimated based on the arithmetic mean [44] of the investment. The heating system investment is $2824/t vapor, and the infrastructure investment is $1303/t vapor. The operation and maintenance costs account for 2.5% [18] of the concentrated heating system investment. i is 0.08, and n is 20 years.

Table 4. Survey of the heating system and infrastructure investment in heating stations.

Heating Station	Establishment Way	CSDF Type	Heating Service Type	Power (t vapor/h)	Equipment Investment (10,000$)	Unit Equipment Investment (10,000$/(t vapor/h))	Infrastructure Investment (10,000$)	Unit Infrastructure Investment (10,000$/(t vapor/h))
A mill factory	Renovation	CSBF	Industrial steam	25	57.9	2.3	10.9	0.4
A feed mill	Newly built	CSBF	Industrial steam	6	25.3	4.2	16.9	2.8
A free trade zone	Newly built	CSPF	Heating	30	118.6	4.0	20.1	0.7
A plant area	Newly built	CSPF	Heating	25	44.7	1.8		
A residential and office area	Renovation	CSBF	Heating	4	7.2	1.8		
Average						2.8		1.3

3.4. Policy Scenario Setting

The policy support to the development of a CSDF-DHS, including subsidies on fixed capital investment, corn straw, CSDF, urban concentrated heating services, carbon trading, and preferential electricity prices, can greatly influence its actual economic performance. To observe the influence of policies on the economic performance of a CSDF-DHS, we designed two policy scenario for the calculation of economic model. One is the benchmark market scenario (BMS) without any policy support, and the other is the current policy scenario (CPS) with some policies currently adopted. The basic settings of the two scenarios are listed in Table 5 and further explained as follows.

- The *benchmark market scenario* (BMS) is a market-oriented scenario without any policy support. Referring to the experiences of the demonstration project in Yangshulin Village, rural residents are willing to freely provide collection enterprises with corn straw with the compensation of a favorable price for CSDF. It is mainly because the manual collection and storage of corn straw by themselves require much more time, manpower, and material resources [34,35].
- The *current policy scenario* (CPS) includes polices current adopted by the government of China [45] and Jilin Province [46] to support the development of a CSDF-DHS. In addition to the tax-free policy for corn straw and CSDF, it also include subsidies for fixed capital investment for CSDF production, stoves, and boilers to increase the competitiveness of the system.

Table 5. Basic settings of the two policy scenarios by stages of a CSDF-DHS.

Scenario	Corn Straw Collection and Bales Transportation	CSDF Production				CSDF Utilization	
		CSDF Production	Selling			Rural Residents, Heating Cooking	Concentrated Heating Enterprise
			Heating Station	Rural Residents			
Bench-mark Market Scenario	$C_{purchase}$: zero [1] / C_{tax}: 6% [1] / C_{profit}: 30% [1] / Current agricultural machinery subsidy [2]	Industrial electricity price: $0.08/kWh [1]	C'_{profit}: 30% [1] / C'_{tax}: 6% [1]	C'_{profit}: 10% [1] / C'_{tax}: 6% [1]		No stove subsidy [1]	C''_{profit}: 20% [1] / C''_{tax}: 6% [1]
Current Policy Scenario	$C_{purchase}$: zero [1] / C_{profit}: 30% [1] / Current agricultural machinery subsidy / C_{tax}: tax-free [3]	Industrial electricity price: $0.08/kWh [1] / Fixed capital investment subsidy: 10% [2]	C'_{profit}: 30% [1] / C'_{tax}: tax-free [3]	C'_{profit}: 10% [1] / C'_{tax}: tax-free [3]		Stove subsidy: 70% [3]	C''_{profit}: 20% [1] / C''_{tax}: 6% [1] / Fixed capital investment subsidy: 100,000/t vapor [3]

Notes: [1] Field survey by authors; [2] The proportion of current agricultural machinery subsidies is 5.3% referring to field survey; [3] Current Government Document and Policy [45,46].

4. Results and Discussion

4.1. The Technical Process of a CSDF-DHS

The technical process of a CSDF-DHS decided by the technical model is illustrated by Table 6 (from corn straw resources to CSDF production) and Table 7 (CSDF utilization in rural and urban areas). The total utilization scale is 58,000 t/a corn straw, and the CSDF produced can serve for 4714 households in rural areas and 1.08 million m^2 of concentrated heating service in urban areas. The CSDF production is either CSPF or CSBF in Table 6.

Table 6. Technical process from corn straw resources to CSDF production.

Category	Corn Straw Resources		Collecting Equipment		CSDF Production	
	Energy Utilization (10,000 t/a)	Collect Area (667 ha)	Packaging Machine (Sets)	Rake (Sets)	CSPF 3.5–4 t/h Production Line	CSBF 7–8 t/h Production Line
Quantities	5.80	11.54	10	5	5	3

Table 7. Technical process of CSDF utilization.

Category	Rural Residents Heating and Cooking			Urban Concentrated Heating			
	Households	CSDF Total Consumption (10,000 t)	CSDF Total Consumption (10,000 t)	Unit Heating Load (MJ/m^2·a)	Heating CSDF Consumption (kg/m^2·a)	20 t vapor/h Boiler Quantity of Heating Station	Heating Area (10,000 m^2)
Quantities	4714	1.65	3.86	425	35.59	4	108

4.2. Economic Performance in the Benchmark Market Scenario

The results of the economic model in the BMS are listed in Tables 8 and 9. The costs of CSPF and heating services provided by CSPF are both higher than those of CSBF because CSPF has higher investment and operation and maintenance costs than CSPF. However, CSPF is more beneficial for the local economy, with higher taxes, profits, and labor income created.

Table 8. Cost of corn straw and CSDF in the BMS.

Category	Corn Straw Cost		CSDF Cost	
	Corn Straw in the Fields ($/t)	Corn Straw Bales in the Factory ($/t)	Rural Residents Heating and Cooking CSDF ($/t)	Urban Concentrated Heating CSDF ($/t)
CSPF	0	30.8	71.2	88.8
CSBF			51.8	73.3

Table 9. Cost of heating service and economic indicators of a CSDF-DHS in the BMS.

Category	Heating Service Price		Economy Indicators		
	Rural Residents, Heating and Cooking ($/household·a)	Urban Concentrated Heating Price ($/m^2·a)	Taxes ($10,000)	Profits ($10,000)	Labor Income ($10,000)
CSPF	275	5.1	59.7	204.3	73.6
CSBF	229	4.4	54.7	177.5	66.3

In this case, CSDF is used to replace traditional energy, mainly coal, LPG, and directly burned firewood and corn straw. According to data collected in Nong'an County, the local price of coal is approximately $101/t and that of LPG is approximately $1.3/kg. Firewood and corn straw have no economic expenses. Therefore, it is estimated that the annual expense of heating and cooking per household is $181/a, which is lower than the expense using CSDF. However, using CSDF can benefit the local economy because it can avoid the payment flowing to other regions to buy coal and LPG. The total payment of 684 households on coal and LPG is 0.85 million USD.

In urban concentrated heating, CSDF mainly replaces coal. The current government-guided price for concentrated heating services mainly by coal is $3.8/m^2, which is lower than that of CSDF and means poor economic performance by a CSDF-DHS in the BMS. However, using CSDF for urban heating can also avoid the payment of coal flowing to other regions. According to the field survey, the concentrated heating coal consumption in Nong'an County is approximately 20 kgce/m^2, the price of purchased coal is approximately $109/tce, and the total expense of coal is approximately 2.35 million USD with a heating area of 1.08 million m^2.

In potential, the total expense of traditional energy flowing out of Nong'an is approximately 3.21 million USD. Using CSDF can avoid this outflow of payments and create taxes, profits, and labor income to help local economic growth, though CSDF has poor economic performance compared to traditional energy in the BMS.

4.3. Economic Performance in the Current Policy Scenario

The results of the economic model in the CPS are listed in Tables 10 and 11. Compared to the BMS, the cost of corn straw and CSPF/CSBF are cut by 5.6% and 8.2%–8.7%/8.7%–9.2%, respectively, because of tax-free policies and subsidies for fixed capital investment. The cost of heating services by CSPF/CSBF for rural residents and urban concentrated heating is reduced by 14.3%/16.1% and 8.8%/9.1%, respectively because of the policies, whereas the taxes and profits created by the CSDF-DHS are also reduced to a certain extent. However, the costs of heating services for rural and urban areas are still higher than those of the traditional energy benchmark ($181/household·a and $3.8/m²), though the economic performance has been improved by policy support totaling 2.0 million USD subsidies and tax-free for corn straw collection and CSPF production.

Table 10. Cost of corn straw and CSDF in the CPS.

Category	Corn Straw Cost		CSDF Cost	
	Corn Straw in the Fields ($/t)	Corn Straw Bales in the Factory ($/t)	Rural Residents, Heating and Cooking CSDF ($/t)	Urban Concentrated Heating CSDF ($/t)
CSPF	0	29.1	65.0	81.5
CSBF			52.7	66.9

Table 11. Cost of heating service and economic indicators of a CSDF-DHS in the CPS.

Category	Heating Service Cost		Economy Indicator			Government Subsidies	
	Rural Residents, Heating and Cooking ($/household·a)	Urban Concentrated Heating Price ($/m²·a)	Taxes ($10,000)	Profits ($10,000)	Labor Income ($10,000)	The First Year Subsidies ($10,000)	Annual Subsidies ($10,000)
CSPF	235	4.7	24.5	196.2	73.6	202.6	0
CSBF	192	4.0	24.5	170.0	66.3	202.6	0

4.4. Influencing Factors of Economic Performance under the CPS

The economic performance of this CSDF-DHS case is poor even with the current policy support. Therefore, we discuss the influence factors of economic performance under the Current Policy Scenario (CPS). These factors includes the subsidy for fixed capital investment, the corn straw price, the CSDF price, the concentrated heating price, carbon trading, and the preferential electricity price, which can be further intervened by government policy in the future.

4.4.1. Influence of Fixed Capital Investment Subsidy

The influence of economic performance on CSDF-DHS heating service costs for a fixed capital investment subsidy is shown in Figures 4 and 5. To discuss the influence of the subsidy proportion changes on heating service cost under the CPS, we set the proportion of fixed capital investment subsidies from 0% to 100%. First, we analyze the urban concentrated heating cost. The results indicate that if the subsidy proportion increased by 10%, the urban concentrated heating cost would decrease by $0.07/m².

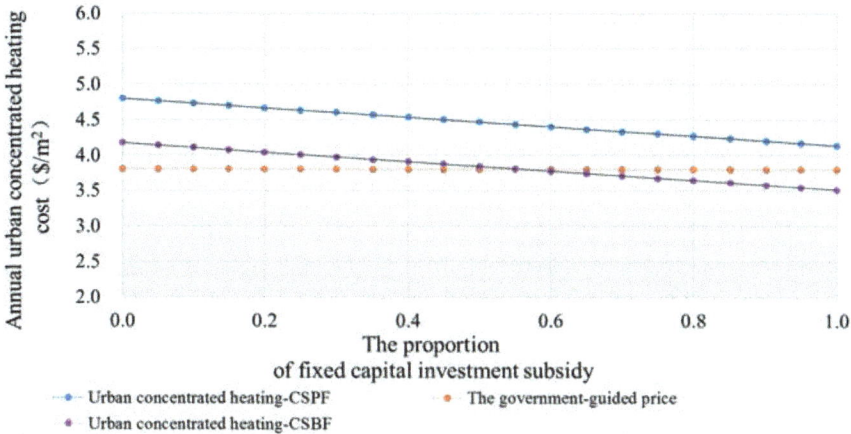

Figure 4. Influence of fixed capital investment subsidy on the economic performance of CSDF for urban concentrated heating.

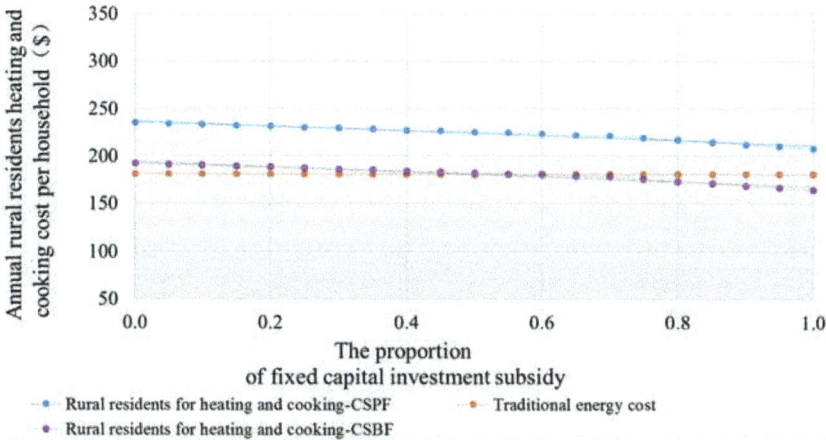

Figure 5. Influence of fixed capital investment subsidy on the economic performance of CSDF for rural residents' heating and cooking.

For CSPF, the urban concentrated heating cost is always higher than the government-guided price. For CSBF, if the subsidy proportion reached 49.0% or higher, the urban concentrated heating cost would be lower than the government-guided price, being competitive in the market. Second, we analyze rural residents' heating and cooking costs. The results show that if the subsidy proportion increased by 10%, the rural residents heating and cooking cost would decrease by $2.6 per household. For CSPF, rural residents' heating and cooking costs are always higher than the cost of traditional energy. For CSBF, if the subsidy proportion reached 62.0% or higher, rural residents' heating and cooking cost would be lower than the cost of traditional energy, being competitive in the market.

4.4.2. Influence of Corn Straw Price

The purchase price of corn straw is set to be zero in both two scenarios. However, when the corn straws are utilized in a large scale as an energy source, the purchase price of corn straw may increase. In addition, to support CSDF-DHS development, the government can also consider corn straw price subsidy to achieve a negative corn straw purchase price, meaning that the collection enterprises can even be paid to obtain corn straw for free, which actually happens in some regions of China because local government tends to reduce field incineration of corn straw to alleviate air pollution by CSDF-DHS.

Under the CPS, we set the purchase price of corn straw to change from −$11.6/t to $11.6/t. The influence of corn straw price on the economic performance of a CSDF-DHS heating service is shown in Figures 6 and 7. The results indicate that if the corn straw price increased by $1.4, the urban concentrated heating cost will rise by $0.12/m². The urban concentrated heating cost for CSPF and CSBF can reach government-guided price when the corn straw price subsidies are at least $10.7/t and $3.3/t, respectively, indicating corn straw prices of −$10.7/t and −$3.3/t, respectively. In terms of rural residents' heating and cooking costs, if the corn straw price increased by $1.4, the rural residents heating and cooking cost will rise by $8.4 per household. The rural residents' heating and cooking costs for CSPF and CSBF can be equal to the average cost of traditional energy per household when the corn straw price subsidies are $9.4/t and $1.9/t, respectively, which indicates that the corn straw price are −$9.4/t and −$1.9/t, respectively.

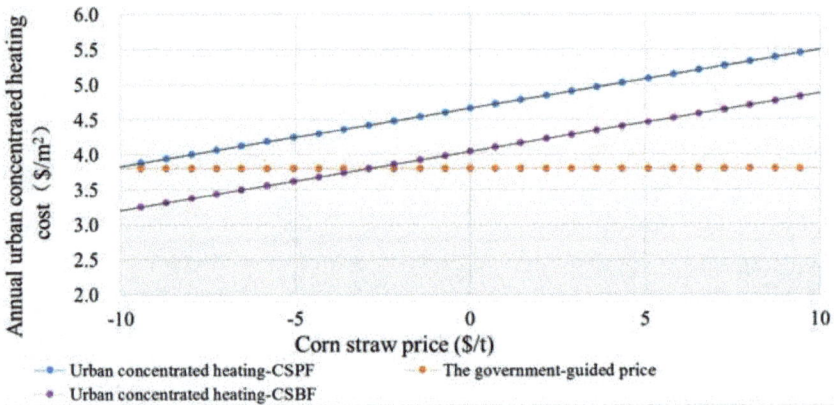

Figure 6. Influence of corn straw price on the economic performance of CSDF for urban concentrated heating.

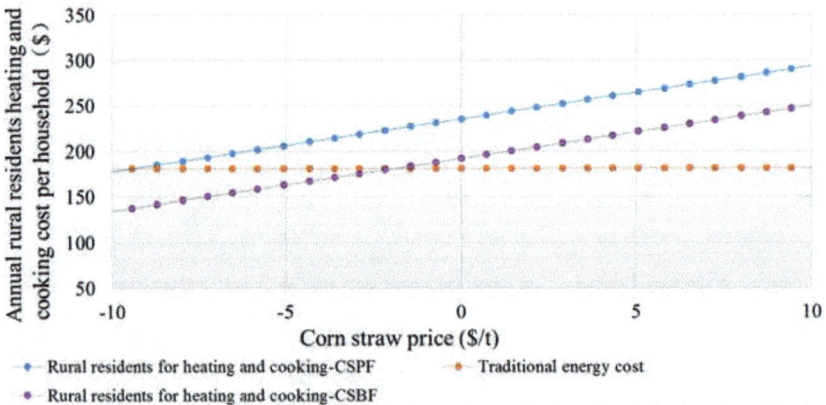

Figure 7. Influence of corn straw price on the economic performance of CSDF for rural residents' heating and cooking.

4.4.3. Influence of Concentrated Heating Fee Subsidy and CSDF Subsidy

Subsidies for urban concentrated heating fees and CSDF subsidies for rural residents can directly reduce final service costs. According to the first law of thermodynamics, an urban concentrated heating fee subsidy can be calculated by the following expression: $a' = \frac{\eta \cdot F_{LHV}}{Q_h^a} \cdot a$, where a' is the CSDF subsidy ($/t), a is the heating fee subsidy ($/m^2), η is the boiler efficiency, F_{LHV} is the low heating value of CSDF (GJ/t) and Q_h^a is the annual heating consumption on space-heating (GJ/m^2·a). The influence of subsidies for urban concentrated heating fees and CSDF subsidies for rural residents on the economic performance of a CSDF-DHS heating service is shown in Figures 8 and 9. The results indicate that for urban concentrated heating cost, when the urban concentrated heating fee subsidy increases by $0.14/m^2, the urban concentrated heating cost decreases by $0.2/m^2 for both CSPF and CSBF. The urban concentrated heating cost reaches the government-guided price when the heating fee subsidies are $0.75/m^2 and $0.23/m^2, respectively. In terms of rural residents' heating and cooking cost, when the rural CSDF subsidy increases by $1.4, the rural residents heating and cooking cost decreases by $5.1 and $4.1 per household for CSPF and CSBF, respectively. Rural residents' heating and cooking costs for CSPF and CSBF can meet average traditional energy costs per household when their subsidy prices are $18.4/t and $3.2/t, respectively.

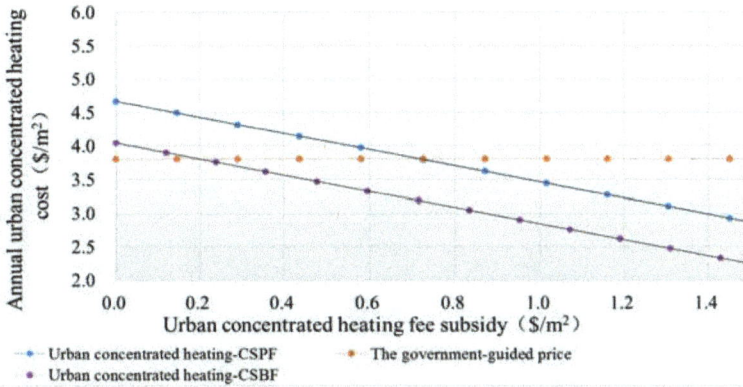

Figure 8. Influence of urban concentrated heating subsidy on economic performance of a CSDF-DHS.

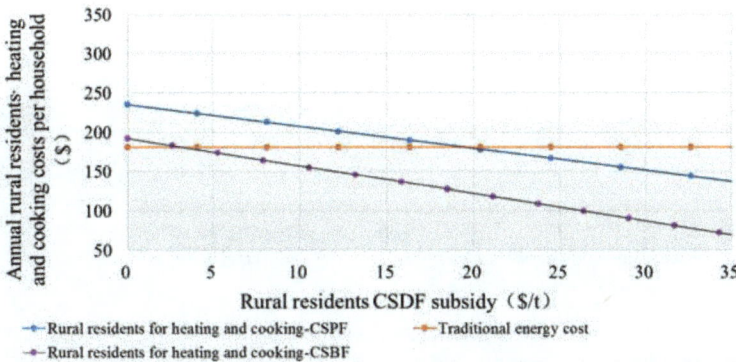

Figure 9. Influence of CSDF subsidy for rural residents' heating and cooking on economic performance of a CSDF-DHS.

4.4.4. Influence of Carbon Trading

Recognized as a renewable energy project, the economic performance of a CSDF-DHS can be improved through carbon trading. Sun [47] studied CDM project development and found that one tonne CSDF can create 1.37 t CO_2,e reduction. Accordingly, we set the carbon trading price from 0 to \$14.5/t CO_2,e. The influence of carbon trading on the economic performance of a CSDF-DHS is illustrated in Figures 10 and 11. The results indicate that, the urban concentrated heating costs of taking CSPF and CSBF reach government-guided prices when carbon trading prices are \$15.3/tCO_2,e and \$4.8/tCO_2,e respectively, and rural residents' heating and cooking costs of taking CSPF and CSBF can meet average traditional energy cost per household when carbon trading prices are \$11.3/tCO_2,e and \$2.3/tCO_2,e respectively.

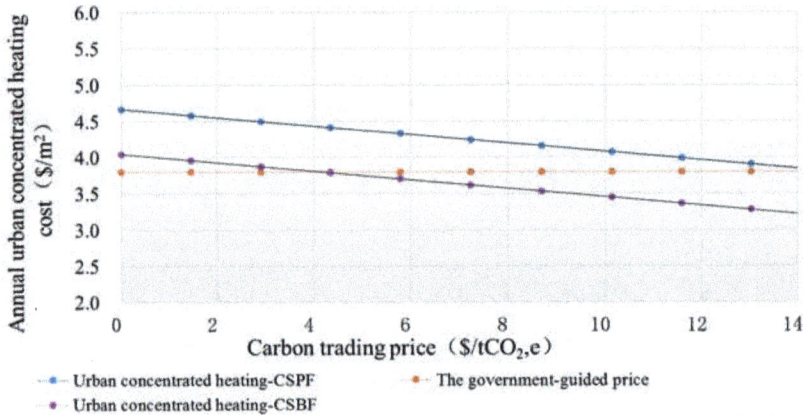

Figure 10. Influence of carbon trading on the economic performance of CSDF for urban concentrated heating.

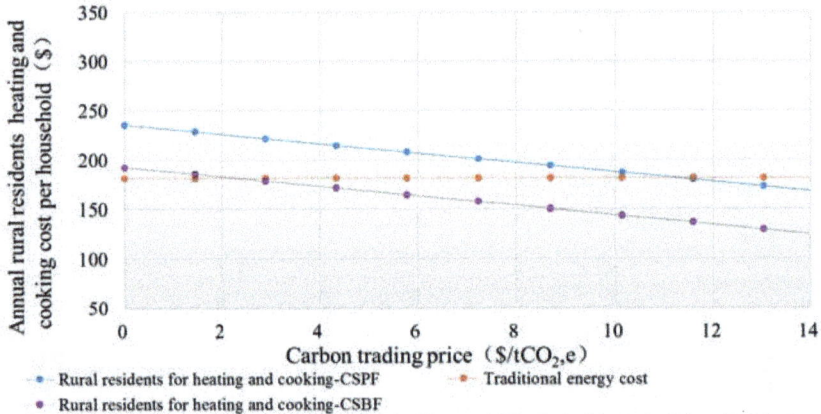

Figure 11. Influence of carbon trading on the economic performance of CSDF for rural residents' heating and cooking.

4.4.5. Influence of Preferential Electricity Price of CSDF Production

China has launched a preferential electricity price policy for the pretreating industry of agricultural products [48]. Some experts argue that CSDF production should be listed in the catalog of the pretreating industry of agricultural products to decrease costs of CSDF production. According to the field survey, the electricity price for agricultural production is 27% lower than that of commerce and industry. The influence of this preferential electricity price for agriculture production on the economic performance of a CSDF-DHS heating service is shown in Table 12.

Table 12. The heating service cost of a CSDF-DHS with preferential electricity price.

Policy Support	CSDF Type	Heating Services Costs	
		Rural Residents, Heating and Cooking Costs ($/household·a)	Urban Concentrated Heating Cost ($/m²·a)
Preferential electricity price	CSPF	225 (reduced by 10.6)	4.5 (reduced by 0.14)
	CSBF	186 (reduced by 6.1)	4.0 (reduced by 0.09)

The heating service cost of a CSDF-DHS under this policy is reduced compared to that under the CPS, but is still higher than that of traditional energy.

4.4.6. Influence of Enterprise Profit Margins

Under the CPS, enterprise profit margins in the three stages of a CSDF-DHS, which are: (1) corn straw collection and bale transportation; (2) CSDF production; and (3) CSDF utilization, are set to be (1) 30%; (2) 30% for urban heating and 10% for rural residents; and (3) 20%, respectively. Based on current enterprise profit margins, enterprises will adjust the profit margins according to the change of market price in CSDF. We set the enterprise profit margins to be increased by 50% or to be reduced by 50% for sensitivity analysis. In the case of increased margins, the profit increase is: (1) 45%; (2) 45% for urban heating and 15% for rural residents; and (3) 30%. In the case of decreased margins, the profit decrease is: (1) 15%; (2) 15% for urban heating and 5% for rural residents; and (3) 10%. The influence of enterprise profit margins in the supply chain on the economic performance of a CSDF-DHS is shown in Figures 12 and 13.

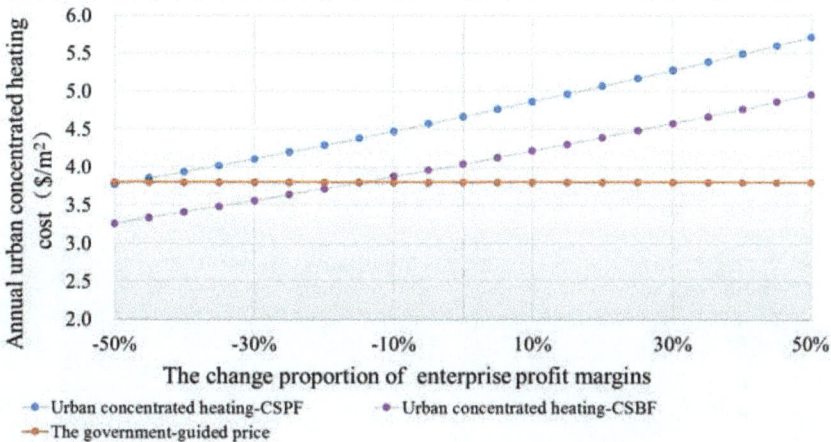

Figure 12. Influence of enterprise profit margin on the economic performance of CSDF for urban concentrated heating.

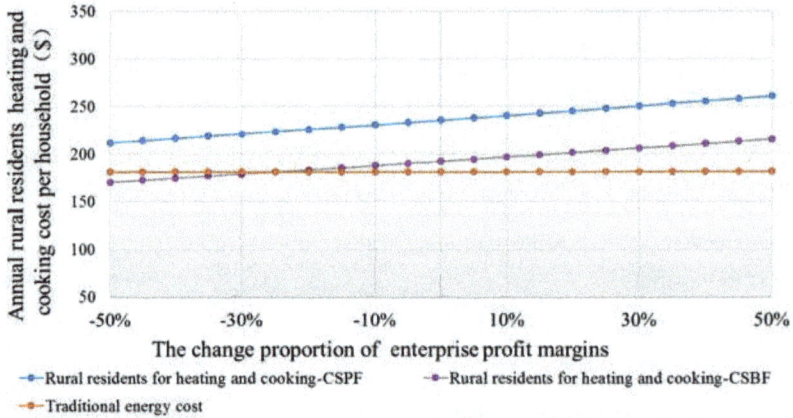

Figure 13. Influence of enterprise profit margin on the economic performance of CSDF for rural residents' heating and cooking.

The results indicate that, in terms of CSPF and CSBF, urban concentrated heating costs reach a government-guided price when enterprise profit margins in the supply chain reduce by 47.8% and 17.8%, respectively. In terms of CSPF and CSBF, rural residents' heating and cooking costs can meet average traditional energy costs when enterprise profit margins in the supply chain reduce by 111.2% and 25.0%, respectively.

4.4.7. Influence of the Boiler Efficiency of Urban Concentrated Heating Station

Improving energy efficiency of CSDF boilers can also reduce heating service cost. How heating station boiler efficiency influences the economic performance of a CSDF-DHS is shown in Figure 14. The results indicate that when CSDF quantity is constant, the urban concentrated heating service area increases with improvement in the energy efficiency of CSDF boilers and the urban concentrated heating cost declines. When the energy efficiency of CSBF boilers is 87.7%, the urban concentrated heating cost can reach the government-guided price. Also, the urban concentrated heating cost for CSPF is always higher than the government-guided price.

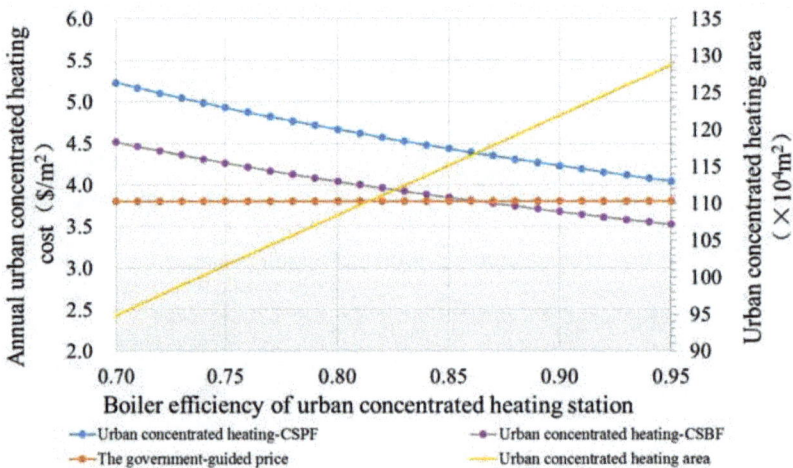

Figure 14. Influence of boiler energy efficiency on the economic performance of a CSDF-DHS.

4.5. Data Uncertainty

Many data used referring to the above in this case study are obtained by a field survey and expert interviews, which are basically constant with the average and rough level of the selected samples but may be different to a certain extent with actual data or statistics accounted by other methods.

5. Conclusions and Policy Implications

This paper established a technical model and economic model of a CSDF-DHS and analyzed its economic performance based on a case study of Nong'an Country in Jilin Province. In addition, this paper analyzes the influence of policy support on its economic performance by scenario analysis. The main conclusions and policy implications are summarized as follows:

- Under the benchmark market scenario oriented by a market without policy support, service prices of CSPF and CSPF as rural residents' heating and cooking fuels are $275/household·a and $229/household·a, which are higher than that of traditional energy of $181/household·a. Service prices of CSPF and CSPF for urban concentrated heating are $5.1/m²·a and $4.4/m²·a, which are also higher than that of traditional energy of $3.8/m²·a.The economic performance measured by the service cost of the CSDF-DHS is poor compared to that of a traditional energy system mainly based on fossil fuels, and producing CSBF is comparatively more economical than producing CSPF. However, a CSDF-DHS has the advantage of extra economic and social benefits created for Nong'an County, including higher taxes ($590,000 for CSPF and $547,000 for CSBF), higher profits ($2,043,000 for CSPF and $1,775,000 for CSBF), and labor income ($736,000 for CSPF and $663,000 for CSBF), compared to a traditional energy system purchasing most of the fossil fuel from other regions. Therefore, it is implicated that current projects of CSDF-DHSs must get policy support, otherwise they will not be competitive in the market, and the policy input can be compensated by its regional economic and environmental benefits to rural areas.
- Under the current policy scenario with current policy support, service prices of CSPF and CSPF as rural residents' heating and cooking fuels are $235/household·a and $192/household·a, which are higher than that of traditional energy. Service prices of CSPF and CSPF for urban concentrated heating are $4.7/m²·a and $4.0/m²·a, which are also higher than that of traditional energy. The economic performance of a CSDF-DHS can be improved considerably but is still not competitive compared to tradition energy system. Therefore, it is implicated more policy support must be considered to make a CSDF-DHS competitive in the market.
- Within the range of influencing factor analysis in this paper, a preferential electricity price still cannot make the CSDF-DHS competitive. Providing a subsidy of fixed capital investment, reducing the profit margins of enterprises by economic compensation, and improving the boiler efficiency can make the CSBF-DHS competitive, but the CSPF-DHS remains expensive. The subsidy of concentrated heating price and of CSDF used for rural residents, carbon trading, and negative corn straw prices can make both CSBF-DHS and CSPF-DHS competitive. Therefore, it is implicated the policy support can be staged to gradually open the market, such as first focus on the commercialization of CSBF-DHSs which are easier to be competitive, and then CSPF-DHSs. Meanwhile, the most powerful policy options are suggested as the subsidy of concentrated heating price and of CSDF used for rural residents, carbon trading, and negative corn straw prices.

In next step of this study, to implement a rapid and healthy development of CSDF-DHSs, the social acceptance of CSDF-DHSs by various operators must be further considered, requiring stakeholder analysis and strategic level research.

Acknowledgments: This work was supported by the Chinese Academy of Engineering (Project No. 2016-ZD-14-2) and the Energy Foundation (Project No. G-1509-23779). The authors also gratefully acknowledge support from BP regarding the Phase II and Phase III Collaboration between BP and Tsinghua University and the support from the Tsinghua-Rio Tinto Joint Research Centre for Resources, Energy and Sustainable Development.

Author Contributions: Shizhong Song, Jing Xu, Chinhao Chong, and Min He coordinated the main theme of this paper and wrote this manuscript. Linwei Ma, Pei Liu, and Xianzheng Huang discussed the research results and commented on the manuscript. Final review was done by Linwei Ma, Weidou Ni and Zheng Li. All the authors read and approved the final manuscript.

Conflicts of Interest: The authors declare no conflict of interest.

Abbreviations

BSDF	biomass solid densified fuel
CSDF	corn straw densified fuel
CSPF	corn straw pellet fuel
CSBF	corn straw briquettes fuel
DHS	district heating system
CSDF-DHS	corn straw densified fuel - district heating system
BMS	Benchmark Market Scenario
CPS	Current Policy Scenario
ha	hectare
kgce	kg coal equivalent
tce	tonne coal equivalent
CO_2,e	CO_2 equivalent

References

1. Widholm, J.; Nagata, T. *Biotechnology in Agriculture and Forestry*; Springer: Heidelberg, Germany, 2010.
2. Nunes, L.J.R.; Matias, J.C.O.; Catalão, J.P.S. Mixed biomass pellets for thermal energy production: A review of combustion models. *Appl. Energy* **2014**, *127*, 135–140. [CrossRef]
3. Zhou, Y.; Zhang, Z.; Zhang, Y.; Wang, Y.; Yu, Y.; Ji, F.; Ahmad, R.; Dong, R. A comprehensive review on densified solid biofuel industry in China. *Renew. Sustain. Energy Rev.* **2016**, *54*, 1412–1428. [CrossRef]
4. Tian, Y.; Zhao, L.; Meng, H.; Sun, L.; Yao, Z. Research on China densified biofuel standards. *Renew. Energy Resour.* **2010**, *28*, 1–5.
5. National Energy Administration of the People's Republic of China. The 13th Five-Year-Plan of Biomass Energy Development. Available online: http://www.gov.cn/xinwen/2016-12/05/content_5143612.htm (accessed on 9 December 2016).
6. National Energy Administration of the People's Republic of China. The 12th Five-Year-Plan of Biomass Energy Development. Available online: http://www.gov.cn/zwgk/2012-12/28/content_2301176.htm (accessed on 9 May 2016).
7. Jiang, D. Improvement plans to strengthen the implementation of effective response to heavy pollution weather—The enlightenment of heavy pollution weather in northeast China in early November 2015. *China Emerg. Manag.* **2015**, *11*, 43–45.
8. Li, R.; Zhang, S.; Wang, Y.; Zhang, X.; Zhao, H.; Zhou, Q.; Chen, W. Mass concentration of atmospheric fine particulates in crop harvesting period in Sanjiang Plain, northeast China. *China Environ. Sci.* **2015**, *3*, 676–682.
9. Chen, J.; Zheng, W.; Gao, H.; Shao, J.; Liu, C. Estimation method of straw burned area based on multi-source satellite remote sensing. *Trans. Chin. Soc. Agric. Eng.* **2015**, *3*, 207–214.
10. China's National Bureau of Statistics. *China Energy Statistical Yearbook 2016*; China Statistics Press: Beijing, China, 2016.
11. Zhou, F. *An Investigation on the Industrialization Development of Biomass Solid Densified Fuel in China*; Energy Foundation China: Beijing, China, 2012.
12. Thomson, H.; Liddell, C. The suitability of wood pellet heating for domestic households: A review of literature. *Renew. Sustain. Energy Rev.* **2015**, *42*, 1362–1369. [CrossRef]
13. Vallios, I.; Tsoutsos, T.; Papadakis, G. Design of biomass district heating systems. *Biomass Bioenergy* **2009**, *33*, 659–678. [CrossRef]
14. Chau, J.; Sowlati, T.; Sokhansanj, S.; Preto, F.; Melin, S.; Bi, X. Techno-economic analysis of wood biomass boilers for the greenhouse industry. *Appl. Energy* **2009**, *86*, 364–371. [CrossRef]
15. Chau, J.; Sowlati, T.; Sokhansanj, S.; Preto, F.; Melin, S.; Bi, X. Economic sensitivity of wood biomass utilization for greenhouse heating application. *Appl. Energy* **2009**, *86*, 616–621. [CrossRef]
16. Michopoulos, A.; Skoulou, V.; Voulgari, V.; Tsikaloudaki, A.; Kyriakis, N.A. The exploitation of biomass for building space heating in Greece: Energy, environmental and economic considerations. *Energy Convers. Manag.* **2014**, *78*, 276–285. [CrossRef]
17. Stolarski, M.J.; Krzyżaniak, M.; Warmiński, K.; Śnieg, M. Energy, economic and environmental assessment of heating a family house with biomass. *Energy Build.* **2013**, *66*, 395–404. [CrossRef]

18. Hendricks, A.M.; Wagner, J.E.; Volk, T.A.; Newman, D.H.; Brown, T.R. A cost-effective evaluation of biomass district heating in rural communities. *Appl. Energy* **2016**, *162*, 561–569. [CrossRef]

19. Stephen, J.D.; Mabee, W.E.; Pribowo, A.; Pledger, S.; Hart, R.; Tallio, S.; Bull, G.Q. Biomass for residential and commercial heating in a remote Canadian aboriginal community. *Renew. Energy* **2016**, *86*, 563–575. [CrossRef]

20. Tabata, T.; Okuda, T. Life cycle assessment of woody biomass energy utilization: Case study in Gifu Prefecture, Japan. *Energy* **2012**, *45*, 944–951. [CrossRef]

21. Ren, L.; Cafferty, K.; Roni, M.; Jacobson, J.; Xie, G.; Ovard, L.; Wright, C. Analyzing and comparing biomass feedstock supply systems in China: Corn stover and sweet sorghum case studies. *Energies* **2015**, *8*, 5577–5597. [CrossRef]

22. Zhao, L.; Zhang, X.; Xu, J.; Ou, X.; Chang, S.; Wu, M. Techno-economic analysis of bioethanol production from lignocellulosic biomass in China: Dilute-acid pretreatment and enzymatic hydrolysis of corn stover. *Energies* **2015**, *8*, 4096–4117. [CrossRef]

23. Toka, A.; Iakovou, E.; Vlachos, D.; Tsolakis, N.; Grigoriadou, A. Managing the diffusion of biomass in the residential energy sector: An illustrative real-world case study. *Appl. Energy* **2014**, *129*, 56–69. [CrossRef]

24. Moiseyev, A.; Solberg, B.; Kallio, A.M.I. The impact of subsidies and carbon pricing on the wood biomass use for energy in the EU. *Energy* **2014**, *76*, 161–167. [CrossRef]

25. Madlener, R. Innovation diffusion, public policy, and local initiative: The case of wood-fuelled district heating systems in Austria. *Energy Policy* **2007**, *35*, 1992–2008. [CrossRef]

26. Gan, L.; Yu, J. Bioenergy transition in rural China: Policy options and co-benefits. *Energy Policy* **2008**, *36*, 531–540. [CrossRef]

27. Shan, M.; Li, D.; Jiang, Y.; Yang, X. Re-thinking china's densified biomass fuel policies: Large or small scale? *Energy Policy* **2016**, *93*, 119–126. [CrossRef]

28. Wang, W.; Wei, O.; Hao, F. A supply-chain analysis framework for assessing densified biomass solid fuel utilization policies in China. *Energies* **2015**, *8*, 7122–7139. [CrossRef]

29. Ministry of Housing and Urban-Rural Development of the People's Republic of China. *Standard of Climatic Regionalization for Architecture, GB50178-93*; China Planning Press: Beijing, China, 1993.

30. Building Energy Research Center, Tsinghua University. *Annual Report on China Building Energy Efficiency*; China Architecture & Building Press: Beijing, China, 2016.

31. Baidu Encyclopedia. Yangshulin Village. Available online: http://baike.baidu.com/link?url=3fT-d2bp2qqqMXeNmCuwJOdnyjNwdcF5aD8BcjbccRz7ipcfnXhD1iAdIc7EChgFcN6v-UWwttDeQ8N_7o9NXq (accessed on 16 March 2016).

32. Bureau of Statistics of Nong'an County. Statistical Bulletin of the National Economic and Social Development of the Nong'an County in 2014. Available online: http://www.nong-an.gov.cn/info/1042/71381.htm (accessed on 20 March 2016).

33. Ministry of Agriculture of the People's Republic of China. *Investigation and Evaluation Report of Crop Straw Resources in China*; Wiley: Hoboken, NJ, USA, 2010.

34. Yu, X.; Wang, L.; Wang, F.; Xiao, J. Research on technology mode of corn straw collection and delivery in northeast Area. *J. Agric. Mech. Res.* **2013**, *5*, 24–28.

35. Xu, Y.; Tian, Y.; Zhao, L.; Yao, Z.; Hou, S.; Meng, H. Comparison on cost and energy consumption with different straw's collection-store-transportation modes. *Trans. Chin. Soc. Agric. Eng.* **2014**, *20*, 259–267.

36. National Development and Reform Commission of People's Republic of China. State Key Low-Carbon Technology Promotion Catalogue. Available online: http://www.sdpc.gov.cn/gzdt/201409/t20140905_625018.html (accessed on 9 December 2016).

37. Energy Bureau of Jilin Province, China. *Construction Guide for "Low Carbon Energy Demonstration Towns" and "Low Carbon Energy Demonstration Counties" in Jilin Province, China*; Energy Bureau of Jilin Province: Changchun, China, 2016.

38. Ministry of Housing and Urban-Rural Development of the People's Republic of China. *Design Code for City Heating Network, CJJ_34-2010*; China Architecture & Building Press: Beijing, China, 2010.

39. National Energy Administration of the People's Republic of China. Notice on the Construction of Heating Demonstration Project of Biomass Solid Densified Fuel Boiler. Available online: http://zfxxgk.nea.gov.cn/auto87/201407/t20140708_1818.htm (accessed on 2 May 2016).

40. Chang, L. *Modeling and Optimization of Hydrogen Supply Chain*; Tsinghua University: Beijing, China, 2008.

41. Financial & Economic Dictionary Editorial Board. *Financial & Economic Dictionary*; China Financial & Economic Publishing House: Beijing, China, 2014.
42. Bureau of Statistics of Jilin Province. Statistical Bulletin of the National Economic and Social Development of Jilin Province in 2012. Available online: http://www.jl.gov.cn/sj/sjcx/ndcx/tjgb/201412/t20141216_1821329.html (accessed on 12 May 2016).
43. New Energy Office of Hebei Province, China. Bid Announcement Enterprises of Clean Stoves Project in Hebei. 2015. Available online: http://news.chinaluju.com/w/a2/9393.html (accessed on 12 May 2016).
44. Zhi, G.; Yang, J.; Zhang, T.; Guan, J.; Du, J.; Xue, Z.; Meng, F. Rural household coal use survey, emission estimation and policy implications. *Res. Environ. Sci.* **2015**, *8*, 1179–1185.
45. General Office of the State Council of the People's Republic of China. The Guiding Opinions on Accelerating the Comprehensive Utilization of Crop Straw. Available online: http://www.chinalaw.gov.cn/article/fgkd/xfg/fgxwj/201003/20100300251211.shtml (accessed on 10 March 2016).
46. The General Office of the People's Government of Jilin Province, China. The Guiding Opinions on Promoting the Comprehensive Utilization of Crop Straw. Available online: http://www.jl.gov.cn/kzgn/nrtj/wj/201605/t20160512_2262625.html (accessed on 10 June 2016).
47. Sun, L.; Tian, Y.; Meng, H.; Zhao, L. Development of clean development mechanism (CDM) project of biomass densified biofuels in China. *Trans. Chin. Soc. Agric. Eng.* **2011**, *8*, 304–307.
48. National Development and Reform Commission of People's Republic of China. Notice on the Adjusted Sale Electricity Price Issues Related to Classification Structure. Available online: http://www.gov.cn/zwgk/2013-06/09/content_2423501.htm (accessed on 10 May 2016).

Article

Data-Reconciliation Based Fault-Tolerant Model Predictive Control for a Biomass Boiler

Palash Sarkar [1,*], **Jukka Kortela** [1], **Alexandre Boriouchkine** [1], **Elena Zattoni** [2] and **Sirkka-Liisa Jämsä-Jounela** [1]

[1]   Department of Biotechnology and Chemical Technology, School of Chemical Engineering, Aalto University, 00076 Aalto, Finland; jukka.kortela@aalto.fi (J.K.); alexandre.boriouchkine@nestejacobs.com (A.B.); sirkka-liisa.jamsa-jounela@aalto.fi (S.-L.J.-J.)
[2]   Department of Electrical, Electronic and Information Engineering "G. Marconi", Alma Mater Studiorum · University of Bologna, 40136 Bologna, Italy; elena.zattoni@unibo.it
*   Correspondence: palash.sarkar@aalto.fi

Academic Editor: Tariq Al-Shemmeri
Received: 26 September 2016; Accepted: 25 January 2017; Published: 9 February 2017

Abstract: This paper presents a novel, effective method to handle critical sensor faults affecting a control system devised to operate a biomass boiler. In particular, the proposed method consists of integrating a data reconciliation algorithm in a model predictive control loop, so as to annihilate the effects of faults occurring in the sensor of the flue gas oxygen concentration, by feeding the controller with the reconciled measurements. Indeed, the oxygen content in flue gas is a key variable in control of biomass boilers due its close connections with both combustion efficiency and polluting emissions. The main benefit of including the data reconciliation algorithm in the loop, as a fault tolerant component, with respect to applying standard fault tolerant methods, is that controller reconfiguration is not required anymore, since the original controller operates on the restored, reliable data. The integrated data reconciliation–model predictive control (MPC) strategy has been validated by running simulations on a specific type of biomass boiler—the KPA Unicon BioGrate boiler.

Keywords: data reconciliation; model predictive control; fault-tolerant contro; BioGrate boiler

---

1. Introduction

Renewable energy production is acknowledged worldwide as a key factor for sustainable growth. As for the energy sources used to replace the fossil fuel sources in power and heat production, local fuel supplies such as biomass fuel, which includes wood chips, bark, and sawdust, have relevant advantages for on-site industries and municipalities, such as, primarily, secure availability and price stability [1]. However, biofuels are inherently affected by highly variable properties, which makes achieving efficient and reliable combustion and low polluting emissions a challenging task in terms of control and monitoring system design [2–4].

In particular, in the operation of a biomass boiler, a crucial role is played by the residual oxygen content in the flue gas. Indeed, this oxygen concentration, which basically depends on the total air supply to the furnace and on the thermal composition of the biofuel, provides the information needed to estimate both the power developed by the combustion process and the level of polluting emissions to the extent that several countries' regulations make explicit reference to this index [5]. Hence, in control systems devised to operate biomass boilers, the oxygen content in the flue gas is directly measured by appropriate sensors and represents a key feedback variable [3,6]. It is therefore obvious that faults occurring in the sensors of the oxygen concentration in the flue gas degrade the performance of the control systems and that, consequently, it is highly advisable to take measures aimed at rendering the control systems tolerant to this type of fault.

A practical way of addressing sensor failures and improving measurement reliability, widely adopted also in biomass boiler operation, includes redundant sensors. However, redundancy implies higher installation and running costs, so that the possibility of implementing more sophisticated fault tolerance techniques is worth being investigated. An evolutionary algorithm for detecting discrepancies in the measurement of the oxygen content in the flue gas, while avoiding hardware redundancy, was developed in [7]. This method, however, did not address the issue of complete sensor breakdown. However, in recent years, several other fault detection methods have been devised in order to handle typical faults occurring in industrial boilers, including boilers exploiting conventional sources of energy. None of them explicitly addresses faults in the oxygen content sensor or even conveniently adapts to detect this kind of fault. In [8], a principal component analysis (PCA) method was developed to detect faults occurring in the sensors of the air flow rate and of the fuel flow rate of an industrial boiler and to consequently reconstruct the faulty measurements. Similarly, in [9], a PCA method was devised to detect malfunctions in the sensors of the air flow rate, fuel flow rate, stack pressure, and wind-box pressure. The methods presented in [8,9] are data-driven methods and, as such, they require a special effort in data preprocessing and validation at the design stage. Furthermore, these methods are required to handle missing values and data outliers, which brings increased complication at their developing phase, as pointed out in [10]. In [11], a generalized likelihood ratio method was devised to detect and accomodate faults occurring in the fuel bed height sensor. Nevertheless, the solution devised in [11] implies an appreciable delay in the reaction to the fault, mainly due to the algorithm run to detect the fault and to the consequent controller reconfiguration.

In [12], model predictive control technology for boiler control was presented to enable the coordination of air and fuel flows during transients. It was shown that this approach can be used to increase the boiler efficiency, and also considerably reduce the production of nitrogen oxides (NO_x). In [4], model predictive control was used as a tool to obtain improved process operation performance for municipal solid waste (MSW) combustion plants. In particular, the conclusion resulting from the comparison with conventional (proportional) P-based control was that the linear model predictive control (LMPC)-based combustion control system outperformed the conventional combustion control system; it was much more capable of handling large temporarily upsets. In [13], an improvement based on the radiation intensity of the flame was presented. The fast response and high sensitivity of the radiation intensity increased the load following capacity of the power plant while keeping the steam pressure stable. In [14], the method to further improve the load change capacity for the water cooled plants through cold source follow adjustment (CSFA) was proposed. Then, an improved control strategy which combines coordinated control system with CSFA was brought forward to be used for the flexible load control in [15]. Still, according to practical tests, the oxygen consumption measurement is the best measure of the heat released in the furnace [3,16].

For these reasons, in this work, a fault tolerant control scheme is proposed, where a data reconciliation algorithm is included in the loop and is activated when the control system has reached the steady state. In particular, the data reconciliation unit acts in such a way that the occurred oxygen content sensor fault is made invisible to the model predictive controller, which is fed with the reconciled data. The idea of exploiting data reconciliation as a means to achieve fault tolerance in a control system is novel. Indeed, data reconciliation methods have been employed in control systems either with the goal of minimizing the noise level in process measurements or with the target of detecting faults in combination with data-based fault detection methods. In particular, in [17,18], data reconciliation was exploited to reduce the noise level in measurements and improve the performance of the control strategy for a continuous stirred tank. In [19], data reconciliation was applied with the same purpose in a distributed control system for a distillation column. Instead, in [20], data reconciliation was used in synergy with PCA for monitoring and sensor fault detection in a modelled ammonia synthesis plant. In [21], data reconciliation improved the estimation of process variables and enabled improved sensor quality control and identification of process anomalies.

As to the effectiveness of the proposed data-reconciliation based fault-tolerant control scheme, this has been shown by integrating data reconciliation with an improved version of the model predictive control (MPC) strategy earlier developed in [2] for a specific type of biomass boiler—the KPA Unicon BioGrate boiler (KPA Unicon Group Oy, Pieksämäki, Finland). Indeed, the MPC strategy presented in this work is based on a more detailed model of the BioGrate boiler, which also includes the dynamics of the oxygen content in the flue gas. The data reconciliation algorithm is based on the constrained optimization of a quadratic cost, which exploits the process static model derived from the dynamic one used for the MPC design. The performance of the overall fault tolerant control system has been evaluated by running simulations on recorded data concerning the measurement of the oxygen content in the flue gas of an actual process under faulty conditions. A schematic diagram of the integrated data reconciliation-model predictive control (DR-MPC) strategy is shown in Figure 1.

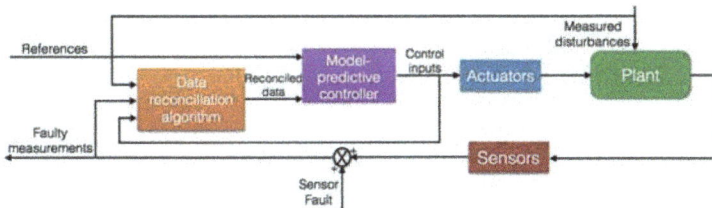

Figure 1. Integrated data reconciliation-model predictive control based fault-tolerant control (FTC) system.

The remainder of the paper is organized as follows. Section 2 describes the process and an improved version of the MPC strategy developed in [2] for the BioGrate boiler. Section 2.1 provides the process description of the BioGrate boiler. Section 2.2 describes the mathematical model of the process. Section 2.3 presents the MPC strategy. Section 3 describes the data reconciliation algorithm implemented on the BioGrate boiler. Section 4 shows the effectiveness of the proposed DR-MPC strategy through the simulation results obtained by using faulty measurements recorded from the real industrial process.

2. Process and Control Description of the BioPower Combined Heat and Power Process

2.1. Process description of the BioGrate boiler

The BioGrate boiler consists of a furnace and a steam-water circulation system as illustrated in Figure 2. The furnace comprises conical-shaped grate rings that are surrounded by heat-insulating refracting brick walls. The grate consists of several rings, half of which rotate while the other half are stationary. Two consecutive rotating rings rotate in opposite directions. This setup ensures a uniform distribution of fuel throughout the grate. Fuel is fed from the bottom of the grate where it spreads towards the outer rings and undergoes combustion. The refracting brick walls radiate the heat generated during combustion back to the grate. The furnace is equipped with air register systems for combustion. Primary air flows from the bottom in a direction perpendicular to the fuel feed movement (cross-flow reactor) to ensure efficient mixing of air and fuel. The secondary air flows from the nozzles in the grate-wall to completely combust volatiles present in the over-bed region [22].

The steam-water circulation system absorbs the heat from the furnace and the flue gas to heat water-steam flowing through it. The important components of the steam-water circulation system include an economizer, a drum and evaporator system and two superheaters. The feedwater is pumped into the economizer where the water absorbs the remaining heat from the flue gas before the flue gas is released into the atmosphere. The heated water is then transferred to the drum and evaporator system. The evaporator consists of downcomers and risers that are located in the walls of the furnace. It absorbs heat from biomass combustion and produces a mixture of water and steam. The steam, separated in the drum, is then passed to the two superheaters where it is further heated with the flue gas to form superheated steam. The superheated steam is transferred to the steam turbine to generate electricity.

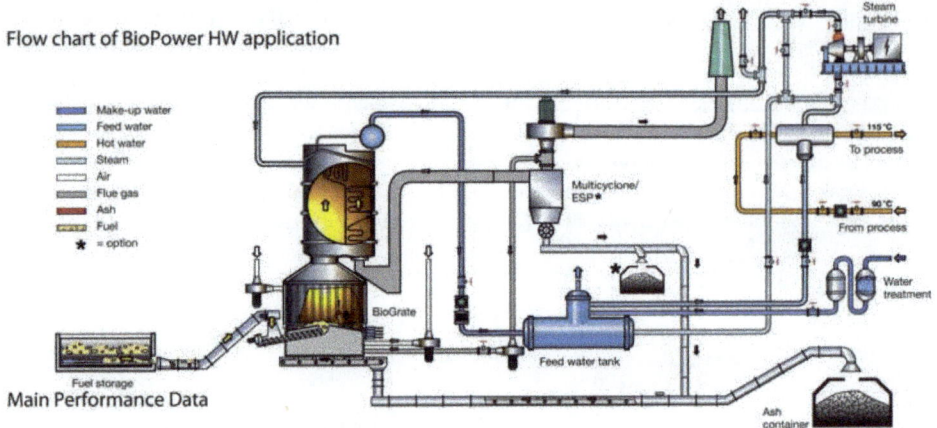

Figure 2. BioPower 5 combined heat and power (CHP) process [23]: 1—primary air; 2—secondary air; 3—economizer; 4—drum; 5—superheaters; 6—evaporator.

The main objective of the BioPower 5 CHP plant is to produce power for the generator and for the hot water network. The split ratio of 2.9 MW electricity and 13.5 MW of heat is defined by the drum pressure and turbine back-pressure. The difference between the consumed and produced power disturbs the pressure in the drum, and the control strategy equalizes the steam production and consumption by controlling the drum pressure, which is achieved by manipulating the fuel and air supply to the furnace.

2.2. Model Description of the BioGrate Boiler

The set of mathematical equations describing the process is given as ([2]):

$$\dot{x}_1(t) = c_{ds}x_1(t) - c_{thd}\beta_{thd}u_2(t) + c_{ds,in}u_1(t), \tag{1}$$

$$\dot{x}_2(t) = -c_{wev}\beta_{wev}x_2(t) + c_{w,in}d_1(t), \tag{2}$$

$$\dot{x}_3(t) = -x_3(t) + q_{wf}(c_{thd}\beta_{thd}u_2(t) - c_{ds}x_1(t)) - 0.0244c_{wev}\beta_{wev}x_2(t), \tag{3}$$

$$\dot{x}_4(t) = -x_4(t) + d_2(t), \tag{4}$$

$$\dot{x}_5(t) = \frac{1}{\alpha_{metal}}(x_3(t) - x_4(t)), \tag{5}$$

$$\dot{x}_6(t) = -x_6(t) + c_{ds,O_2}x_1(t) + c_{wev,O_2}x_2(t) + c_{thd,O_2}u_2(t) + c_{sa,O_2}u_3(t) + c_{ds,in,O_2}u_1(t), \tag{6}$$

where $x_1(t)$, $x_2(t)$, $x_3(t)$, $x_4(t)$, $x_5(t)$ and $x_6(t)$ are the fuel bed height, the moisture content in the furnace, the power generated from the biomass combustion, the filtered steam demand, the drum pressure and the oxygen content in flue gas, respectively; $u_1(t)$, $u_2(t)$ and $u_3(t)$ are the fuel flow rate, the primary air flow rate and the secondary air flow rate, respectively; $d_1(t)$ and $d_2(t)$ are measured disturbances and consist of the moisture content in the fuel and the steam demand, respectively; c_{ds}, c_{thd}, $c_{ds,in}$, c_{wev} and $c_{w,in}$ are model coefficients identified from process data; β_{thd} describes the dependence on the position of the moving grate; β_{wev} is the coefficient for a dependence on the position from the centre to the periphery of the moving grate; α_{metal} is a coefficient that depends on the material type of the metal tubes of the evaporator system; and c_{ds,O_2}, c_{wev,O_2}, c_{thd,O_2}, c_{sa,O_2} and c_{ds,in,O_2} are the parameters for the linearized model of the oxygen content.

Equation (1) describes the dynamics of the amount of dry fuel in the furnace. The first two terms on the right side represent the thermal decomposition rate, whereas the last term describes the fuel feed to the furnace. Equation (2) represents the dynamics of water evaporation. The energy for the water evaporation is mainly provided by the combustion of biomass near the surface of the grate. The temperature near the bottom of the biomass layer is almost independent of the primary air flow; thus, the water evaporation rate is assumed to be independent of the primary air flow as well. Equation (3) is the equation representing the power generated from the biomass combustion. The power is a filtered difference of the power released by the fuel thermal decomposition and the power consumed by water evaporation. The second term on the right-hand side represents the power of the fuel thermal decomposition and the last term describes the power needed for water evaporation. Equation (4) represents the filtered steam demand. Equation (5) represents the dynamics of the drum pressure. The drum level is kept constant by its controller, and, therefore, the variations in the steam volume are neglected. Equation (6) describes the dynamics of the residual oxygen in the flue gas after the complete combustion.

The process models have been identified in [24], and the continuous-time state-space model of the process is:

$$\dot{x}(t) = A_c\,x(t) + B_c\,u(t) + E_c\,d(t),$$
$$y(t) = C_c\,x(t),$$

where:

$$A_c = \begin{bmatrix} c_{ds} & 0 & 0 & 0 & 0 & 0 \\ 0 & -c_{wev}\beta_{wev} & 0 & 0 & 0 & 0 \\ -q_{wf}c_{ds} & -0.0244c_{wev}\beta_{wev} & -1 & 0 & 0 & 0 \\ 0 & 0 & 0 & -1 & 0 & 0 \\ 0 & 0 & \frac{1}{\alpha_{metal}} & -\frac{1}{\alpha_{metal}} & 0 & 0 \\ c_{ds,O_2} & c_{wev,O_2} & 0 & 0 & 0 & -1 \end{bmatrix},$$

$$B_c = \begin{bmatrix} c_{ds,in} & -c_{thd}\beta_{thd} & 0 \\ 0 & 0 & 0 \\ 0 & q_{wf}c_{thd}\beta_{thd} & 0 \\ 0 & 0 & 0 \\ 0 & 0 & 0 \\ c_{ds,in,O_2} & c_{thd,O_2} & c_{sa,O_2} \end{bmatrix},$$

$$E_c = \begin{bmatrix} 0 & 0 \\ c_{w,in} & 0 \\ 0 & 0 \\ 0 & 1 \\ 0 & 0 \\ 0 & 0 \end{bmatrix},$$

$$C_c = \begin{bmatrix} 1 & 0 & 0 & 0 & 0 & 0 \\ 0 & 0 & 1 & 0 & 0 & 0 \\ 0 & 0 & 0 & 0 & 1 & 0 \\ 0 & 0 & 0 & 0 & 0 & 1 \end{bmatrix}.$$

2.3. Model Predictive Control for the BioGrate Boiler

The prime objective of the MPC strategy for the BioGrate boiler is to produce the desired amount of power for the electricity generator and for the hot water network, while maintaining the drum pressure, the fuel bed height and the oxygen content in flue gas at constant values. These variables are affected by exogenous factors such as variation in the moisture content of the fuel and the steam demand. The objective of the control strategy is achieved by utilizing the process model defined by (1)–(6) and by manipulating the air flow rates and the fuel flow rates. In order to design the MPC, a sample data model of the process has been obtained by zero-order hold discretization with sampling time $T_s = 1\,\text{s}$. Hence, from now on, we will refer to the following discrete-time system:

$$
\begin{aligned}
x(k+1) &= Ax(k) + Bu(k) + Ed(k), \\
y(k) &= Cx(k).
\end{aligned}
\tag{7}
$$

According to (7), the k-step ahead prediction is formulated as:

$$
y(k) = CA^k x(0) + \sum_{j=0}^{k-1} H(k-j)u(j),
\tag{8}
$$

where $H(k-j)$ contains the impulse response coefficients. Therefore, using (8), the MPC optimization problem consists in minimizing:

$$
\phi = \frac{1}{2}\sum_{k=1}^{N_p} \|y(k) - r(k)\|_{Q_z}^2 + \frac{1}{2}\|\Delta u(k)\|_{Q_u}^2
\tag{9}
$$

under the constraints:

$$
\begin{aligned}
x(k+1) &= Ax(k) + Bu(k) + Ed(k), & k &= 0,1,\ldots,N_p - 1, \\
y(k) &= Cx(k), & k &= 0,1,\ldots,N_p, \\
u_{\min} &\leq u(k) \leq u_{\max}, & k &= 0,1,\ldots,N_p - 1, \\
\Delta u_{\min} &\leq \Delta u(k) \leq \Delta u_{\max}, & k &= 0,1,\ldots,N_p - 1, \\
y_{\min} &\leq y(k) \leq y_{\max}, & k &= 1,2,\ldots,N_p,
\end{aligned}
$$

where $r(k)$ is the target value and $\Delta u(k) = u(k) - u(k-1)$. The predictions by (8) are formulated as presented in [11] and the MPC regulator problem of (9) is then solved by convex quadric programming algorithms.

The original system (7) is augmented with disturbance dynamics to achieve the offset-free tracking in the presence of model-plant mismatch or unmeasured disturbances [25]. Hence, the extended system is the following:

$$
\begin{bmatrix} x(k+1) \\ \eta(k+1) \end{bmatrix} = \begin{bmatrix} A & B_d \\ 0 & A_d \end{bmatrix}\begin{bmatrix} x(k) \\ \eta(k) \end{bmatrix} + \begin{bmatrix} B \\ 0 \end{bmatrix} u(k) + \begin{bmatrix} E \\ 0 \end{bmatrix} d(k) + \begin{bmatrix} w(k) \\ \xi(k) \end{bmatrix},
\tag{10}
$$

$$
y(k) = \begin{bmatrix} C & C_\eta \end{bmatrix}\begin{bmatrix} x(k) \\ \eta(k) \end{bmatrix} + v(k),
\tag{11}
$$

where $w(k)$ and $v(k)$ are white noise disturbances with zero mean. Thus, the disturbances and the states of the system are estimated as follows:

$$
\begin{bmatrix} \hat{x}(k|k) \\ \hat{\eta}(k|k) \end{bmatrix} = \begin{bmatrix} \hat{x}(k|k-1) \\ \hat{\eta}(k|k-1) \end{bmatrix} + \begin{bmatrix} L_x \\ L_\eta \end{bmatrix}(y(k) - C\hat{x}(k|k-1) - C_\eta \hat{\eta}(k|k-1)),
\tag{12}
$$

and the state predictions of the augmented system (10) are obtained by:

$$\begin{bmatrix} \hat{x}(k+1|k) \\ \hat{\eta}(k+1|k) \end{bmatrix} = \begin{bmatrix} A & B_d \\ 0 & A_d \end{bmatrix} \begin{bmatrix} \hat{x}(k|k) \\ \hat{\eta}(k|k) \end{bmatrix} + \begin{bmatrix} B \\ 0 \end{bmatrix} u(k) + \begin{bmatrix} E \\ 0 \end{bmatrix} d(k). \tag{13}$$

Since $\eta(k)$ are observable, their estimates are used to remove their influence from the controlled variables. The disturbance model is defined by choosing the matrices B_d and C_η. Since the additional disturbance modes introduced by disturbance are unstable, it is necessary to check the detectability of the augmented system. The augmented system (10) is detectable if and only if the nonaugmented system (7) is detectable, and the following condition holds:

$$\text{rank} \begin{bmatrix} I - A & -B_d \\ C & C_\eta \end{bmatrix} = n + n_\eta, \tag{14}$$

where n is the dimension of A and n_η is the dimension of A_d. In addition, if the system is augmented with a number of integrating disturbances n_η equal to the number of the measurements p ($n_\eta = p$) and, if the closed-loop system is stable and constraints are not active at a steady state, there is zero offset in controlled variables.

3. The Data Reconciliation Algorithm for the BioGrate Boiler

The aim of this section is to describe the data reconciliation algorithm and how it works as a fault tolerant component in the steady-state closed-loop system.

By processing the control inputs, the measured disturbances, and the measured outputs of the process—including the faulty one—the data reconciliation algorithm provides the reconciled measurements as the solution of a problem consisting in the minimization of a quadratic cost under a linear equality constraint. In addition to the measurement of the flue gas oxygen content, the other measurements benefit from corrected values. The quadratic cost is a function of the difference between the faulty measurement and the actual measurement (i.e., the fault-free value of the measurement). The linear constraint is the steady-state equation of the regulated plant, where the states which are not directly measurable are replaced by the respective soft-sensors.

In order to formally state the constrained optimization problem, the vector $h(k)$ is defined as:

$$h(k) = \begin{bmatrix} y(k)^\top & \hat{x}_2(k) & \hat{x}_4(k) & u(k)^\top & d(k)^\top \end{bmatrix}^\top, \tag{15}$$

where $y(k)$, $\hat{x}_2(k)$, $\hat{x}_4(k)$, $u(k)$, and $d(k)$, respectively, are the measured outputs, the estimates of $x_2(k)$ and $x_4(k)$, the control inputs, and the measurable disturbances. Likewise, the vector $h^f(k)$ is defined as:

$$h^f(k) = \begin{bmatrix} y^f(k)^\top & \hat{x}_2(k) & \hat{x}_4(k) & u(k)^\top & d(k)^\top \end{bmatrix}^\top, \tag{16}$$

where $y^f(k)$ denotes the faulty measured output or, more precisely, the measured output, where, in particular, the measurement of the oxygen content is affected by the fault occurred in the related sensor: i.e.,

$$y^f(k) = \begin{bmatrix} y_1(k) & y_2(k) & y_3(k) & y_4^f(k) \end{bmatrix}^\top. \tag{17}$$

With the notation introduced in (15)–(17), the cost functional is defined as follows:

$$J = (h(k) - h^f(k))^\top W (h(k) - h^f(k)), \tag{18}$$

where W denotes a positive definite diagonal matrix, whose entries are identified from the indstrial data from a BioPower CHP power plant.

The linear constraint is derived from the process steady-state equations:

$$(A - I) x(k) + B u(k) + E d(k) = 0, \tag{19}$$

where I denotes the identity matrix, through the following reasoning. While the control input $u(k)$ and the disturbance $d(k)$ are directly available, the states must be replaced by their measurement—fault-free or actual ones, respectively—or by their estimates. In particular, the states $x_1(k)$, $x_3(k)$, and $x_5(k)$ are replaced by the fault-free measurements $y_1(k)$, $y_2(k)$, and $y_3(k)$. The state $x_6(k)$ is replaced by the actual measurement $y_4(k)$. Moreover, the states $x_2(k)$ and $x_4(k)$ are replaced by their estimates $\hat{x}_2(k)$ and $\hat{x}_4(k)$. Hence, the linear constraint (19) can be restated in compact form as a function of $h(k)$, provided that a state-space basis transformation T is applied with the aim of re-ordering the state variables consistently with the definition of $h(k)$ and the replacements described above. Namely, with T such that:

$$x'(k) = T^{-1} x(k) = \left[\begin{array}{cccccc} x_1(k) & x_3(k) & x_5(k) & x_6(k) & x_2(k) & x_4(k) \end{array} \right]^{\mathsf{T}},$$

and M is defined by:

$$M = \left[\begin{array}{ccc} (A' - I) & B' & E' \end{array} \right],$$

where $A' = T^{-1} A T$, $B' = T^{-1} B$, and $E' = T^{-1} E$, (19) can be rewritten as:

$$M h(k) = 0. \tag{20}$$

In order to minimize the quadratic cost (18) with respect to the difference $h(k) - h^f(k)$, under the linear equality constraint (20), one can straightforwardly apply the Lagrange multiplier method. Then, the Lagrangian function is:

$$\mathcal{L}((h(k) - h^f(k)), \lambda) = (h(k) - h^f(k))^{\mathsf{T}} W (h(k) - h^f(k)) + \lambda^{\mathsf{T}} M h(k),$$

where λ denotes the vector of the Lagrangian multipliers. Therefore, solving the constrained optimization problem reduces to solving the system of equations:

$$\begin{cases} 2 \left(h(k) - h^f(k) \right)^{\mathsf{T}} W + \lambda^{\mathsf{T}} M = 0, \\ M \left(h(k) - h^f(k) \right) + M h^f(k) = 0, \end{cases}$$

which provides, for the reconciled variable:

$$h^{rec}(k) = h^f(k) - W^{-1} M^{\mathsf{T}} (M W^{-1} M^{\mathsf{T}})^{\dagger} M h^f(k). \tag{21}$$

4. Results

Two case studies are presented to demonstrate the effectiveness of the integrated DR-MPC based strategy. Simulation studies utilise industrial data from a BioPower 5 CHP power plant and the system is identified and implemented in the MATLAB (R2016b, MathWorks Inc., Natick, MA, USA) environment, as described in Section 2.2. The first case study discusses the performance of the DR-MPC during an intermittent oxygen content sensor fault. The second case study analyses the performance of the system during complete failure of the sensor.

4.1. Case Study I

This case study describes the effectiveness of the integrated DR-MPC system during an additive intermittent fault that occurs in the flue gas oxygen content sensor from 501 s to 5400 s. The occurrence of this fault is presented in Figure 3. The effectiveness of the system is demonstrated by comparing the performances of the DR-MPC and an MPC strategy. The faultless "normal" operation of the flue gas oxygen content sensor is also demonstrated with the MPC strategy for both case studies.

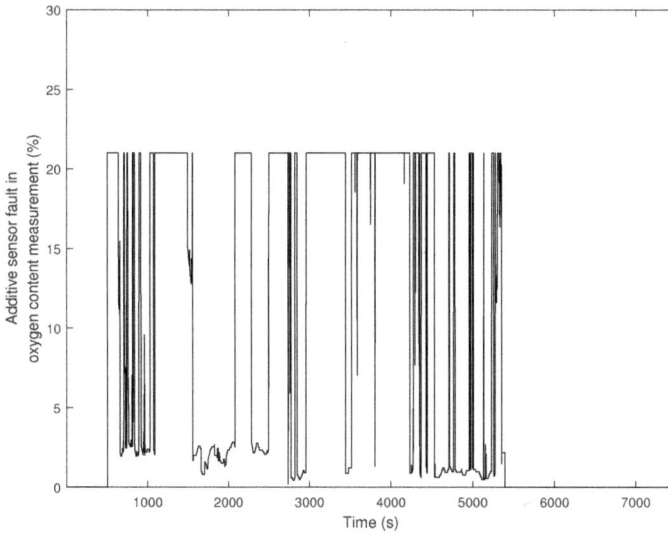

Figure 3. Additive intermittent sensor fault in flue gas oxygen content measurement.

In both simulation setups, the weighting matrices of the MPC controller are $Q_z = \text{diag}\{0.001, 0.001, 0.1, 0.01\}$ and $Q_u = \text{diag}\{0.1, 0.1, 0.1\}$, while the boundary conditions are $u_{min} = [0\ 0\ 0]^\top$, $u_{max} = [5\ 20\ 20]^\top$, $\Delta u_{min} = [-0.8\ -0.8\ -0.8]^\top$, $\Delta u_{max} = [0.8\ 0.8\ 0.8]^\top$, $y_{min} = [0\ 0\ 0]^\top$ and $y_{max} = [1\ 35\ 55\ 30]^\top$. The setpoint values for combustion power, drum pressure, fuel bed height and oxygen content are 15 MW, 50 bar, 0.5 m and 4%, respectively. The matrix W in the data reconciliation algorithm is defined as $W = \text{diag}\{2.9914, 3.0297, 0.0605, 3.0216, 2.9852, 2.9808, 3.0000, 3.0000, 3.0000, 3.0000, 3.0000\} \times 10^6$.

The simulation is performed for a period of 7500 s. Figures 4 and 5 present the resulting behaviour of controlled outputs and control inputs, respectively. The results demonstrate that the closed-loop performance of the MPC strategy (black lines) is significantly deteriorated by the fault as it introduced undesirable oscillations to the control inputs and controlled outputs: The calculated soft-sensor value of the combustion power [24] greatly differs from its nominal value, due to the faults in the flue gas oxygen content measurement. In contrast, the integrated DR-MPC (red lines) was able to notably reduce the effects of this fault on the considered process inputs and outputs. In other words, with the data reconciliation, the fault does not affect the steady-state behavior of the system. The integrated DR-MPC works as follows: Firstly, Equation (21) is used to calculate the reconciled measurements shown in Figure 4. Secondly, Equation (12) is utilized to estimate the states and the disturbances of the system. Thirdly, the optimization problem (9) is solved, giving the new inputs as illustrated in Figure 5. Finally, new state predictions are given by Equation (13). The oscillation of the fuel bed height is caused by the smaller weight of 0.001 used by the MPC for the fuel flow and primary air flow rates in comparison to 0.1 of the drum pressure. Moreover, there is the small degradation of the measurements with the DR-MPC in comparison to the 'normal' faultless situation.

The flue gas oxygen content measurement is still needed: in its normal operation, it measures oxygen content fast (less than second) that is needed for fast dynamics of excess air control and to prevent pollution.

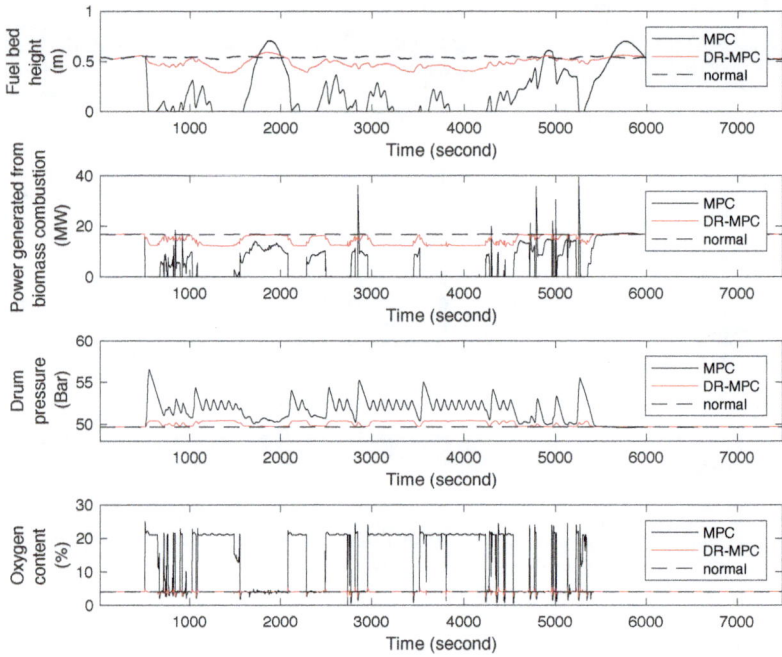

Figure 4. Case I—Closed-loop simulation results: Controlled outputs.

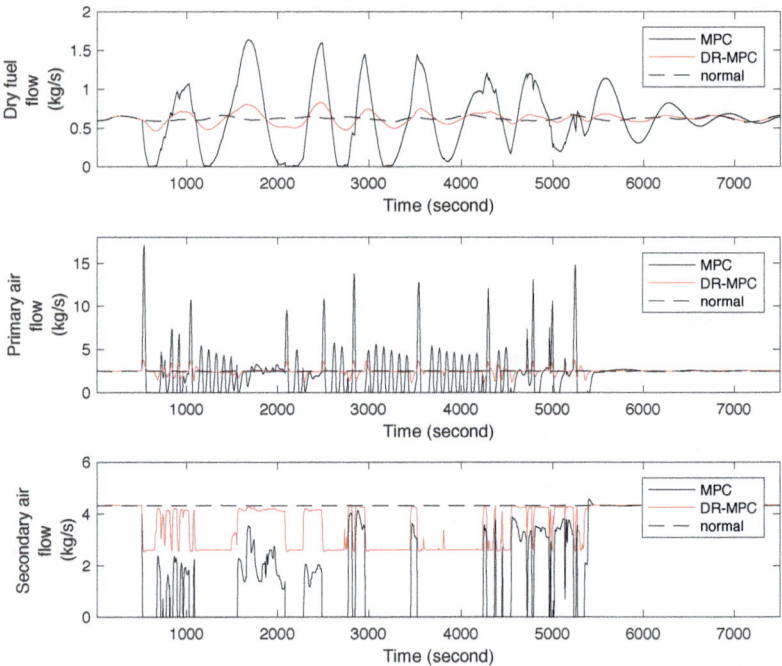

Figure 5. Case I—Closed-loop simulation results: Control inputs.

4.2. Case Study II

In this case study, the performance of the integrated DR-MPC strategy is evaluated during the occurrence of an oxygen content sensor failure. The failure occurs between 500 s and 1000 s. The tuning matrices of the MPC controller, the weighting matrix of the data reconciliation algorithm, and the input and output constraints are given the same values of the previous case study. Figures 6 and 7 show the performance of the integrated DR-MPC.

The MPC strategy cannot perform under the absence of measurement data from the oxygen content sensor due to the failure. Instead, when the DR-MPC strategy is active, the reconciliation algorithm provides a reconstructed signal for the oxygen content measurement. As a result, the closed-loop system is prevented from being unstable. When comparing DR-MPC with the other available methods, usually the controller reconfiguration is needed.

In summary, the case studies demonstrated that the fault effects on the BioPower process, originating from the oxygen content sensor malfunction, can be effectively minimised by integrating the data reconciliation method into the MPC strategy.

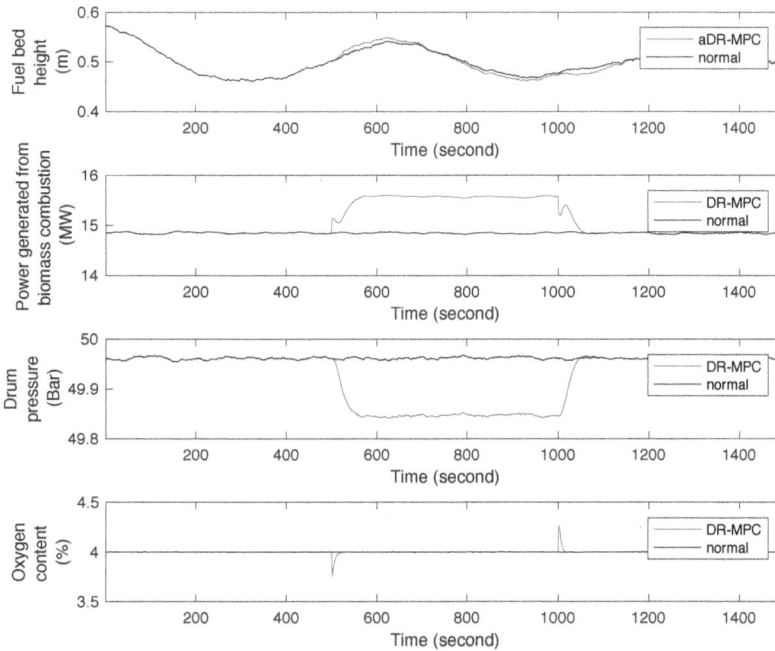

Figure 6. Case II—Closed-loop simulation results: Controlled outputs.

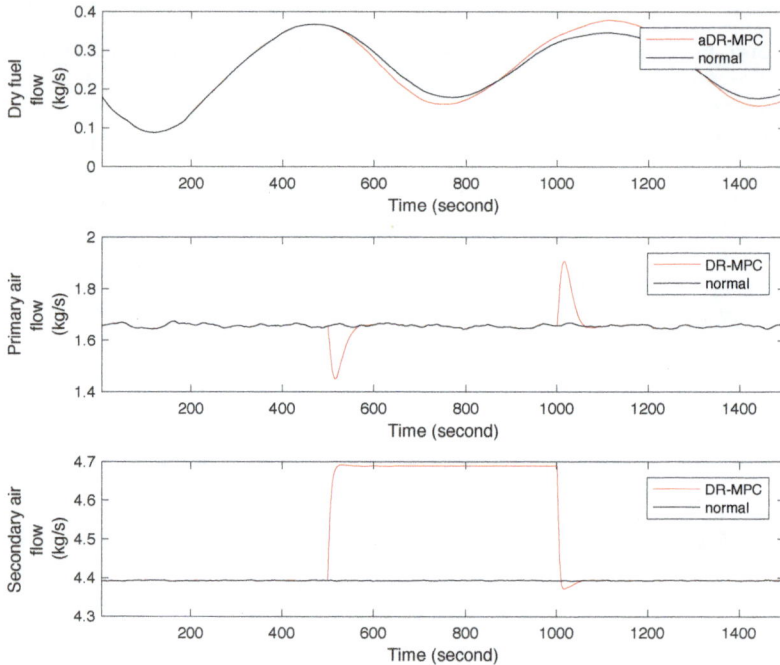

Figure 7. Case II—Closed-loop simulation results: Control inputs.

5. Conclusions

In this paper, we have introduced an integrated DR-MPC strategy to achieve a control system for the BioPower 5 CHP plant with the property of being tolerant to faults or a complete breakdown of the flue gas oxygen content sensor.

The DR-MPC strategy presented in this work was based on a more detailed model of the BioGrate boiler, which also includes the dynamics of the oxygen content in the flue gas. The data reconciliation algorithm provided the reconciled measurement as the solution of a problem consisting of the minimization of a quadratic cost under a linear equality constraint in the MPC strategy.

The effectiveness of the DR-MPC strategy has been shown by considering both an intermittent fault in the flue gas oxygen content sensor and its failure. The results showed that the closed-loop performance of the MPC strategy was significantly deteriorated by the fault as it introduced undesirable oscillations to the control inputs and controlled outputs. In contrast, the integrated DR-MPC was able to notably reduce the effects of this fault on the considered process inputs and outputs. In the case of the complete breakdown of the oxygen content sensor, the reconciliation algorithm provided a reconstructed signal for the oxygen content measurement .

Acknowledgments: This work was done as a part of the "Integrated condition-based control and maintenance" (ICBCOM) consortium project of the Finnish Funding Agency for Technology and Innovation (TEKES).

Author Contributions: Palash Sarkar conceived and designed the data reconciliation algorithm; Jukka Kortela identified the models and developed the model predictive control; All authors reviewed the work and improved the paper continuously.

Conflicts of Interest: The authors declare no conflict of interest.

Abbreviations

CHP	Combined heat and power
CSFA	Cold source flow adjustment
DR-MPC	Data-reconciliation model predictive control
LMPC	Linear model predictive control
MPC	Model predictive control
MSW	Municipal solid waste
NOx	Nitrogen oxides
PCA	Principal component analysis

References

1. Aslani, A.; Helo, P.; Naaranoja, M. Role of renewable energy policies in energy dependency in Finland: System dynamics approach. *Appl. Energy* **2014**, *113*, 758–765.
2. Kortela, J.; Jämsä-Jounela, S.L. Model predictive control utilizing fuel and moisture soft-sensors for BioPower 5 combined heat and power (CHP) plant. *Appl. Energy* **2014**, *131*, 189–200.
3. Gölles, M.; Reiter, S.; Brunner, T.; Dourdoumas, N.; Obernberger, I. Model based control of a small-scale biomass boiler. *Control Eng. Pract.* **2014**, *22*, 94–102.
4. Leskens, M.; van Kessel, L.B.M.; Bosgra, O.H. Model predictive control as a tool for improving the process operation of MSW combustion plants. *Waste Manag.* **2005**, *25*, 788–798.
5. United States Environmental Protection Agency. *National Emission Standards for Hazardous Air Pollutants for Major Sources: Industrial, Commercial, and Institutional Boilers and Process Heaters*; United States Environmental Protection Agency: Washington, DC, USA, 2015.
6. Benyó, I. Cascade Generalized Predictive Control—Applications in Power Plant Control. Ph.D. Thesis, University of Oulu, Oulu, Finland, 2006.
7. Klimánek, D.; Sulc, B. Evolutionary detection of sensor discredibility in control loops. In Proceedings of the 31st Annual Conference of IEEE Industrial Electronics Society (IECON 2005), Raleigh, NC, USA, 6–10 November 2005; pp. 1–6.
8. Dunia, R.; Qin, S.J. Joint diagnosis of process and sensor faults using principal component analysis. *Control Eng. Pract.* **1998**, *6*, 457–469.
9. Qin, S.J.; Li, W. Detection, identification, and reconstruction of faulty sensors with maximized sensitivity. *AIChE J.* **1999**, *45*, 1963–1976.
10. Kadlec, P.; Gabrys, B.; Strandt, S. Data-driven soft sensors in the process industry. *Comput. Chem. Eng.* **2009**, *33*, 795–814.
11. Kortela, J.; Jämsä-Jounela, S.L. Fault-tolerant model predictive control (FTMPC) for the BioGrate boiler. In Proceedings of the 2015 IEEE 20th Conference on Emerging Technologies & Factory Automation (ETFA), Luxembourg, 8–11 September 2015.
12. Havlena, V.; Findejs, J. Application of model predictive control to advanced combustion control. *Control Eng. Pract.* **2005**, *13*, 671–680.
13. Wang, F.; Huang, Q.; Liu, D.; Yan, J.; Cen, K. Improvement of load-following capacity based on the flame radiation intensity signal in a power plant. *Energy Fuels* **2008**, *22*, 1731–1738.
14. Wang, W.; Zeng, D.; Liu, J.; Niu, Y.; Cui, C. Feasibility analysis of changing turbine load in power plants using continuous condenser pressure adjustment. *Energy* **2014**, *64*, 533–540.
15. Wang, W.; Liu, J.; Zeng, D.; Niu, Y.; Cui, C. An improved coordinated control strategy for boiler-turbine units supplemented by cold source flow adjustment. *Energy* **2015**, *88*, 927–934.
16. Kortela, U.; Lautala, P. A new control concept for a coal power plant. *Control Sci. Technol. Prog. Soc.* **1982**, *6*, 3017–3023.
17. Leibman, M.J.; Edgar, T.F.; Lasdon, L.S. Efficient data reconciliation and estimation for dynamic process using nonlinear programming techniques. *Comput. Chem. Eng.* **1992**, *16*, 963–986.
18. Abu-el-zeet, Z.H.; Roberts, P.D.; Becerra, V.M. Enhancing model predictive control using dynamic data reconciliation. *AIChE J.* **2002**, *48*, 324–333.
19. Schladt, M.; Hu, B. Soft sensors based on nonlinear steady-state data reconciliation in the process industry. *Chem. Eng. Process. Process Intensif.* **2007**, *46*, 1107–1115.

20. Amand, T.; Heyen, G.; Kalitventzeff, B. Plant monitoring and fault detection: Synergy between data reconciliation and principal component analysis. *Comput. Chem. Eng.* **2001**, *25*, 501–507.
21. Korpela, T.; Suominen, O.; Majanne, Y.; Laukkanen, V.; Lautala, P. Robust data reconciliation of combustion variables in multi-fuel fired industrial boilers. *Control Eng. Pract.* **2016**, *56*, 101–115.
22. Yin, C.; Rosendahl, L.A.; Kær, S.K. Grate-firing of biomass for heat and power production. *Prog. Energy Combust. Sci.* **2008**, *34*, 725–754.
23. Wärtsilä Biopower. *Wärtsilä Power Plants: Bioenergy Solutions*; Wärtsilä: Vaasa, Finland, 2005.
24. Kortela, J.; Jämsä-Jounela, S.L. Modeling and model predictive control of the BioPower combined heat and power (CHP) plant. *Electr. Power Energy Syst.* **2015**, *65*, 453–462.
25. Pannocchia, G.; Rawlings, J.B. Disturbance models for offset-free model-predictive control. *AIChE J.* **2003**, *49*, 426–437.

![energies logo] *energies*

MDPI

Article

Computational Model of a Biomass Driven Absorption Refrigeration System

Munyeowaji Mbikan [†] **and Tarik Al-Shemmeri** [*,†]

Department of Mechanical Engineering, Staffordshire University, College Road, Stoke-On-Trent ST4 2DE, UK; munyeowaji.mbikan@staffs.ac.uk
* Correspondence: T.T.Al-Shemmeri@staffs.ac.uk; Tel.: +44-178-535-3335
† These authors contributed equally to this work.

Academic Editor: Vasily Novozhilov
Received: 12 December 2016; Accepted: 9 February 2017; Published: 16 February 2017

Abstract: The impact of vapour compression refrigeration is the main push for scientists to find an alternative sustainable technology. Vapour absorption is an ideal technology which makes use of waste heat or renewable heat, such as biomass, to drive absorption chillers from medium to large applications. In this paper, the aim was to investigate the feasibility of a biomass driven aqua-ammonia absorption system. An estimation of the solid biomass fuel quantity required to provide heat for the operation of a vapour absorption refrigeration cycle (VARC) is presented; the quantity of biomass required depends on the fuel density and the efficiency of the combustion and heat transfer systems. A single-stage aqua-ammonia refrigeration system analysis routine was developed to evaluate the system performance and ascertain the rate of energy transfer required to operate the system, and hence, the biomass quantity needed. In conclusion, this study demonstrated the results of the performance of a computational model of an aqua-ammonia system under a range of parameters. The model showed good agreement with published experimental data.

Keywords: absorption refrigeration; modelling; aqua-ammonia; biomass; renewable energies

---

1. Introduction

Refrigeration and cooling demand, either for indoor climate control or preservation and chilling purposes, cuts across every region of the globe. Conventional refrigeration and air-conditioning systems are high energy intensive systems [1,2]. In cases where a vapour compression refrigeration cycle (VCRC) applies, and a conventional energy source is used, the resulting climatic impact is two-fold—the effect of carbon dioxide emission from fossil fuel, and the ozone depletion effect from Hydrochlorofluorocarbons (HCFCs) and Hydrofluorocarbons (HFCs) refrigerants used in these systems. Research and development of technologies, aimed at reducing both the energy consumption and environmental impact of these systems, is on-going [3,4]. Ground source heating and cooling is one such emerging technology [5]. This involves exploring the ground's huge capacity to receive and supply heat for space heating and cooling. Others are solar powered refrigeration systems (mechanical compression or thermal sorption) [6,7], ground coupled solar panel cooling [8], and passive heating and cooling [1]. Passive heating and cooling involves strategic control of thermal conditions in buildings to improve indoor atmospheric conditions with near-zero energy consumption. The technology promises a huge energy benefit. However, its impact falls outside heavy refrigeration demands, and its application to existing buildings will be of high capital intensity, where possible. The vapour absorption refrigeration cycle (VARC) is a thermally driven system with very low electrical energy demand compared with the VCRC. VARC uses ozone depletion-free and environmentally friendly refrigerants. The thermal requirement of the VARC system can be met by inexpensive sources such as waste heat from industrial processes, and exhaust heat from engines, solar and biomass.

As part of a trigeneration system, VARC is driven by hot water (Figure 1) generated in the boiler and split in three ways by three separate valves leading to the three applications. Operation of the absorption refrigeration systems with solar heat was studied by [7,9–14], and operation with waste heat recovery was studied by [15–18]. However, literature on biomass driven absorption cooling is scarce. In this paper, a computer model for the evaluation of a single-stage aqua-ammonia vapour absorption refrigeration system is developed. An important and new novelty aspect of this model is that the performance of the refrigeration cycle is coupled with the biomass fuel necessary to achieve it, and also, the determination of the actual value of the refrigerant vapour at the generator exit (state point 7, Figure 2) using the vapour split Rachford–Rice equation [19] which previous researchers have assumed to be 100%; whereas, in this paper, it is actually determined using a combination of Equations (17)–(19).

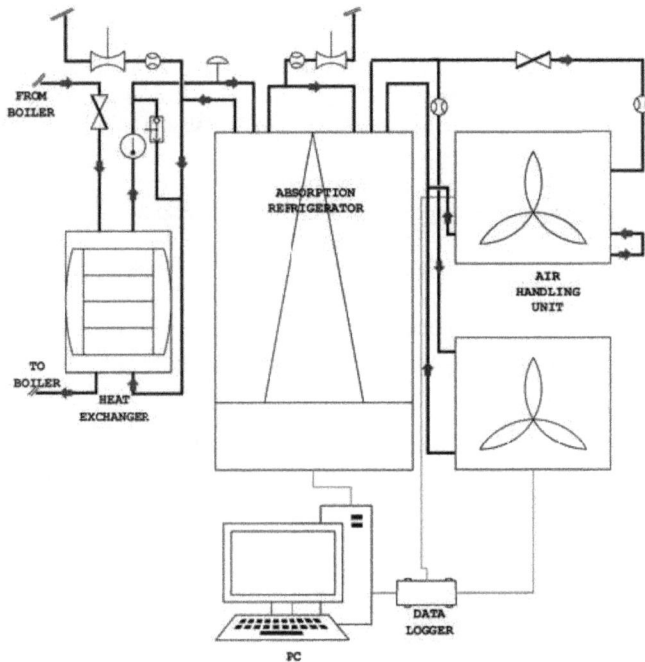

Figure 1. Absorption refrigeration test rig.

2. Biomass Source to Drive Absorption Refrigeration

Biomass refers to fuels produced directly or indirectly from organic materials. This includes plants, agricultural, domestic and industrial wastes. Biomass is carbon neutral and has been in use for energy production since the beginning of time; for many parts of the world, it is still the main source of heat. The growing production of biomass in recent times falls on the back of fossil fuel market volatility, environmental considerations, security concerns and energy supply diversification. Modern technologies have improved the efficiency of biomass fuels and enhance its cleanness far more than the traditional uses which are mainly open fires. As the fourth largest energy source, it is widely recognised as a potential sustainable global source of energy. The main annual global production of biomass is estimated to be equivalent to the 4500 EJ of the annual global energy capture contributing about 10% of the global primary energy demand mainly in the form of traditional non-commercial biomass [20]; but now, biomass is known to be applied on a large scale with biomass boilers of over 500 MWth [21]; the potential to produce 50% of Europe's total energy requirement from purposefully grown biomass such as dedicated energy crops and short rotation coppice (SRC) is proposed [22]. Most renewable energy sources depend on back-up power or a battery bank to forestall intermittent supply [23]. Biomass is one renewable source that can be used in many applications with hardly any need for back-up. This versatility is further enhanced by the emerging conversion technologies and the improvement of the existing ones, making it a flexible energy source [24]. However, the economic and environmental gains of biomass fuels are influenced by the proximity of the field-storage, and mode of transportation to the point of utilsation [25]. Biomass used for fuels is categorized into two main groups [26].

Woody Biomass:

- Forest residues;
- Wood waste;
- Crop residues;
- Wood crops (SRC, Willow, and Miscanthus).

Non-Woody Biomass:

- Animal wastes;
- Industrial and municipal wastes;
- High energy crops;
- Algae—a huge aquatic biomass source, with water covering about 75% of the earth.

Woody biomass is considered in this paper. The demand for biomass will vary for different end users but will generally depend on:

- Boiler capacity;
- Boiler efficiency;
- Operating hours;
- Type and availability of biomass.

3. Modelling of the Aqua-Ammonia Refrigeration System

Figure 2 is a schematic of the VARC. The refrigerant vapour leaving the evaporator at state 10 (low temperature and pressure side) is absorbed by a weak solution of ammonia coming from a pressure reduction valve at state point 6. After absorbing ammonia vapour, it becomes a strong solution with respect to the concentration of ammonia in solution, and leaves the absorber at state 1 as a saturated liquid—the enthalpy of condensation being rejected and hence, the heat rejection Q_a in the absorber. A liquid pump increases the pressure of the strong solution from the evaporator pressure (p_e) to the condenser pressure (p_c) through the solution heat exchanger where it is preheated by the warm weak solution and enters the generator at state 3. The strong solution is heated in the generator to liberate the refrigerant. The weak solution leaves the generator at state 4 through the solution heat exchanger and leaves the solution expansion valve (pressure reduction valve) at state 6 back to the absorber. The refrigerant enters the condenser at state 7 at high pressure p_c and the heat Q_c is rejected to the surroundings by condensation at higher temperature. The liquid refrigerant leaves the condenser at state 8 and undergoes isentropic expansion, and enters the evaporator at state 9 where heat Q_e is added from the space being cooled at low pressure, causing the refrigerant to boil, and the vapour leaves to the absorber at state 10, where it is absorbed by the weak solution.

Based on Figure 2, $\dot{m}_1$, $\dot{m}_6$ and $\dot{m}_{10}$ (kg/s) are respectively, the flow rates of the strong solution, weak solution and the refrigerant, and x_1, x_6 and x_{10} are ammonia mole fraction in the weak solution, strong solution and the refrigerant. h_i (kJ/kg) is the state point enthalpy $i = 1, 2, \ldots, n$ is the state point. Q_i (kW) is the energy rate of change, where $i = A, C, E, G$ for the absorber, condenser, evaporator and generator, respectively. T_{jk} is the flow temperature where j is the component initial and k indicates flow direction—i for inlet and o for outlet. COP_{ref}—the performance coefficient of the refrigeration cycle—is the ratio of the heat supplied in the generator to the rate of heat transfer to the system from the surroundings (refrigeration capacity).

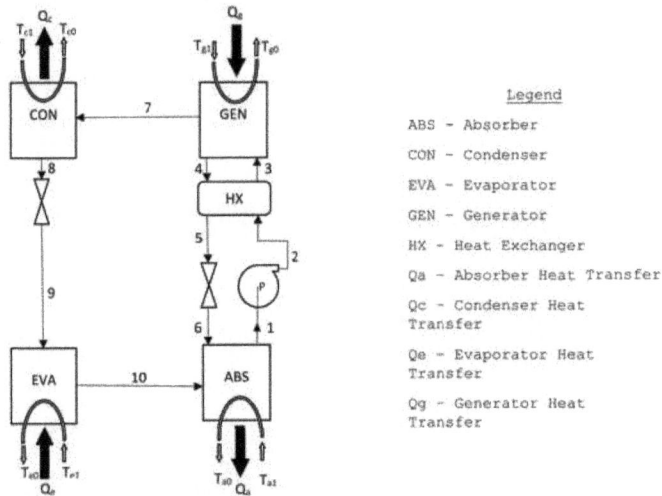

Figure 2. Schematic of the single-stage aqua-ammonia vapour absorption refrigeration system.

Mass conservation—overall:

$$\dot{m}_1 = \dot{m}_6 + \dot{m}_{10} \tag{1}$$

Mass conservation—ammonia:

$$\dot{m}_1 x_1 = \dot{m}_6 x_6 + \dot{m}_{10} x_{10} \tag{2}$$

Energy conservation:

$$\dot{m}_1 h_1 = \dot{m}_6 h_6 + \dot{m}_{10} h_{10} \tag{3}$$

Combining Equations (2) and (3):

$$x_1 = x_6 + (\dot{m}_{10}/\dot{m}_1)(x_{10} - x_6)$$

Circulation ratio (f):

$$f = \dot{m}_1/\dot{m}_{10} \tag{4}$$

$$x_1 = x_6 + (1/f)(x_{10} - x_6)$$

Similarly,

$$h_1 = h_6 + (1/f)(h_{10} - h_6)$$

Energy balance—absorber:

$$Q_A = \dot{m}_{10} h_{10} + \dot{m}_6 h_6 - \dot{m}_1 h_1 \tag{5}$$

From Equations (4) and (5),

$$Q_A = \dot{m}_{10}[h_{10} + (f-1)h_6] - \dot{m}_1 h_1 \tag{6}$$

Energy balance—heat exchanger:

$$Q_{hx} = \dot{m}_6(h_4 - h_6) = \dot{m}_1(h_3 - h_2) - w_p \tag{7}$$

where $\dot{m}_r = \dot{m}_{10} = \dot{m}_7$—refrigerant mass flow, $x_1 = x_2 = x_3 = X_{ss}$—strong solution (with respect to the ammonia concentration in solution) and $x_4 = x_5 = x_6 = X_{ws}$—the weak solution.

Assume pump work $w_p \approx 0$ (negligible), the enthalpy at the generator inlet:

$$h_3 = h_2 + (^1/_f)(h_4 - h_5) \tag{8}$$

Energy balance—generator:

$$Q_G + \dot{m}_1 h_3 = \dot{m}_{10} h_7 + \dot{m}_6 h_4$$

$$Q_G = \dot{m}_{10}[h_7 + (f-1)h_4] - \dot{m}_1 h_1 \tag{9}$$

Energy balance—condenser:

$$Q_C = \dot{m}_{10}(h_7 - h_8) \tag{10}$$

Energy balance—evaporator:

$$Q_E = \dot{m}_{10}(h_{10} - h_8) \tag{11}$$

Coefficient of performance:

$$COP_{ref} = \frac{Q_E}{Q_G + w_p} = \frac{Q_E}{Q_G} \tag{12}$$

3.1. Governing Equations

In order to fix the thermodynamic state for a compressible binary solution, the composition is required in addition to the two independent properties—temperature and pressure. The bubble point (T_b) and dew point (T_d) temperatures are calculated from the correlations developed by Patek and Klomfar [27]—Equation (13).

$$T_b(p, x) = T_0 \sum_i a_i (1-x)^{m_i} [ln\frac{p_0}{p}]^{n_i} \tag{13}$$

$$T_d(p, y) = T_0 \sum_i a_i (1-y)^{\frac{m_i}{4}} [ln\frac{p_0}{p}]^{n_i} \tag{14}$$

$$x = f(T, p) \tag{15}$$

$$y = f(T, p) \tag{16}$$

where T is temperature (K), T_0, T_b and T_d are the reference, bubble and dew points temperatures, respectively. The coefficients are shown in Tables 1 and 2. x and y are the ammonia mole fraction in the liquid and vapour phases, respectively.

$p_0 = 2$ MPa and $T_0 = 100$ K. The reduced thermodynamic properties are as follows:

$$T_0 = 100 \text{ K}; p_b = 10 \text{ bar}; T_r = \frac{T}{T_0}; p_r = \frac{p}{p_b}; R = 8.314 \text{ J/kmol}.$$

The composition of the solution phase is estimated from Equation (15), developed from Equation (13). To estimate the composition in the vapour phase, the vapour split, V/F, from Equation (18) is calculated from the Rachford–Rice [19] flash Equation (17) numerically using Matlab (fzero) function. The vapour composition is obtained from Equation (19).

$$\sum_{i=1}^{S}(y_i - x_i) = \sum_{i=1}^{S}\frac{(K_i - 1)z_i}{((K_i - 1)V + 1)} = 0, 0 \leqslant \frac{V}{F} \leqslant 1 \tag{17}$$

$$x_i = \frac{z_i}{1 + \frac{V}{F(K_i-1)}} \tag{18}$$

$$y_i = K_i x_i \tag{19}$$

K-value defines the ratio of each component:

$$K_i = \frac{y_i}{x_i} \tag{20}$$

F in the above Equations represents the feed with composition z_i. V (with composition y_i) is the vapour product and L (with composition x_i) is the liquid product.

Table 1. Coefficients for Equation (13).

i	m_i	n_i	a_i
1	0	0	$+0.322302 \times 10^1$
2	0	1	-0.384206
3	0	2	$+0.460965 \times 10^{-1}$
4	0	3	-0.378945×10^{-2}
5	0	4	$+0.135610 \times 10^{-3}$
6	1	0	$+0.487755$
7	1	2	-0.120108
8	1	2	$+0.106154 \times 10^{-1}$
9	2	3	-0.533589×10^{-3}
10	4	0	$+0.785041 \times 10^1$
11	5	0	-0.115941×10^1
12	5	1	-0.523150×10^2
13	6	0	$+0.489596 \times 10^{-1}$
14	13	1	0.421059×10^{-1}

Table 2. Coefficients for Equation (14).

i	m_i	n_i	a_i
1	0	0	$+0.324004 \times 10^1$
2	0	1	-0.395920
3	0	2	-0.434624×10^{-1}
4	0	3	-0.218943×10^{-2}
5	1	0	-0.143526×10^1
6	1	1	$+0.105256 \times 10^1$
7	1	2	-0.719281×10^{-1}
8	2	0	$+0.122362 \times 10^2$
9	2	1	-0.224368×10^1
10	3	0	-0.210780×10^2
11	3	1	0.110834×10^1
12	4	0	$+0.145399 \times 10^2$
13	4	2	$+0.644312$
14	5	2	-0.221264×10^1
15	5	2	-0.756266
16	6	0	-0.135529×10^1
17	7	2	$+0.183541$

The Equations for the enthalpy and the Gibbs free energy equation [28–30] are as follows:

$$h = -RT_b T_r{}^2 [\frac{\partial}{\partial T_r}\left(\frac{G_r}{T_r}\right)] \tag{21}$$

$$G = h_0 - TS_0 + \int_{T_0}^{T} C_p dT + \int_{p_0}^{p} v dp - T \int_{T_0}^{T} \left(\frac{C_p}{T}\right) dT \tag{22}$$

The correlation for the heat capacity at constant pressure Cp and volume v are given below:

$$C_{pl} = b_1 + b_2 T + b_3 T^2 \tag{23}$$

$$C_{pg} = d_1 + d_2 T + d_3 T^2 - T \int_{p_0}^{p} \left(\frac{\delta^2 v}{\delta T^2}\right) dP \tag{24}$$

$$v_l = a_1 + a_2 p + a_3 T^2 + a_4 T^2 \tag{25}$$

$$v_g = \frac{RT}{p} + c_1 + \frac{c_2}{T^3} + \frac{c_3}{T^{11}} + \frac{c_4 p^2}{T^{11}} \tag{26}$$

Equations (21)–(22) were solved with Mathcad using the correlations (23)–(26), to obtain (27) and (28). Equations (29) and (30) are the liquid (h_{liquid}) and vapour (h_{vapour}) phase enthalpy. The coefficients [28] are shown in Tables 3 and 4.

Table 3. Coefficients for Equations (27) and (28).

Coefficients	Ammonia	Water
a_1	3.971423×10^{-2}	2.748796×10^{-2}
a_2	-1.790557×10^{-5}	-1.016665×10^{-5}
a_3	-1.308905×10^{-2}	-4.452025×10^{-3}
a_4	3.752836×10^{-3}	8.389264×10^{-4}
b_1	1.634519×10^{1}	1.214557×10^{1}
b_2	-6.508119	-1.898065
b_3	1.448937	2.911966×10^{-2}
c_1	-1.049377×10^{-2}	2.136131×10^{-2}
c_2	-8.288224	-3.169291×10^{1}
c_3	-6.647257×10^{2}	-4.631611×10^{4}
c_4	-3.045352×10^{3}	0.0
d_1	3.673647	4.019170
d_2	9.989629×10^{-2}	-5.175550×10^{-2}
d_3	3.617622×10^{-2}	1.951939×10^{-2}
h_0^l	4.87853	21.821141
h_0^v	26.468879	60.965058
T_{ro}	3.2252	3.0705
p_{ro}	2.0000	3.0000

Table 4. Coefficients for Equation (30).

Coefficients		Coefficients	
e_1	-41.733398	e_9	0.387983
e_2	0.02414	e_{10}	0.004772
e_3	6.702285	e_{11}	-4.648107
e_4	-0.11475	e_{12}	0.836376
e_5	63.608968	e_{13}	-3.553627
e_6	-62.490768	e_{14}	0.000904
e_7	1.761064	e_{15}	21.361723
e_8	0.008626	e_{16}	-20.736547

$$h_{liquid} = -RT_b \left(-h_0^l + b_1 (T_{r0} - T_r) + \frac{b_2}{2}\left(T_{r0}^2 - T_{r0}^3\right) + (a_4 T_r^2 - a_1)(p_r - p_{r0}) - \frac{a_2}{2}\left(p_r^2 - p_{r0}^2\right)\right) \tag{27}$$

$$h_{vapour} = - RT_b \left[- h_0^v + d_1 T_{r0} + \frac{d_2}{2} \left(T_r^2 - T_{r0}^2 \right) + \frac{d_3}{3} \left(2T_r^3 - T_{r0}^3 \right) - d_1 T_r - d_2 T_{r0}^2 - \frac{d_3}{2} \left(T_r^2 - T_{r0}^2 \right) \right.$$
$$\left. - c_1 (p_r - p_{r0}) + c_2 \left(\frac{-4p_r}{T_r^3} + \frac{4p_{r0}}{T_{r0}^3} \right) + c_3 \left(\frac{-12p_r}{T_r^{11}} + \frac{12p_{r0}}{T_{r0}^{11}} \right) + \frac{c_4}{3} \left(\frac{-12p_r}{T_r^3} + \frac{12p_{r0}}{T_{r0}^3} \right) \right] \tag{28}$$

Equations (29) and (31):

$$h^l = (1 - x)h_{liquid,H_2O} + x h_{liquid,NH_3} + h^E \tag{29}$$

h^E is the energy of mixing as shown in Equation (30).

$$h^E = e_1 + e_2 p + (e_3 + e_4 p) T + \frac{e_5}{T} + \frac{e_6}{T^2} + (2x - 1)$$
$$\left(e_7 + e_8 p + (e_9 + e_{10} p) T + \frac{e_{11}}{T} + 3\frac{e_{12}}{T^2} + (2x - 1)^2 \left(e_{13} + e_{14} p + \frac{e_{15}}{T} + \frac{e_{16}}{T^2} \right) \right) \tag{30}$$

Similarly, the vapour phase enthalpy is given as shown in (31) where y' is the ammonia vapour fraction.

$$h^v = (1 - y')h_{vapour,H_2O} + y' h_{vapour,NH_3} \tag{31}$$

The computation procedure for the analysis was implemented in Matlab. The coefficient of performance (*COP*), and the rate of energy due to the absorber, condenser, and generator are computed. The refrigeration capacity (Q_E) is calculated (if it is not given as an input parameter) from the refrigerant mass flow; and the refrigerant mass flow is calculated if the Q_E is given instead.

3.2. Boiler Heat Exchanger Description

Unlike the VCRC which relies on the power of the compressor—a high electrical energy consuming device—the VARC combines the affinity of the binary solution, a comparatively low energy consuming solution pump and thermal energy for the regeneration of the refrigerant and transport of fluid around the refrigeration cycle. The regenerating unit of the VARC can be operated by low grade heat from a variety of thermal sources including solar collectors, waste process heat from an exhaust and hot water or steam generating processes. Biomass solid fuels can be used to provide the required energy via the production of hot water [17]. The quantity of biomass required to operate a VARC for a given load is estimated based on the calorific value of the species. The heat generated in the boiler, from combusting biomass fuel, is used to heat water flowing through its circuitry and provides energy that is transferred through a heat exchanger to the generator circuit. A counter flow heat exchanger configuration (Figure 3) is used to enhance heat transfer of the hot water from the boiler to the regenerating unit.

The amount of heat required to raise the temperature of a substance is given as:

$$q = mC_p \frac{dT}{t} \tag{32}$$

For a continuous flow rate of change,

$$q = \dot{m} C_p \frac{dT}{t} \tag{33}$$

For a set temperature of hot water coming from the boiler and assuming that for the heat exchanger:

- Steady operating conditions apply;
- The heat exchanger is well insulated so that the heat loss to the surroundings is negligible;
- The kinetic and potential energy changes of the fluid are very small and negligible;
- No fouling conditions apply;
- The properties of the fluid remain the same throughout the process.

the rate of heat required for the generator, for a set inlet and exit temperature, is estimated, and hence, the biomass quantity required for the load and operating duration:

$$\dot{Q}_{hx} = \dot{m}C_p\,(T_{hi} - T_{ho}) \tag{34}$$

and

$$\dot{Q}_G = \dot{m}C_p\,(T_{co} - T_{ci}) \tag{35}$$

For a heat exchanger effectiveness of E,

$$\dot{Q}_{hx} = \dot{m}C_p E\,(T_{hi} - T_{ho}) \tag{36}$$

the rate of heat for Equations (34)–(36) is assumed to be same.

For boiler efficiency of η_B and full load heating hours' equivalent (FLHE), the energy required E_R and the biomass fuel quantity required B_Q, is estimated thus:

$$E_R = \frac{Q_{hx} * FLHE}{\eta_B} \tag{37}$$

where Q_{hx} is the rate of heat transfer from the heat exchanger.

$$B_Q = Q_{hx} * F_d * E_R \tag{38}$$

F_d is the fuel density of the biomass species.

The quality of combustion in the boiler depends on several factors including design, air velocity, air temperature, etc. The influence of the design factor which can be evaluated by Computational Fluid Dynamics (CFD) analysis [31], is not directly considered but assumed to be built into the boiler efficiency and the effectiveness of the heat exchanger.

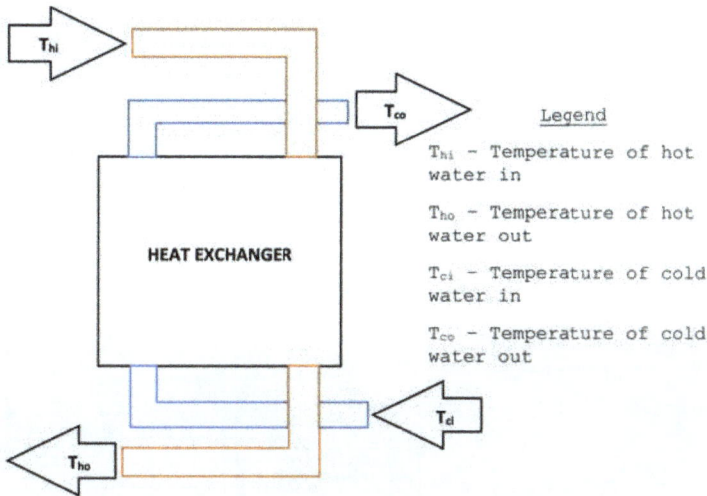

Figure 3. Heat exchanger.

4. Result and Discussion

For a given operating condition, the quantity of biomass in tonnes/kW required would vary in accordance with the energy density of the species. To perform an estimation calculation, the rate of heat transfer to the generator is first calculated. This is done by the computer simulation as described above.

4.1. Biomass Quantity Required

The rate of energy transfer due to the heat exchanger (Figure 3), the generator and the heat-loss is calculated as follows [13]:

$$Q_G = \dot{m}_c C_{pc} (T_{ci} - T_{co}) \tag{39}$$

$$Q_{hx} = \dot{m}_h C_{ph} (T_{ho} - T_{hi}) \tag{40}$$

$$\dot{q}_{loss} = U A_{hx} * (T_{hx} - T_{amb}) + U A_G * (T_G - T_{amb}) \tag{41}$$

where $T_{hx} = T_{hi} - T_{ci}$.

$$Q_{hx} = Q_G + \dot{q}_{loss} \tag{42}$$

$$B_Q = Q_G + \dot{q}_{loss} * \left(\frac{FLHE}{F_d * \eta_B} \right) \tag{43}$$

Equation (41) is the total rate of energy loss for the system apart from the boiler, and Equation (42) relates the energy transfer rate due to the heat exchanger to that of the generator. The biomass quantity required is given by Equation (43), where Q_G is the rate of energy transfer at the generator of the absorption refrigeration system. The equation shows that the quantity of biomass required for a given load over a time period would depend on the density of the biomass species and the boiler efficiency. Four biomass solid fuel types were compared. The energy densities are given in Table 5. It is observed that the higher the energy content of the fuel (kWh/tonne), the less the quantity required. Figure 4. shows that the higher the temperature of the generator, the higher the quantity of fuel required, provided all other operating conditions are held constant. The mass of fuel required increased as the flow rate of hot water from the boiler is increased (Figure 5). Woodchips with the least dense energy content of the fuels under consideration, require nearly twice the mass of wood, with the highest energy content, for the same operation.

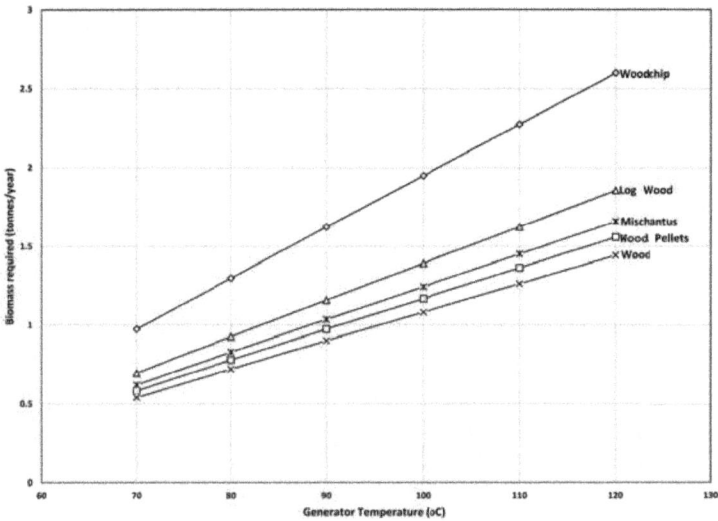

Figure 4. Comparison of fuel consumed with varied generator temperature.

Table 5. Average densities of biomass solid fuels [32]. MC: moisture content.

Fuel Type	Average Energy Density (kJ/tonne)
Wood chip (30% MC)	3000
Wood pellets	5000
Log wood (stacked-air dried; 20% MC)	4200
Wood	5400
Mischantus	4700

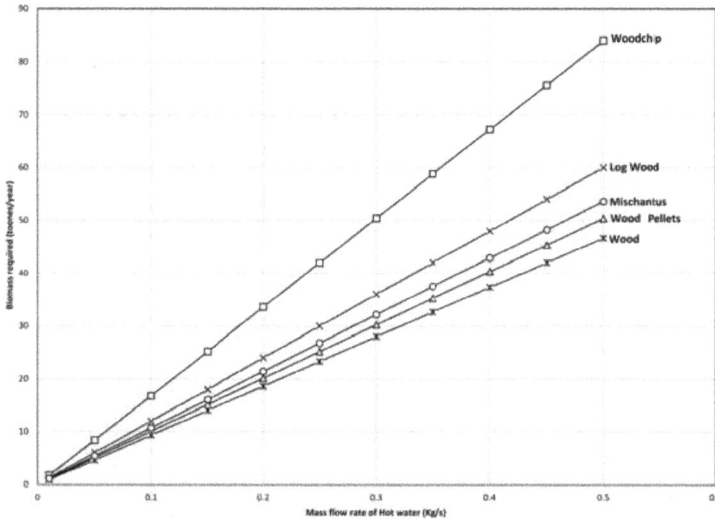

Figure 5. Effect of mass flow rate of hot water on fuel consumption.

4.2. Absorption Refrigeration

In calculating the COP, it is assumed that no pressure drops except through the expansion valves and the flow pumps. Pumping is isentropic; at state points 1, 4, and 8, the liquid is saturated; the composition of the solution at state points 1, 2, and 3 (strong solution), and 4, 5, and 6 (weak solution) remain unchanged throughout the process. The condenser pressure (high pressure) is the same as that at state points 3, 4 and 7, while the evaporator pressure (low pressure) holds for points 1, 6 and 10. The pump work is neglected, the refrigerant heat exchanger is not considered, and the generator is considered a single unit. simulated results from the absorption refrigeration cycle model developed were compared with published dated from the referenced authors [33–35]. The comparison was selected to reflect conditions of both light ($T_E = 2.5\,°C, T_A = 20\,°C, T_C = 20\,°C, T_G = 60 \rightarrow 90\,°C$) and heavy cooling ($T_G = 100\,°C, T_C = 40\,°C, T_A = 30\,°C, T_E = -5\,°C, \dot{m}_r = 1\ \text{kg/s}, E = 0.80, p_e = 354.42\ \text{kPa}, p_c = 1166.92\ \text{kPa}$) requirements. The results, show close similarity with a variation of ±0.02. Table 6 shows the comparison of the variation of the solution composition and refrigerant circulation ratio with generator temperature. The variation of the energy rate of change in the various heat-exchanger units and COP with the generator temperature is presented in Tables 7 and 8. The slight divergence in the result is to be expected partly due to the assumptions made. The temperature and pressure of the solution at the heat exchanger are computed by the model. Also, the composition of the refrigerant vapour at the exit of the generator is computed and not assumed; these parameters are normally estimated. In the calculation of the COP, the solution pump work was considered in the references. However, it is neglected in this work (Figure 6).

Table 6. Comparison of the flows of solution and refrigerant ratio at varied generator temperature (T_E = 2.5 °C, T_C = 20 °C, and T_A = 20 °C).

Literature	Generator Temp (°C)	Weak Sol (%)	Strong Sol (%)	Circulation Ratio (f)
Sun [34]	60	47.1	62.4	3.41
Present work	60	47.1	62.5	3.40
Sun [34]	70	41.4	62.6	2.70
Present work	70	41.4	62.6	2.76
Sun [34]	80	36.2	62.5	2.4
Present work	80	36.4	62.5	2.4
Sun [34]	90	31.4	62.5	2.2
Present work	90	31.5	62.5	2.2

Table 7. Comparison of the energy flows and coefficient of performance (*COP*) at varied generator temperature (T_E = 2.5 °C, T_C = 20 °C, and T_A = 20 °C).

Literature	Generator Temp (°C)	Q_G (kW)	Q_C (kW)	Q_A (kW)	Q_E (kW)	*COP*
Sun [34]	60	25.73	21.8	24.43	20.52	0.80
Present work	60	25.84	21.6	26.5	21.75	0.84
Sun [34]	70	26.14	21.94	24.73	20.52	0.78
Present work	70	26.19	21.47	23.80	20.76	0.79
Sun [34]	80	26.55	21.99	25.10	20.52	0.77
Present work	80	26.65	20.13	25.70	20.65	0.77
Sun [34]	90	26.92	21.97	25.48	20.52	0.76
Present work	90	27.19	20.31	24.40	20.81	0.76

Table 8. Comparison of the energy flows and COP (T_E = −5 °C, T_C = 40 °C, T_A = 30 °C and T_G = 100 °C) [33,35].

Literature	Q_g (kW)	Q_E (kW)	Q_A (kW)	Q_C (kW)	*COP*
Sun [35]	30.131	18.597	30.327	18.461	0.617
Present work	29.781	20.936	29.838	20.079	0.703

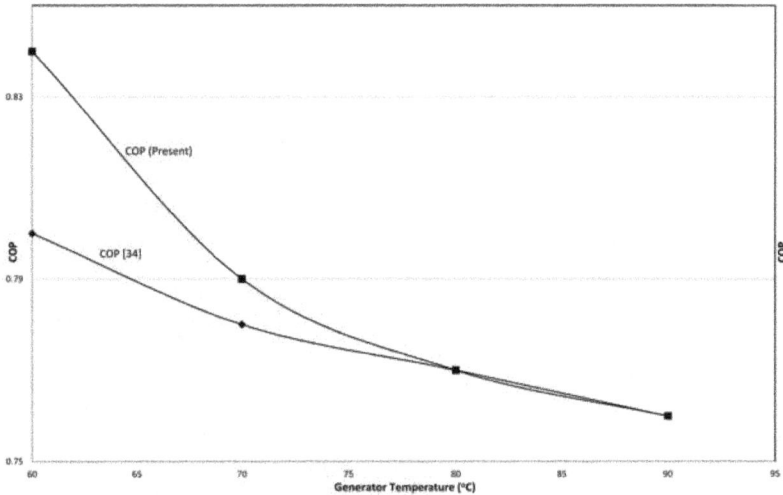

Figure 6. Comparison of the present work with published data in [34].

5. Conclusions

A routine for the analysis of a single stage aqua-ammonia refrigeration system has been developed. It estimates the concentration of the strong and the weak solutions as well as calculating the concentration of ammonia in the refrigerant vapour. The calculated *COP*, properties and rate of energy transfer in all components are reasonably approximated. Simulation results were produced for a practical range of heat inputs. Results were compared with published data and found good agreement. The use of biomass thermal energy which is renewable is proved to provide a CO_2 free refrigeration option. However, the economic viability and carbon footprint will depend on availability, proximity to source, system size, full load operating hours and storage capacity required. Prediction of the biomass quantity required could be useful in the economic evaluation of the feasibility of operating an absorption refrigeration system with biomass, particularly for areas with access to biomass.

Acknowledgments: This paper presents a summary of some the results from the project "ARBOR" sponsored by the European Union, INTERREG IVB, NWE to promote the conversion of Biomass for Energy.

Author Contributions: Munyeowaji Mbikan carried out the numerical computation. Tarik Al-Shemmeri is the principal supervisor of the research.

Conflicts of Interest: The authors declare no conflict of interest in the publication of this paper.

Nomenclature

A_G	Area of generator heat transfer surface (m²)
A_{hx}	Area of heat transfer surface of heat exchanger (m²)
B_Q	Biomass Quantity (Tonnes)
C_p	Specific heat capacity (kJ/kgK)
E_R	Energy Required (kW/h)
f	Circulation ratio
F_d	Fuel density (kJ/m³)
FLHE	Full load heating hours equivalent (h)
h_{liquid}	Enthalpy of liquid component (kJ/kg)
h_{vapour}	Enthalpy of vapour component (kJ/kg)
h^l	Liquid phase enthalpy (kJ/kg)
h^v	Vapour phase enthalpy (kJ/kg)
MC	Moisture content (%)
$\dot{m}_r$	Refrigerant mass flow (kg/s)
p_c	Condenser pressure (MPa)
p_e	Evaporator pressure (MPa)
Q_A	Absorber heat transfer (kW)
Q_C	Condenser heat transfer (kW)
Q_E	Evaporator heat transfer (kW)
Q_G	Generator heat transfer (kW)
Q_{hx}	Heat exchanger heat transfer (kW)
$\dot{q}_{loss}$	Rate of heat loss (kW)
T_{amb}	Ambient temperature (°C)
T_{ci}	Temperature of cold water at inlet (°C)
T_{co}	Temperature of cold water at outlet (°C)
T_{hi}	Temperature of hot water at inlet (°C)
T_{ho}	Temperature of hot water at outlet (°C)
U	Overall heat transfer coefficient
η_B	Boiler Efficiency (%)

References

1. Santamouris, M.; Argiriou, A. Renewable energies and energy conservation technologies for buildings in southern Europe. *Int. J. Sol. Energy* **1994**, *15*, 69–79.
2. Santamouris, M.; Papanikolaou, N.; Livada, I.; Koronakis, I.; Georgakis, C.; Argiriou, A.; Assimakopoulos, D. On the impact of urban climate on the energy consumption of buildings. *Sol. Energy* **2001**, *70*, 201–216.

3. Ullah, K.; Saidur, R.; Ping, H.; Akikur, R.; Shuvo, N. A review of solar thermal refrigeration and cooling methods. *Renew. Sustain. Energy Rev.* **2013**, *24*, 499–513.
4. Florides, G.A.; Tassou, S.A.; Kalogirou, S.A.; Wrobel, L.C. Review of solar and low energy cooling technologies for buildings. *Renew. Sustain. Energy Rev.* **2002**, *6*, 557–572.
5. Banks, D. *An Introduction to Thermogeology: Ground Source Heating and Cooling*, 2nd Ed.; Wiley-Blackwell: Oxford, UK, 2012.
6. Kim, D.S.; Infante Ferreira, C.A. Solar refrigeration options—A state-of-the-art review. *Int. J. Refrig.* **2008**, *31*, 3–15.
7. Fan, Y.; Luo, L.; Souyri, B. Review of solar sorption refrigeration technologies: Development and applications. *Renew. Sustain. Energy Rev.* **2007**, *11*, 1758–1775.
8. Sahay, A.; Sethi, V.K.; Tiwari, A.C.; Pandey, M. A review of solar photovoltaic panel cooling systems with special reference to Ground coupled central panel cooling system (GC-CPCS). *Renew. Sustain. Energy Rev.* **2015**, *42*, 306–312.
9. Siddiqui, M.U.; Said, S.A.M. A review of solar powered absorption systems. *Renew. Sustain. Energy Rev.* **2015**, *42*, 93–115.
10. Desideri, U.; Proietti, S.; Sdringola, P. Solar-powered cooling systems: Technical and economic analysis on industrial refrigeration and air-conditioning applications. *Appl. Energy* **2009**, *86*, 1376–1386.
11. Said, S.A.M.; El-Shaarawi, M.A.I.; Siddiqui, M.U. Alternative designs for a 24-h operating solar-powered absorption refrigeration technology. *Int. J. Refrig.* **2012**, *35*, 1967–1977.
12. Sarbu, I.; Sebarchievici, C. General review of solar-powered closed sorption refrigeration systems. *Energy Convers. Manag.* **2015**, *105*, 403–422.
13. Bales, C.; Ayadi, O. Modelling of commercial absorption heat pump with integral storage. In Proceedings of the 11th International Conference on Energy Storage, Stockholm, Sweden, 14–17 June 2009.
14. Mathkor, R.Z.; Agnew, B.; Al-Weshahi, M.A.; Latrsh, F. Exergetic analysis of an integrated tri-generation organic rankine cycle. *Energies* **2015**, *8*, 8835–8856.
15. Kalinowski, P.; Hwang, Y.; Radermacher, R.; Al Hashimi, S.; Rodgers, P. Application of waste heat powered absorption refrigeration system to the LNG recovery process. *Int. J. Refrig.* **2009**, *32*, 687–694.
16. Manzela, A.A.; Hanriot, S.M.; Cabezas-Gómez, L.; Sodré, J.R. Using engine exhaust gas as energy source for an absorption refrigeration system. *Appl. Energy* **2010**, *87*, 1141–1148.
17. Lin, L.; Wang, Y.; Al-Shemmeri, T.; Zeng, S.; Huang, X.; Li, S.; Yang, J. Characteristics of a diffusion absorption refrigerator driven by waste heat from engineexhaust. *Proc. Instit. Mech. Eng. Part E J. Process Mech. Eng.* **2006**, *220*, 139–149.
18. Lin, L.; Wang, Y.; Al-Shemmeri, T.; Ruxton, T.; Turner, S.; Zeng, S.; Huang, J.; He, Y.; Huang, X. An experimental investigation of a household size trigeneration. *Appl. Therm. Eng.* **2007**, *27*, 576–585.
19. Rachford, H.H.; Rice, J.D. Procedure for use of electrical digital computers in calculating flash vaporization hydrocarbon equilibrium. *Trans. AIME* **1952**, *4*, 19–20.
20. Ladanai, S.; Vinterbäck, J. *Global Potential of Sustainable Biomass for Energy*; Department of Energy and Technology, Institutionen för Energi Och Teknik Swedish University of Agricultural Sciences: Uppsala, Sweden, 2009; p. 32.
21. Caillat, S.; Vakkilainen, E. Chapter 9—Large-scale biomass combustion plants: An overview. In *Biomass Combustion Science, Technology and Engineering*; Elsevier: Amsterdam, The Netherlands, 2013; pp. 189–224.
22. Bridgwater, A.V. The technical and economic feasibility of biomass gasification for power generation. *Fuel* **1995**, *74*, 631–653.
23. Treado, S. The effect of electric load profiles on the performance of off-grid residential hybrid renewable energy systems. *Energies* **2015**, *8*, 11120–11138.
24. Long, H.; Li, X.; Wang, H.; Jia, J. Biomass resources and their bioenergy potential estimation: A review. *Renew. Sustain. Energy Rev.* **2013**, *26*, 344–352.
25. Sopegno, A.; Rodias, E.; Bochtis, D.; Busato, P.; Berruto, R.; Boero, V.; Sørensen, C. Model for energy analysis of Miscanthus production and transportation. *Energies* **2016**, *9*, 392.
26. Bekele, K.; Hager, H.; Mekonnen, K. Woody and non-woody biomass utilisation for fuel and implications on plant nutrients availability in the mukehantuta watershed in ethiopia. *Afr. Crop Sci. J.* **2013**, *21*, 625–636.
27. Pátek, J.; Klomfar, J. Simple functions for fast calculations of selected thermodynamic properties of the ammonia-water system. *Int. J. Refrig.* **1995**, *18*, 228–234.

28. Ziegler, B.; Trepp, C. Equation of state for ammonia-water mixtures. *Int. J. Refrig.* **1984**, *7*, 101–106.
29. Kherris, S.; Makhlouf, M.; Zebbar, D.; Sebbane, O. Contribution study of the thermodynamics properties of the ammonia-water mixtures. *Therm. Sci.* **2013**, *17*, 891–902.
30. Ganesh, N.S.; Srinivas, T. Evaluation of thermodynamic properties of ammonia- water mixture up to 100 bar for power application systems. *J. Mech. Eng. Res.* **2011**, *3*, 25–39.
31. Collazo, J.; Porteiro, J.; Míguez, J.L.; Granada, E.; Gómez, M.A. Numerical simulation of a small-scale biomass boiler. *Energy Convers. Manag.* **2012**, *64*, 87–96.
32. Martin, R.; Mark, M. *Biomass Heating: A Practical Guide for Potential Users*; In-Depth Guide CTG012; Carbon Trust: London, UK, 2007.
33. Cai, W.; Sen, M.; Paolucci, S. Dynamic simulation of an ammonia-water absorption refrigeration system. *Ind. Eng. Chem. Res.* **2012**, *51*, 2070–2076.
34. Sun, D.W. Thermodynamic design data and optimum design maps for absorption refrigeration systems. *Appl. Therm. Eng.* **1997**, *17*, 211–221.
35. Sun, D.W. Comparison of the performances of NH_3-H_2O, NH_3-$LiNO_3$ and NH_3-$NaSCN$ absorption refrigeration systems. *Energy Convers. Manag.* **1998**, *39*, 357–368.

energies

MDPI

Article

Potential of Livestock Generated Biomass: Untapped Energy Source in India

Gagandeep Kaur [1,*] (iD), Yadwinder Singh Brar [1] and D.P. Kothari [2]

[1] I. K. Gujral Punjab Technical University, 144603 Punjab, India; braryadwinder@yahoo.co.in
[2] Ex-Vice Chancellor, VIT, Vellore, 632014 Tamil Nadu, India; dpkvits@gmail.com
*   Correspondence: simicgagan@gmail.com or gaganee@ptu.ac.in; Tel.: +91-947-809-8118

Received: 9 May 2017; Accepted: 20 June 2017; Published: 25 June 2017

Abstract: Modern economies run on the backbone of electricity as one of major factors behind industrial development. India is endowed with plenty of natural resources and the majority of electricity within the country is generated from thermal and hydro-electric plants. A few nuclear plants assist in meeting the national requirements for electricity but still many rural areas remain uncovered. As India is primarily a rural agrarian economy, providing electricity to the remote, undeveloped regions of the country remains a top priority of the government. A vital, untapped source is livestock generated biomass which to some extent has been utilized to generate electricity in small scale biogas based plants under the government's thrust on rural development. This study is a preliminary attempt to correlate developments in this arena in the Asian region, as well as the developed world, to explore the possibilities of harnessing this resource in a better manner. The current potential of 2600 million tons of livestock dung generated per year, capable of yielding 263,702 million m^3 of biogas is exploited. Our estimates suggest that if this resource is utilized judiciously, it possesses the potential of generating 477 TWh (Terawatt hour) of electrical energy per annum.

Keywords: biogas; cows dung; electrical energy; India; livestock

---

1. Introduction

Indiscriminate consumption of fossil fuels to meet energy demand for the burgeoning human population world-wide, especially in high population density nations, is primarily responsible for three quarters of the total world figures for greenhouse gas (GHG) generation [1,2]. The role of GHG as the main contributor to global warming poses a challenge to all life forms on the planet and it is now recognized as a distinct obstacle in the sustainability and living conditions of future generations [3,4]. The high growth rate of the human population and tendencies towards urbanization during the last few decades have raised energy demand which is likely to grow by 25% (700 Quadrillion British Thermal Units) of the present figure, by the year 2040 [5]. In order to fulfill this expected rise in energy demand, alternative sources need to be identified and tapped due to the now recognized likelihood of the exhaustion of fossil fuels. It therefore becomes essential that the best available source of green energy be identified, with a premise that it should be sustainable, should possess the capability of being replenished consistently, and the process should be both natural and bio-friendly [6,7]. If the focus is shifted to a regional level in the Asia-Pacific region, the Indian sub-continent is a likely candidate to explore such possibilities due to the qualifying criteria of being a land mass of varied flora and fauna which can generate enough biomass for the investigation being conducted. India is the world's seventh largest land-mass spreading over 328 million hectares and is expected to become the most populous nation of the world by the year 2025 [5,8]. Politically, India is divided into 35 provinces and union-territories (UTs). India has six main climatic sub-types, ranging from deserts in the west, glaciers in the north, tropical humid climate in the southwest, and has a huge coastline because the sub-continent is a geographic peninsular. India's climate is influenced by its unique geographical and geological features conducive to varied life forms. These features facilitate the existence of a wide variety of vegetable and animal life forms which contribute to, as well as generate, huge biomass, which can be tapped as potential source for energy. The last two decades have witnessed an economic surge in India and the Gross Domestic Product (GDP) is expected to rise three-fold by the year 2040 [5]. With considerable demand for energy in rural and urban Indian communities, there is a discernible trend of a gradual shift from non-renewable to renewable energy sources, the latter being identifiable as ideal and a better alternative for guaranteed energy access to consumers [9,10]. Agriculture, forest, and livestock-based bio waste/ byproducts are called biomass or bio-residues, and are available in plenty in India [11]. Biomass-based energy generation is popular in rural areas due to infrastructural constraints in delivery of conventional energy fuels [12]. Bio energy has merit as it is renewable and extractable from organic matter by utilizing simple and economical techniques and processes of anaerobic digestion (AD). Such techniques possess the ability of yielding sufficient amounts of practically usable biogas. [13]. Biogas is an already identified eco-friendly energy source with main components being methane (60%–70%) and carbon-dioxide (35%–40%) [14,15]. Another innovative technique developed in recent years to utilize available biomass is development of the Microbial Fuel Cell (MFC) [16]. MFCs are currently under intensive research and thus far researchers have been able to obtain a maximum power density of 3600 MW/m^2 with glucose fed as substrate using common available raw biomass components [17]. A typical MFC is a bio-reactor which converts chemical energy in bio-convertible substrates directly into electricity by action of specific microorganisms which facilitate the conversion of substrate directly into electrons [18]. It is not that biomass available in India is so far being wasted. Since ancient times, biomass generated by domesticated livestock has been the primary source of domestic energy in India. It meets the kitchen energy requirements of most rural, and approximately one-half of urban house-holds [11]. Livestock dung, comprising of excreta and urine still finds usage as a vital, renewable, and sustainable precursor for producing bio-fertilizer (farm yard manure), biogas, and a unique product specific to South Asia i.e., cow-dung cakes, caked and dried dairy animal feces which are used for burning as fuel in traditional, earthen ovens [19]. To promote usage of biomass, efficient energy conversion technologies have been developed by Government of India and many biogas plants are successfully running in progressive rural sectors. Unfortunately, despite exponential economic growth, energy production in India remains deficient, as approximately one fourth of the country's population, and 44% of the rural population, do not have access to grid-based electricity. It is pertinent to mention here that per capita electricity consumption in India is just 814 kWh, which is moderate (24%) when compared with the global average consumption [20,21]. Annual energy demand rose from 830,594 MU (million units) to 1,142,092 MU as per 2009 assessment of the energy sector. Subsequently, although electricity generation was hiked from 770 MUs to 1160 MUs, deficiency persists in the rural sector. Current renewable source energy generation stands at just 15% of total energy generated from all known sources. Of this, biomass-based power generation stands at just 2%, despite it being capable of contributing more [22]. India thus has ample room to enhance energy generation by utilizing renewable energy sources, especially biomass-based bio energy. India's livestock population is one of the largest in world with current total population standing at 512.05 million [23]. This sector occupies second rank after agriculture in energizing and boosting rural economic trends. Contribution of livestock to GDP is 4.11% at current price index [24,25]. In the majority of Indian provinces, two-thirds of the population is directly or indirectly involved in dairy and poultry farming. Livestock dung therefore plays a vital role in rural economy and its judicious and proper utilization by introducing modern processing technologies may contribute in a significant manner to enhance rural incomes by utilizing this huge biomass for power generation [25,26]. Probing livestock excreta as a likely potential green energy source and therefore needs in-depth analysis. If this potential source of energy can be harnessed in an economical manner, it is likely to contribute in a significant manner to reducing stress on the conventional coal and hydro-electric-based national electric grid.

2. Indian Scenario

2.1. Energy Scenario in India

Despite world-wide recession, the Indian economy is growing at a tremendous rate compared to rest of the world. Ironically, the margin of economic growth in rural and urban regions is separated by a great chasm [27]. Further, current estimates suggest that around 25% of the national population, and 44% of the rural population, are bereft of access to grid-based electricity distribution, the primary reason for this gap [28]. All regions continue to experience shortage in electrical energy year round and peak demand rises with each passing year. Surplus power in some provinces is generally sold to deficit provinces but the national situation remains dismal. However, over the last few years India has been able to squeeze the gap to a certain extent [29]. Table 1 and Figure 1a summarize the updated overall national energy scenario. Despite confronting consistent deficit, installed energy generation capacity in India has improved from 145,755 MW in the year 2006 to 319,606 MW in year 2017, which can be considered a fair level of improvement [30]. Since 2009, energy generation was enhanced from 771 BU to current level of 1160 BU, which is significant [22]. Figure 1b illustrates the year to year growth rate of installed energy generation capacity. In 2017, from the total installed energy generation capacity share of provinces, central, and private sectors stood at 35%, 25%, and 42% respectively. Presently, energy is generated primarily from thermal power units [(68%) (218,330 MW)] by utilizing conventional sources like coal, gas, and oil as fuel. The remaining requirement is met from hydroelectric [(14%) (44,478 MW)], nuclear [(2%) (6780 MW)], and renewable energy source [(16%) (50,018 MW)] based power plants [30]. For sustainable development, efforts to promote bio energy from available biomass resources have been ongoing [31]. However, the full potential has not been exploited for the generation of electricity from biomass. It still stands at a poor 2% of all potential alongside other exploitable renewable energy resources [32]. Current renewable energy source utilization stands at 16% of the total energy produced, of which 83.54% is from solar power, 11.46% from wind power, 2.2% from small hydro-electric power plants, 1.96% from biomass, and 0.56% from bagasse burning [22]. Current estimates suggest that it is possible to generate approximately 17,538 MW of power from presently available biomass. The Ministry of New and Renewable Energy (MNRE) of India actively promotes biomass-based power plants. Such plants are primarily located in rural areas where abundant biomass exists. As per available data, India has 5940 MW biomass based plants, out of which 4946 MW are grid connected and 994 MW are off-grid [33].

Table 1. Energy and peak demand scenario in India.

Year	Energy Requirements (MUs)	Energy Achieved (MUs)	Energy Surplus(+)/ Deficit(−) (%)	Peak Energy Demand (MW)	Peak Energy Achieved (MW)	Peak Energy Surplus(+)/ Deficit(−) (%)
2009–2010	830,594	746,644	−10.1	119,166	104,009	−12.7
2010–2011	861,591	788,355	−8.5	122,287	110,256	−9.8
2011–2012	937,199	857,886	−8.5	130,006	116,191	−10.6
2012–2013	995,557	908,652	−8.7	135,453	123,294	−9.0
2013–2014	1,002,257	959,829	−4.2	135,918	129,815	−4.5
2014–2015	1,068,923	1,030,785	−3.6	148,166	141,160	−4.7
2015–2016	1,114,408	1,090,850	−2.1	153,366	148,463	−3.2
2016–2017	1,142,092	1,134,633	−0.7	159,542	156,934	−1.6

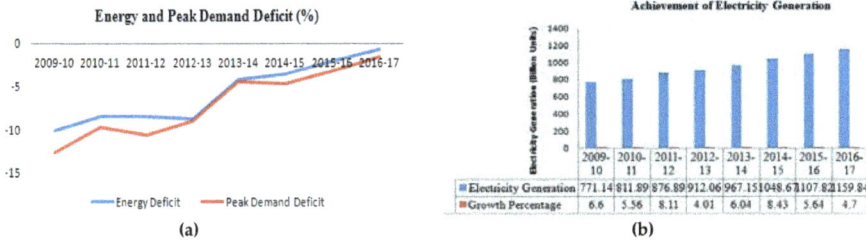

Figure 1. (**a**) Energy and peak demand deficit in India; (**b**) Achievement of electricity generation in India.

Off-grid capacity contributes 652 MW from captive power plants, 18 MW from biomass gasifier systems, of which 164 MW equivalent biomass gasifier systems are deployed for thermal applications. Leading provinces for biomass projects are Maharashtra, Uttar Pradesh, Karnataka, Tamil Nadu, Andhra Pradesh, Chhattisgarh, and Punjab [33]. Provincial figures of energy generation utilizing biomass are depicted in Figure 2a.

2.2. Livestock Scenario

The livestock sector contributes significantly to generating employment in rural India, especially for landless people and marginal farmers, providing nutritious food to millions of people. Livestock practice has been a lifesaving asset during catastrophes like floods and drought. According to national survey, about 16.44 million people are directly or indirectly involved in the livestock sector [34]. According to the 19th Livestock census report, national livestock population stands at 512 million, comprising mainly of cows, buffaloes, sheep, goats, pigs, and other species including poultry. Figure 2b presents the percentage population share of these species in total livestock. Consistent increase in herd sizes over the years has led to generation of enormous livestock waste. This waste not only causes air pollution, but contributes to atmospheric and ground water contamination as well. Generation of obnoxious and potent gases, like methane and nitrous oxide, contributes to environmental pollution leading to serious public health issues. Scientific measures to stop this environmental degradation are therefore absolutely essential [35–37]. Safe disposal of such gigantic proportions of livestock waste otherwise rich in organic matter poses a serious hurdle to development [38]. Indian traditional lifestyles have a very efficient means of disposal of dung in which dung is collected from individual houses in a large village heap at a strategic location in the village. Bacterial degradation results in the heap turning into compost which is later bought by farmers to fertilize their fields. Some portion of the cow-dung is shaped into cakes used as fuel in earthen ovens (Figure 2c).

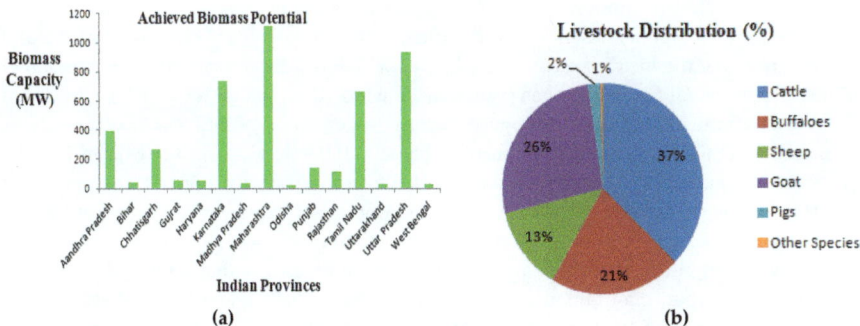

Figure 2. *Cont.*

Step 1: Fresh Cow-Dung from farm	
Collected and shaped into 8-10 inches cakes	Collected and thrown on previous day heap

Step 2: Cakes dried in Sun & heap allowed to ferment	
Dried, tough cakes	Heap ferments & degrades into nitrogenous compounds

Step 3: Utilization	
Cakes ignited & burnt to cook meals	Heap collected after 4 to 6 weeks and mixed in Soil in agricultural fields

(c)

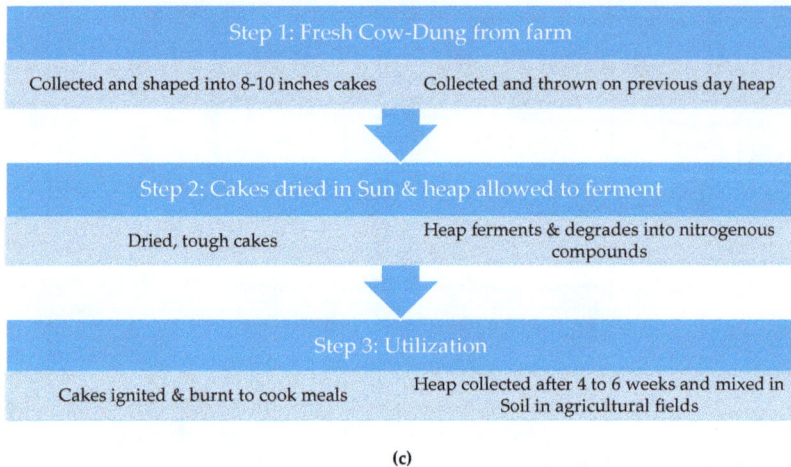

Figure 2. (a) Province-wise commulative achieved biomass and cogeneration power; (b) Species-wise livestock distribution in Indain provinces; (c) Flow-chart showing traditional use of cow-dung as kitchen fuel and manure.

Dried cow-dung (calorific value 14 MJ/kg) has been used to generate power in other countries as it yields the same amount of combustible energy as dried firewood (calorific value 16–20 MJ/kg) [39]. 150 million tons of cow-dung is presently being used in the world as fuel and 40% of this total usage occurs in India. The most widely used practices in ascending order for dung management are composting, dumping in landfills, disposal via common effluent treatment plant, or by dumping in open waste spaces. Animal dung and waste water collectively generate waste on the order of 3 million to 6000 million tons per day. It is estimated that just 9% of the total biomass generated by livestock is utilized for biogas recovery [40,41]. As a remedial measure, the latest cost-effective and eco-friendly energy techniques for utilization of livestock waste biomass to generate energy are needed.

2.3. Energy Conversion Techniques

Among all energy conversion techniques, the most efficient and appropriate technique for dung conversion to electrical energy is anaerobic digestion. Anaerobic digestion (AD) converts energy stored in dung into biogas, which can be utilized to generate electrical energy. Anaerobic digestion is degradation of organic material by microbial action in absence of air, transforming it into biogas, a mixture of methane, carbon dioxide, and some other trace gases. In this system dung is collected and mixed with an optimum amount of water. Microbes are already present or special cultures can be added to enhance microbial degradation. Pre-treatments like screening, grit removal, mixing, and flow equalization enhances the yield of recoverable biogas which is later utilized in a combustion engine to generate electricity. Generated biogas primarily consists of methane (50%–75%), carbon dioxide (25%–50%), and minute quantities of other gases like hydrogen-sulfide (H_2S) and ammonia (NH_4) [42]. Biogas generated in this manner has 'High Heating Value' (HHV) ranging from 16 MJ/m^3 to 25 MJ/m^3. The electrical energy content of typical biogas is 5–7 kWh/m^3 of biogas produced if a standard biogas yield of 0.04 m^3/kg is recoverable [43,44]. Methane gas generated by this methodology has traditionally been used as fuel for lighting and other electricity-dependent machines [45]. Current research reveals that 1 kg of cow-dung mixed with an equal quantity of water, and with a total hydration retention time of 55–60 days, when maintained at ambient temperature of 24–26 °C and yields 35–40 liters of biogas [45]. As compared to other renewable energy generation raw materials, livestock dung is a better alternative as it is economical, requires less capital investment, and has the least per unit cost of production [46,47]. Researchers are consistently working for other alternative means for generation of sustainable energy from livestock dung and MFC is presumed to be the front runner as an alternative to just using the methane gas to fuel generators. A MFC is an assembly which can capture the electrons generated by the metabolism of microorganisms and hold the charge to maintain a stable, continuous source of energy [48]. Special strains of microorganisms in dung promote conversion of substrate into electrons in a better manner [49]. Cow-dung contains myriad organic carbon sources which are subject to oxidative processes by action of microorganisms to yield electric energy [50]. The maximum achievable MFC voltage is theoretically on the order of 1.1 Volt [48,49,51]. This has raised an immense curiosity in cows dung as a probable contender for sustainable energy generation. At this time, bio-energy is being looked at as a prospective alternative to fulfill the requirements of an ideal renewable energy source of the future. The livestock, dairy, and poultry industries generate enormous amounts of waste and the annual amount can be staggering even for a small-scale farm. It is crystal clear that in the Indian scenario, inexhaustible potential for conversion of livestock biomass waste to energy exists, and possesses the likelihood of contributing significantly to the total energy demand [32]. An attempt, therefore, needs to be made to explore the available livestock waste and its potential for conversion to biogas for generation of electricity by all means possible with currently available and feasible technology. Any constraints to this effect need to be identified and thoroughly investigated before a venture in this direction can be made. In India, many agencies and private commercial ventures practicing conversion of a variety of biomass resources to energy, such as bagasse utilization in the sugar industry, to generate electricity exist. However, their contribution is miniscule if compared to the requirement [32]. A comprehensive region-wise study needs to be conducted to ascertain the potential of biogas production by utilizing livestock dung as the primary source so that essential guidelines and recommendations for relevant biomass management technology can be formulated. The outcome of this exercise may serve as a guiding force for the application of this venture as a likely, still untapped source of energy which holds promise as a bio-friendly and sustainable means to the envisaged end.

3. Materials and Methods

To evaluate the extent and potential of effectively utilizing biogas as a source of power generation, the amount of livestock and poultry waste generated in the country was obtained from figures published by the Ministry of Agriculture and Farmer Welfare, India [23]. Out of 35 provinces and union territories (UTs) of India, 18 major provinces had significant amounts of livestock waste generated as per latest figures, while 17 provinces/UTs (labeled "Other Provinces" in above figures) contribution was miniscule. Data from the major livestock waste-generating provinces will therefore only be considered. The statistical data includes population of livestock assets comprising mainly cows, buffaloes, sheep, goats, pigs, and poultry. Species like horses, ponies, camel, yak, and mithun were excluded from this study due to their being less than 1% of the total livestock population in India [23]. By considering the average body weight of each participating animal species, annual livestock dung and biogas generation were estimated. Significant parameters like total solids and animal dung availability were considered as they had implications and inter-dependence on biogas generation potential. Relevant calculations are depicted in following sub sections:

3.1. Gross Estimate of Amount of Livestock Waste Generated in India

To determine the livestock and poultry waste generated in India, state-wise statistical livestock data published by DADF, Ministry of Agriculture was accessed (Figure 3). In terms of population, India's current livestock population figures include 191 million cows, 109 million buffaloes, 65 million sheep, 135 million goats, 10 million pigs, and about 729 million poultry. The quantity of livestock dung generated invariably depends on multiple factors: such as type of animal, animal husbandry practices in operation, quantity, and ingredients of feed, whether confined or range animals, and general features like weight and size of typical breeds of each animal species [52,53]. To calculate the average amount of dung produced by livestock, enlisted species were categorized by size as large animals (cows and buffaloes), small animals (sheep and goats), pigs, or poultry (broiler, layer, and other). Past research pinpoints that the average production of dung ranged between 10–20 kg/day (5 to 6% of body weight), 2 kg/day (4 to 5% of body weight), 4 kg/day (5 to 7% of body weight) and 0.1 kg/day (3 to 4% of body weight) for cows and buffalo, sheep and goat, pig and poultry respectively in Asia [53,54]. In the present study, the quantity of dung generated by enlisted species was calculated by assuming livestock weight as 250, 40, 80, and 1.5 kg for Bovine, Caprine, Porcine, and avian species respectively, measured in kilograms for the average domesticated breeds in India. Accordingly, average standard values of 22.5 kg/day (9% body weights), 1.6 kg/day (based on 4% of body weight), 2.7 kg/day (based on 9% of body weight), and 0.045 kg per day (3% of the body weight) were considered for the above-mentioned livestock categories [2,55,56].

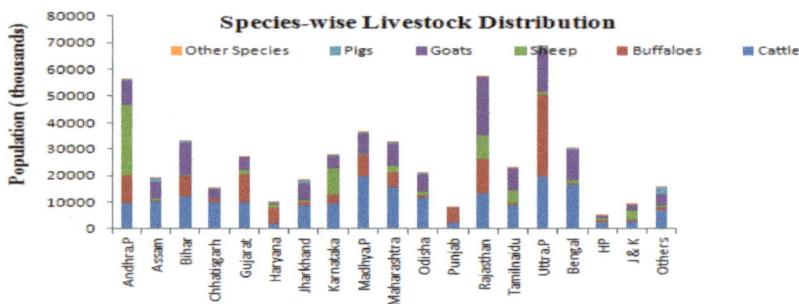

Figure 3. Species-wise livestock distribution in Indian provinces.

3.2. Determination of the Potential of Biogas and Power Generation

Cumulative biogas generated from livestock waste depends on many key factors: The type of animal and its breed, typical body weight, type of feed intake, and typical content of total solids excreted in dung waste [53]. However, uniform means of techniques for collection of the dung need to be standardized in order to meet the requirements for generation of ideal quantities of biogas per unit [53,55,57]. Hence only from a standardized cumulative biogas volume can the availability coefficient factor can be calculated and considered for determination of the expected and likely biogas yield, as per Equation (1):

$$TPB = M \times TS \times AC \times EBTS, \tag{1}$$

where TPB is the value of 'theoretical potential of biogas' (m^3 per year), M indicates the cumulative amount of procurable/collectable dung for specific province per year, measured in kilograms, TS denotes percentage of total solids with respect to the gross mass of animal dung, AC denotes the availability coefficient and EBTS is the biogas quantity generated per kilogram of TS (m^3 per kg TS). In this study, TS value was assumed to be 25% from large animal generated waste. As far as small animal waste is concerned, TS was assumed at 29% for pigs and poultry birds each, with the EBTS value calculated as 0.6 and 0.4, 0.8 m^3 per kg TS, respectively. The availability coefficient was assumed as 70% for large dairy animals, 20% for small ruminants, and 60% each in case of waste generated from commercial piggery and poultry industries [55]. It is quite evident that the methane content of biogas depends upon the type of dung used in the anaerobic digestion. Many studies indicate that, during the AD of cow-dung, average biogas recovered constitutes 50%–70% methane. Methane generated from small animals' dung is lesser at 40% to 50% [18,19,48,58]. Methane content recoverable from poultry and swine dung on average ranges between of 50%–70% and 60% [55,56]. In present study, 60% and 45% methane content was considered as likely to be generated from large animals, pigs, poultry, and small animals, respectively, as per expected Indian norms. While calculating heating value, heat conversion was calculated as 90% of total harvested methane, which was denoted as heat conversion efficiency in the boiler, assuming that 36 MJ per m^3 of methane is the calorific value [56]. For calculation of potential electricity generation from biogas, the following Equation (2) was utilized:

$$e_{biogas} = E_{biogas} \times \eta, \tag{2}$$

here e_{biogas} is the quantity of electricity generated (kWh per year), E_{biogas} is the yet to be converted raw energy in the biogas (kWh per year) and the η is the overall conversion efficiency of the biogas to electricity. η is variable and largely depends on power plant technology utilized. The η value is considered between 35%–42% in large turbine-system-operated power plants and 25% in small generators. In the current study, the η value was assumed as 30% [58]. By using the following Equation (3), the quantity of E_{biogas} is calculated

$$E_{biogas} = \text{Energy content biogas} \times m_{biogas}, \tag{3}$$

here, energy content biogas denotes biogas in terms of calorific value of (kWh per m^3) and m_{biogas} indicates annual quantity of biogas produced (m^3 per year). Energy content biogas is assumed to be 6 kWh per m^3 by considering 21.5 MJ per m^3 biogas as the calorific value [58]. For the present study, many factors like national policy, administrative technical guidelines, and financial barriers were not taken into account which can affect future implications of the usefulness of this technique. The key objective of this detailed assessment is to determine the electrical potential likely to be generated by converting dung into biogas. The following section outlines and analyzes some of the essential and pertinent parameters for harnessing this prospective inexhaustible source of energy.

4. Data Reporting and Results

4.1. Potential Quantifiable Sources of Livestock Dung

As per the latest 19th livestock census, prominent provinces with huge generation of livestock and poultry-based dung are listed in Table 2. From the figures, it is evident that annual generation of dung stands at around 2600 million tons (MT), which is stupendous in terms of amount generated, making it a vital untapped energy source. The largest proportion (95%) is generated by large dairy animals and the remaining 5% by small animals including sheep, goats, pigs, and poultry. Prominent among the regional provinces contributing to maximal generation include Uttar Pradesh, Rajasthan, Madhya Pradesh, Andhra Pradesh, Maharashtra, Gujarat, Bihar, and Bengal, of which the dung generation ranges between 149.5 MT to 423.7 MT per annum in an ascending order (Considering a minimum of 150 MT per annum as criterion for grouping the provinces on the basis of dung yield). In terms of percentage, the extent of dung generation has been calculated as 16.74% in Uttar Pradesh, 9.26% in Madhya Pradesh, 8.78% in Rajasthan, and 7% each in Maharashtra, Gujarat, Bihar, and Andhra Pradesh. If small animal dung generation is considered, Andhra Pradesh emerges as the front-runner with 18% and 23% annual yield from small animals and poultry respectively. Assam occupies the exclusive status of being the state with the highest generation of pig dung which stands at 1.5 MT (15%) per annum.

Table 2. Potential quantifiable sources of livestock dung.

Provinces/Union Territories	Large Animal Dung (MT)	Small Animal Dung (MT)	Pigs Dung (MT)	Poultry Dung (MT)	Total Dung (MT)
Andhra P	166	21	0.4	2.7	190.1
Assam	88	4	1.5	0.5	94
Bihar	163	7	0.6	0.2	170.8
Chhattisgarh	92	2	0.4	0.3	94.7
Gujurat	167	4	0	0.2	171.2
Haryana	65	1	0.2	0.7	66.9
Jarkhand	81	4	0.9	0.2	86.1

Table 2. *Cont.*

Provinces/Union Territories	Large Animal Dung (MT)	Small Animal Dung (MT)	Pigs Dung (MT)	Poultry Dung (MT)	Total Dung (MT)
Karnataka	107	8	0.3	0.9	116.2
Madhya P	228	5	0.2	0.2	233.4
Maharashtra	173	6	0.3	1.2	180.5
Odisha	101	5	0.3	0.3	106.6
Punjab	62	0.4	0	0.3	62.7
Rajasthan	216	18	0.3	0.1	234.4
Tamil N	79	7.6	0.2	1.9	88.7
Uttra P	412	10	1.4	0.3	423.7
W. Bengal	141	7	0.6	0.9	149.5
HP	24	1	0	0.1	25.1
J & K	29	3	0	0.2	32.2
Others	66	3	2.5	0.7	72.2
Total	2460	117	10.1	11.9	2599

4.2. Potential for Biogas Production

Livestock and poultry dung has already been identified as a potential raw material for biogas production and considered as a vital aspect in the sustainable development of developing countries in South Asia. The tentative theorized estimate of this untapped source for biogas production is shown in Table 3. Total potential biogas production from all dung sources was calculated in terms of annual yield measured in million m^3 per year.

Table 3. Potential for biogas generation (TPB) in provinces.

Provinces/Union Territories	Large Animal Biogas (Million m^3)	Small Animal Biogas (Million m^3)	Pigs Biogas (Million m^3)	Poultry Biogas (Million m^3)	Total Biogas (Million m^3)
Andhra P	17,430	420	55.68	375.84	18,281.5
Assam	9240	80	208.8	69.6	9598.4
Bihar	17,115	140	83.52	27.84	17,366.3
Chhattisgarh	9660	40	55.68	41.76	9797.4
Gujurat	17,535	80	0	27.84	6970.28
Haryana	6825	20	27.84	97.44	66.9
Jarkhand	8505	80	125.28	27.84	8738.1
Karnataka	11,235	160	41.76	125.28	11,562
Madhya P	23,940	100	27.84	27.84	24,095.6
Maharashtra	18,165	120	41.76	167.04	18,493.8
Odisha	10,605	100	41.76	41.76	10,788.5
Punjab	6510	8	0	41.76	6559.7
Rajasthan	22,680	360	41.76	13.92	23,095.6
Tamil N	8295	152	27.84	264.48	8739.32
Uttra P	43,260	200	194.88	41.76	43,696.6
W. Bengal	14,805	140	83.52	125.28	15,153.8
HP	2520	20	0	13.92	2553.9
J & K	3045	60	0	27.84	3132.8
Others	6930	60	348	97.44	7435.44
Total	258,300	2340	1405.9	1656.48	263,702.4

Considering figures of expected yield of biogas per annum, Uttar Pradesh at 43,696 million m^3 leads the pack, followed by Madhya Pradesh at 24,095 million m^3, Rajasthan at 23,095 million m^3, Andhra Pradesh at 18,281 million m^3, Maharashtra at 18,493 million m^3, Gujarat at 17,642 million m^3, Bihar at 17,366 million m^3, and Bengal at 15,153 million m^3. Of this, 98% of total dung utilizable for biogas production is obtainable from large (dairy) animals and the remaining 2% from smaller domesticated animals as well as poultry. Andhra Pradesh leads other provinces with respect to the maximal exploitable biomass being generated from small animal farms and the poultry industry, with a total estimated yield of 795 million tons of biogas per annum. In terms of dung obtainable from pig-rearing provinces, the expected biogas yield stands at 1400 million m^3 per annum. Assam leads with an annual potential of 208 million m^3 (15%), followed by Uttar Pradesh at 195 million m^3 (14%). Biogas potential from animal dung has already been calculated and exploited with annual figures from some of the prominent countries of the world in this arena showing at 481.7 million m^3 for Sweden, 550.7 million m^3 for Denmark, 461.9 million m^3 for Finland, 8599.8 million m^3 for Iran, 1477.4 million m^3 for Turkey, and 4589.5 million m^3 for Malaysia [55,56].

4.3. Estimates of Methane Yield and Corresponding Heating Value Obtainable from Biomass

The methane content of biogas was calculated by assuming standard content of methane produced from the variety of dung sources under consideration. Total estimate of methane yield from all sources of animal dung for the country was calculated as 157,870 million m^3 per annum. Further calculations suggest that this methane gas if utilized as an energy source can yield heat energy of about 5,146,576 million MJ per annum.

4.4. Calculation of Electrical Potential

Detailed calculations were carried out to estimate the exploitation of methane from biogas generated from animal dung as fuel for generating electricity. Table 4 shows the calculated results compiled on the basis of the resource exploited, and the potential in each state. From the illustration and Figure 4, it is evident that highest potential, in terms of recoverable electrical energy in Terawatt hour (TWh) on an annual basis, exists in Uttar Pradesh at 16.6% (79 TWh), followed by Madhya Pradesh 9.15% (43.5 TWh), Rajasthan 8.75% (41.6 TWh), Maharashtra 7% (33.4 TWh), Andhra Pradesh 6.9% (32.9 TWh), Gujarat 6.7%(31.9 TWh), Bihar 6.6%(33.1 TWh), and Bengal 5.6% (27.3 TWh). In India, total dung available stands at 2600 million tons per annum which has the potential to generate 263,702 million m^3 of biogas each year. As shown in calculations above, the methane yield from biogas stands at 157,870 million m^3 per annum, which has heating value of 5,146,576 million MJ that possesses the capacity to generate an additional 477 TWh of electrical energy per annum, which by all means is a significant figure. A similar study carried out in the United Provinces reported that animal dung generated across all provinces in the continent has a potential of 928 trillion BTU in terms of raw energy which can be utilized to generate heat and electricity, meeting approximately one percent of energy demand [58]. Previous studies carried out in India claim a biogas potential of 16,030 million m^3 from collectable dung considering only dairy animal dung generated in the country. According to this study, as per the figures in Year 1997, dung collected from cows and buffaloes stand at 458 million tons per annum, with an energy potential of 336 PJ [52]. Small ruminants, pigs, and poultry were not considered in this potential estimate [52].

Table 4. Estimated electrical energy potential of Indian provinces.

Provinces/Union Territories	Large Animal Potential (TWh)	Small Animal Potential (TWh)	Pigs Potential (TWh)	Poultry Potential (TWh)	Total Eletrical Potential (TWh)
Andhra P	31.57	0.57	0.1	0.68	32.92
Assam	16.73	0.11	0.38	0.13	17.35
Bihar	31	0.19	0.15	0.05	31.4
Chhattisgarh	17.5	0.054	0.10	0.076	17.8
Gujarat	31.76	0.11	0	0.050	31.92
Haryana	12.36	0.027	0.050	0.18	12.61
Jarkhand	15.4	0.11	0.23	0.050	15.79
Karnataka	20.35	0.22	076	0.23	20.87
Madhya P	43.36	0.14	0.050	0.050	43.6
Maharashtra	32.9	0.16	0.076	0.3	34.34
Odisha	19.2	0.14	0.076	0.076	19.5
Punjab	11.8	0.011	0	0.076	11.88
Rajasthan	41.08	0.49	0.075	0.025	41.67
Tamil N	15.02	0.20	0.05	0.48	15.76
Uttra P	78.35	0.27	0.35	0.076	79.05
Bengal	26.81	0.19	0.15	0.23	27.38
HP	4.56	0.027	0	0.025	4.62
J & K	5.51	0.082	0	0.050	5.65
Others	12.55	0.082	0.63	0.18	13.44
Total	467.81	3.18	2.55	3	476.53

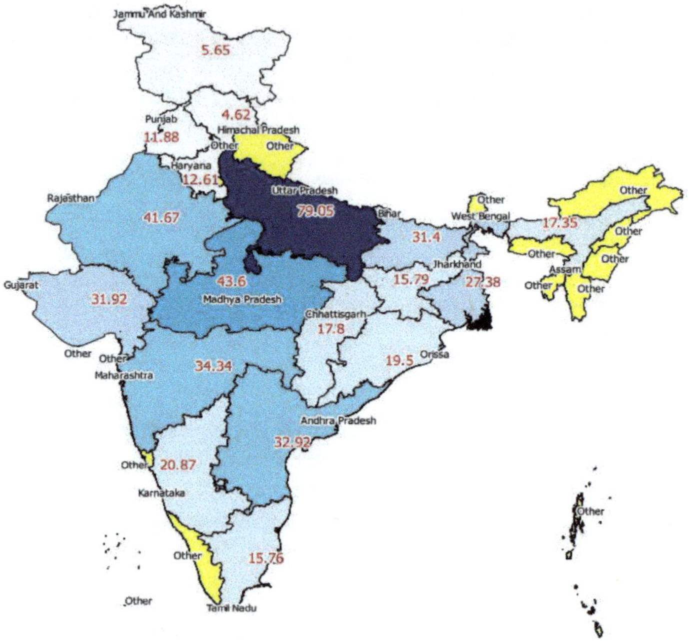

Figure 4. Potential of electrical energy generation (TWh per year) in different provinces of India.

The observations in this study with regard to the potential for energy generation are divergent from our calculations primarily because of variations inherent in the timing of the study, growth in animal population, higher recoverable dung mass per animal at present, availability of better techniques of energy extraction due to improved methods of extraction as per latest available technology, and consideration of all species of domesticated animals as generators of biomass in our study. The present annual electric energy demand in India has been calculated at 98,734 MU units. By our calculations, 477 TWh of electrical energy generation is possible from the identified biogas resources in our study, even if the lowest energy conversion efficiency at 25% is considered. If higher energy conversion efficiency is achievable by some means, an even greater amount of electricity can be generated. Our study suggests that livestock and poultry-based dung biomass utilization for electricity production can contribute to decreasing electric power generation costs in a consistent and sustainable manner, if exploited judiciously. This can contribute to reducing the expenditure on conventional fuels presently being employed to generate electricity in India. Coal, a non-renewable and heavily-mined raw material used for electricity generation at present is a major financial burden on the exchequer, in addition to its implications on environmental pollution. Biomass-based sustainable green energy production, by utilizing the huge biomass generated as animal waste can go a long way to reduce this burden.

5. Conclusions

India is largely an agrarian society with the majority of the population thriving in rural, environs enriched with natural resources. Livestock and poultry dung waste is a vital biomass resource that is largely wasted. Current utilization of this precious biomass involves conversion of dung into biogas by effective utilization of organic wastes through recycling operations and sometimes conversion to electrical energy in modern large-scale dairies. A robust analysis of livestock and poultry biomass generated in India and its potential as a renewable energy source has been presented. It can be safely expected that the vast biomass generated as livestock dung in India at 2600 million tons per annum has huge potential if exploited as an energy source. Our calculations suggest that it may be possible to generate 477 TWh of electrical energy per annum if the current potential of livestock dung to generate 263,702 million m³ of biogas is exploited. Over the years, anaerobic digestion technology has improved substantially to maximize the production of harvestable biogas from livestock dung. This means that modern AD technology may effectively contribute to higher yields of biogas, particularly methane, which can serve as a green and renewable source of energy. However, there exist plenty of bottlenecks in initiating and actually making use of such technologies. Some of the identifiable bottlenecks include economic constraints in building the infrastructure, logistic issues, outdated and inefficient mechanisms of waste disposal, collection and management, and a general lack of acceptability to new ideas in favor of time-tested traditional methods of waste disposal. Despite constraints, even a fraction of this dung-based sustainable energy source, if utilized judiciously, can revolutionize rural lifestyles and open the door for establishing modern industries requiring electricity in the rural sector. Moreover, as traditional mechanisms of waste disposal are already operational, this new venture will hardly make a dent in the present rural lifestyle, as the technology needs to be made operational in a step-wise manner. Potential ventures in this arena need to be introduced in regions with heavy dairy presence as a means of livelihood. Pilot projects of this nature are already in operation and need to be strengthened to harness this green and sustainable source of energy.

Acknowledgments: The work is based on the doctorate thesis of the first author, who wants to affectionately express gratitude to her mentors. We deeply indebted to Rupinder Singh Kanwar and Charanjeet Sarangal for their invaluable contribution in english editing and livestock data handling. We are very grateful to I.K. Gujral Punjab Technical University for academic and technical support.

Author Contributions: Gagandeep Kaur contributed comprehensively to the preparation of manuscript. Yadwinder Singh Brar gave the technical advice and improved the structure of manuscript and, D. P. Kothari confer the scientific expert guidance for this study. All the authors have approved the submission of manuscript.

Conflicts of Interest: No potential conflict of interest reported by authors.

Abbreviations

Units Used

MU	Million Units
BU	Billion Units
MW	Mega Watts
MJ	Mega Joules
MT	Million Tones
kWh	Kilowatt hours
TWh	Terawatt hours
PJ	Peta Joules

References

1. IPCC. Climate Change 2014 Synthesis Report Summary for Policymaker's Summary. Available online: https://www.ipcc.ch/pdf/assessment-report/ar5/syr/SYR_AR5_FINAL_full_wcover.pdf (accessed on 28 March 2016).
2. Scheftelowitz, M.; Thran, D. Unlocking the energy potential of dung—An assessment of the biogas production potential at the farm level in Germany. *Agriculture* **2016**, *6*, 20. [CrossRef]
3. Hosseini, S.E.; Wahid, M.A.; Aghili, N. The scenario of greenhouse gases reduction in Malaysia. *Renew. Sustain. Energy Rev.* **2013**, *28*, 400–409. [CrossRef]
4. Hosseini, S.E.; Wahid, M.A. Development of biogas combustion in combined heat and power generation. *Renew. Sustain. Energy Rev.* **2014**, *40*, 868–875. [CrossRef]
5. Exxonmobil 2017 the Outlook for Energy: A View to 2040. Available online: http:/cdn.exxonmobil.com/~/media/global/files/outlook-for-energy/2017/2017-outlook-for-energy.pdf (accessed on 15 April 2017).

6. Abdeshahian, P.; Al-Shorgani, N.K.N.; Salih, N.K.M.; Shukor, H.; Kadier, A.; Hamid, A.A.; Kalil, M.S. The production of biohydrogen by a novel strain Clostridium sp. YM1 in dark fermentation process. *Int. J. Hydrog. Energy* **2014**, *39*, 12524–12531. [CrossRef]
7. Atlas of Sustainable Development Goals 2017 from World Bank Development Indicators. Available online: http://openknowledge.worldbank.org/handle/10986/26306 (accessed on 28 March 2016).
8. World Development Indicator. Available online: http://wdi.worldbank.org/table/WV.1 (accessed on 28 March 2016).
9. Mukhopadhyay, K. An assessment of biomass gasification based power plant in the sunderbans. *Biomass Bioenergy* **2004**, *27*, 253–264. [CrossRef]
10. Sinha, C.S.; Ramana, P.V.; Joshi, V. Rural energy planning in India: Designing effective intervention strategies. *Energy Policy* **1994**, *22*, 403–414. [CrossRef]
11. Suresh, C. Biomass resource assessment for power generation: A case study from Haryana state, India. *Biomass Bioenergy* **2010**, *34*, 1300–1308.
12. Bhattacharyya, S.C. Energy access problem of the poor in India: Is rural electrification a remedy? *Energy Policy* **2006**, *34*, 3387–3397. [CrossRef]
13. Murphy, J.D.; McKeogh, E. Technical, economic and environmental analysis of energy production from municipal solid waste. *Renew. Energy* **2004**, *29*, 1043–1057. [CrossRef]
14. Chasnyk, O.; Solowski, G.; Shkarupa, O. Historical technical and economic aspects of biogas development: Case of Poland and Ukraine. *Renew. Sustain. Energy Rev.* **2015**, *52*, 227–239. [CrossRef]
15. Sun, Q.; Li, H.; Yan, J.; Liu, L.; Yu, Z.; Yu, X. Selection of appropriate biogas upgrading technology—A review of biogas cleaning, upgrading and utilization. *Renew. Sustain. Energy Rev.* **2015**, *51*, 521–532. [CrossRef]
16. Logan, B.E.; Regan, J.M. Microbial challenges and fuel cell applications. *Environ. Sci. Technol.* **2010**, *40*, 5172–5180. [CrossRef]
17. Rahimnejad, M.; Adhami, A.; Darvari, S. Microbial fuel cell as new technology for bioelectricity generation: A review. *Alex. Eng. J.* **2015**, *54*, 745–756. [CrossRef]
18. Rabaey, K.; Lissens, G.; Siciliano, S.D.; Verstraete, W. A microbial fuel cells capable of converting glucose to electricity at high rate and efficiency. *Biotechnol. Lett.* **2003**, *25*, 1531–1535. [CrossRef] [PubMed]
19. Raj, A.; Jhariya, M.K.; Toppo, P. Cow-dung for eco-friendly and sustainable productive farming. *Int. J. Sci. Res.* **2014**, *3*, 201–202.
20. Understanding Energy Challenges in India. Available online: http://www.iea.org/publications/freepublications/publications/publication/india_study_FINAL_WEB.pdf (accessed on 28 March 2017).
21. National Electricity Plan. Volume 1, Generation. Available online: http://climateobserver.org/wp-content/uploads/2015/01/National-Electricity-Plan.pdf (accessed on 15 February 2017).
22. Energy Statistics 2016. Twenty Third Issue by Central Statistics Office. Available online: http://mospiold.nic.in/Mospi_New/upload/Energy_statistics_2016.pdf (accessed on 26 December 2016).
23. 19th Livestock Census, All India Report. Ministry of Agriculture and Farmer welfare, Department of Animal Husbandry, Dairying & Fisheries, Government of India. Available online: http://www.indiaenvironmentportal.org.in/content/399839/19th-livestock-census-2012-all-india-report/ (accessed on 15 September 2016).
24. Kumar, A.A. A study on renewable resources in India. In Proceedings of the IEEE International Conference on Environmental Engineering & Application (ICEEA), Singapore, 10–12 September 2010; pp. 49–53.
25. Saha, S.; Biswas, S.; Pal, S. Survey analysis scope and applications of biomass energy in India. In Proceedings of the 2014 1st International Conference on Non Conventional Energy (ICONCE 2014), Kalyani, West Bengal, India, 16–17 January 2014; pp. 136–141.
26. Varshney, R.; Bhagoria, J.L.; Mehta, C.R. Small scale biomass gasification technology in India—An overview. *J. Eng. Sci. Manag. Educ.* **2010**, *3*, 33–40.
27. Energy for All: Financial Access for the Poor. Available online: http://www.iea.org/media/weowebsite/energydevelopment/presentation_oslo_oct11.pdf (accessed on 14 September 2015).
28. Balachandra, P. Dynamics of rural energy access in India: An assessment. *Energy* **2012**, *36*, 5556–5567. [CrossRef]
29. Power Sector at a Glance all India. Available online: http://powermin.nic.in/en/content/power-sector-glance-all-India (accessed on 14 April 2017).

30. Report on Power Sector 2017. Available online: http://cea.nic.in/reports/monthly/executivesummary/2017/exe_summary-01.pdf (accessed on 3 April 2017).
31. Pachauri, S.; Jiang, L. The household energy transition in India and China. *Energy Policy* **2008**, *36*, 4022–4035. [CrossRef]
32. Kumar, A.; Kumar, N.; Baredar, P.; Shukla, A. A review on biomass energy resources, potential, conversion and policy in India. *Renew. Sustain. Energy Rev.* **2015**, *45*, 530–539. [CrossRef]
33. Biomass Knowledge Portal. Available online: http://biomasspower.gov.in/ (accessed on 7 May 2017).
34. DADF 2016-17. Annual Report. Ministry of Agriculture and Farmer welfare, Department of Animal Husbandry, Dairying & Fisheries, Government of India. Available online: http://dahd.nic.in/sites/default/files/Annual%20Report%202016-17.pdf (accessed on 17 April 2017).
35. Review of Emission Factors and Methodologies to Estimate Ammonia Emissions from Animal Waste Handling. Available online: https://cfpub.epa.gov/si/si_public_record_Report.cfm?dirEntryID=55098 (accessed on 28 March 2016).
36. The US Inventory of Greenhouse Gas emission and Sinks: Fast Facts. Available online: https://www.epa.gov/ghgemissions/fast-facts-inventory-us-greenhouse-gas-emissions-and-sinks-1990-2014 (accessed on 17 January 2016).
37. Martinez, J.; Dabert, P.; Barrington, S.; Burton, C. Livestock waste treatment systems for environment quality, food safety and sustainability. *Bioresour. Technol.* **2009**, *100*, 5527–5536. [CrossRef] [PubMed]
38. Nasir, I.M.; Ghazi, T.I.M.; Omar, R.; Idris, A. Anaerobic digestion of cows dung: Influence of inoculums concentration. *Int. J. Eng. Technol.* **2013**, *10*, 22–26.
39. Fraenkel, P.L. Biomass and Coal (the Non-Petroleum Fuel). In *Water lifting Device, FAO Irrigation and Drainage paper 43*; Food and Agriculture Organization of the United Nations: Rome, Italy, 1986; pp. 286–306.
40. Sunil, B.; Mathews, E.B. Management of livestock waste. *J. Indian Vet. Assoc.* **2015**, *13*, 13–15.
41. Resource Assessment for Livestock and Agro-Industrial Wastes–India. Available online: https://www.globalmethane.org/documents/ag_india_res_assessment.pdf (accessed on 15 February 2017).
42. Surendra, K.C.; Takar, D.; Andrew, G.; Hashimoto; Khandal, S.K. Biogas sustainable energy source for developing countries: Opportunities and challenges. *Renew. Sustain. Energy Rev.* **2014**, *31*, 846–859. [CrossRef]
43. Maithal, S. Biomass Energy Resources Assessment Handbook. Available online: http://apctt.org/recap/sites/all/themes/recap/pdf/Biomass.pdf (accessed on 4 April 2016).
44. Fact Sheet "Biomass—An Important Renewable Energy Sources". Available online: http://www.wmaa.asn.au/lib/pdf/07_publications/1306_biogas_factsheet.pdf (accessed on 6 May 2017).
45. Gupta, K.K.; Aneja, K.R.; Rana, D. Current status of cow as a bioresource for sustainable development. *Bioresour. Bioprocess* **2016**, *3*. [CrossRef]
46. Rao, P.V.; Baral, S.S.; Day, R.; Mutnuri, S. Biogas generation potential by anerobic digestion for sustainable energy development in India. *Renew. Sustain. Energy Rev.* **2010**, *14*, 2086–2094. [CrossRef]
47. Buragohain, B.; Mahanta, P.; Moholkar, V.S. Biomass gasification for decentralized power generation: Indian perspective. *Renew. Sustain. Energy Rev.* **2010**, *14*, 73–92. [CrossRef]
48. Lovely, D. Microbial fuel cells: Novel microbial physiologies and engineering approaches. *Curr. Opin. Biotechnol.* **2006**, *17*, 327–332. [CrossRef] [PubMed]
49. Franks, A.E.; Nevin, K.P. Microbial fuel cells: A current review. *Energies* **2010**, *3*, 899–919. [CrossRef]
50. Tharali, A.D.; Sain, N.; Osborme, W. Microbial fuel cells in bioelectricity production. *Front. Life Sci.* **2016**, *9*, 252–266. [CrossRef]
51. Lovely, D.R.; Juice, B. Harvesting electricity with microorganism. *Nat. Rev. Microbiol.* **2006**, *4*, 497–508. [CrossRef] [PubMed]
52. Ravindranatha, N.H.; Somashekara, H.I.; Nagarahjaa, M.S.; Sudhaa, P.; Sangeethaa, G.; Bhattacharya, S.C.; Salam, P.A. Assessment of sustainable non plantation biomass resource potential for energy in India. *Biomass Bioenergy* **2005**, *29*, 178–190. [CrossRef]
53. Avcioglu, A.O.; Turker, U. Status and potential of biogas energy from animal wastes in Turkey. *Renew. Sustain. Energy Rev.* **2012**, *16*, 1557–1561. [CrossRef]
54. Kaygusuz, K. Renewable and sustainable energy use in Turkey: A review. *Renew. Sustain. Energy Rev.* **2002**, *6*, 339–366. [CrossRef]

55. Afazeli, H.; Jafari, A.; Rafiee, S.; Nosrati, M. An investigation of biogas production potential from livestock and slaughter house wastes. *Renew. Sustain. Energy Rev.* **2014**, *34*, 380–386. [CrossRef]
56. Abdeshhahian, P.; Lee, J.S.; Ho, W.S.; Hashim, H.; Lee, C.T. Potential of biogas production from farm animal waste in Malaysia. *Renew. Sustain. Energy Rev.* **2016**, *60*, 714–723. [CrossRef]
57. Deublein, D.; Steinhauser, A. *Biogas from Waste and Renewable Resources: An Introduction*; Wiley-VCH-Verl: Weinheim, Germany, 2011.
58. Cu'ellar, A.D.; Webber, M.E. Cow power: The energy and emissions benefits of converting dung to biogas. *Environ. Res. Lett.* **2008**, *3*. [CrossRef]

energies

MDPI

Article

Development of the IBSAL-SimMOpt Method for the Optimization of Quality in a Corn Stover Supply Chain

Hernan Chavez [1], Krystel K. Castillo-Villar [1,*] and Erin Webb [2]

1 Mechanical Engineering Department, The University of Texas at San Antonio, One UTSA Circle, San Antonio, TX 78249, USA; hernan.77@gmail.com
2 Environmental Sciences Division, Oak Ridge National Laboratory, One Bethel Rd., Oak Ridge, TN 37831, USA; webbeg@ornl.gov
* Correspondence: Krystel.Castillo@utsa.edu; Tel.: +1-210-458-6504

Received: 28 June 2017; Accepted: 31 July 2017; Published: 3 August 2017

Abstract: Variability on the physical characteristics of feedstock has a relevant effect on the reactor's reliability and operating cost. Most of the models developed to optimize biomass supply chains have failed to quantify the effect of biomass quality and preprocessing operations required to meet biomass specifications on overall cost and performance. The Integrated Biomass Supply Analysis and Logistics (IBSAL) model estimates the harvesting, collection, transportation, and storage cost while considering the stochastic behavior of the field-to-biorefinery supply chain. This paper proposes an IBSAL-SimMOpt (Simulation-based Multi-Objective Optimization) method for optimizing the biomass quality and costs associated with the efforts needed to meet conversion technology specifications. The method is developed in two phases. For the first phase, a SimMOpt tool that interacts with the extended IBSAL is developed. For the second phase, the baseline IBSAL model is extended so that the cost for meeting and/or penalization for failing in meeting specifications are considered. The IBSAL-SimMOpt method is designed to optimize quality characteristics of biomass, cost related to activities intended to improve the quality of feedstock, and the penalization cost. A case study based on 1916 farms in Ontario, Canada is considered for testing the proposed method. Analysis of the results demonstrates that this method is able to find a high-quality set of non-dominated solutions.

Keywords: renewable energy; bioenergy; biofuels; biomass; supply chain network design; logistics; discrete-event simulation; optimization; simulation-based optimization

---

1. Introduction

Biofuel has been recognized as an alternative source of renewable energy [1]. This research focuses on the use of energy crops, agricultural and forest residues to produce second-generation biofuels. First-generation biofuels are produced using edible products such as corn and soybean. These types of biomass feedstock have raised the debate of food versus fuel and, as a result, second-generation biofuels were developed. Second generation biomass feedstock exhibit more variability on its physical properties (e.g., higher ash and moisture contents) than first generation biomass. The variability of these physical properties has an important effect on the cost to deliver biomass feedstock to the reactor throat, which is known as delivered cost [2]. This delivered cost includes the cost of production, harvest, storage, handling, preprocessing, and transportation operations. Particularly, the cost of production includes the cost of feedstock, which accounts for 40–60% of the operating cost of conversion facilities [3,4].

The term *"biorefinery"* refers to the facility that processes biomass coming from plant, animal and food wastes into energy, fuels, chemicals, polymers, and food additives, among others [5]. At a constant operation rate, biorefinery processing facilities have lifetimes of 20 to 25 years [6]. Considering such a limited time for biorefineries to obtain profit on their investment, it becomes extremely important that all factors that have a direct effect on the delivered cost are considered when developing a bioenergy project. However, it is not possible to maintain delivered cost at a level that supports the profitability of the bioenergy projects and allows reduction of fuel cost when it is not possible to maintain the required supply of convertible biomass (i.e., biomass with uniform physical characteristics at the specification values determined by the conversion process).

The purpose of this paper is to propose and test a Discrete Events Simulation (DES) model coupled with a metaheuristic-based optimization method to design and manage a biomass supply chain (SC). The SC spans from the collection of biomass to the delivery at the gate of the biorefinery for its conversion to biofuel. The objective is to minimize the cost of having imperfect biomass quality and total dry matter loss.

The proposed model is based on an extension of the IBSAL (Integrated Biomass Supply Analysis and Logistics) model developed by Sokhansanj et al. [7]. This time-dependent DES model with activity-based costing has been extensively used by Oak Ridge National Laboratory (ORNL) to estimate the collection, storage, and transportation costs in biomass logistic systems. These estimates have been considered by the U.S. Department of Energy (DoE) and the bioenergy industry for making decisions related to the supply of biomass for the production of biofuel. In this paper, the IBSAL model is extended. The extended version estimates the cost of imperfect quality of feedstock and evaluates its effect on the performance measures of the SC. The inherent variability in biomass quality is added as well as several preventive pre-processing operations. Moreover, the extended IBSAL is coupled with a novel SimMOpt (Simulation-based Multi-Objective Optimization) method that allows finding near-optimal solutions. This method uses a Simulated Annealing (SA) approach for multi-objective problems with stochastic elements included in its formulations. In this way, the IBSAL (previously used as a manual what-if scenario tool) is enhanced to find near-optimal designs while considering multiple competing objectives.

This paper is structured as follows: Section 2 presents a literature review on models for biomass SCs. Section 3 describes the methodology: (a) the impact of feedstock quality; (b) interaction between the extended IBSAL and the SimMOpt optimization method; (c) detailed description of the SimMOpt optimization method; and (d) extended IBSAL with quality related blocks. Section 4 describes the case study and provides the geographical and operational characteristics of the implementation. The SA schedules (algorithmic parameters) considered during the experimentation are also discussed in this section. Section 5 presents the results of the case study using different SA schedules. Results are evaluated by computing a hyper volume indicator to determine the ability of the model of producing Pareto-optimal fronts. Section 6 discuss the quality of the set of non-dominated solutions for three SA schedules and future algorithmic improvements, which are intended to reduce the computational cost.

2. Literature Review

The delivered cost is highly sensitive to the variability on the attributes behavior (i.e., the properties of the elements intervening in the SC), and the interactions among them. Therefore, it is important to build models that realistically represent this variability and complex interactions, so that any conclusion based on the results from the experimentations with these models can be validated before design and management decisions are made to the real system. Reliable models, correct assumptions and accurate cost estimates are vital to appropriately design the system that will supply the feedstock to the conversion process.

Success in designing a biomass SC that optimizes the delivered cost depends on the model's ability to: (1) represent the behavior of the elements intervening in the SC; (2) consider plausible assumptions; and (3) perform the computations within a reasonable computational burden.

These models are commonly based on approaches such as simulation-based optimization or mathematical programming (MP). Simulation-based optimization models may require a considerable amount of execution time and data. Alternatively, in some cases, it is possible to formulate simpler time-independent models using methods such as metaheuristics or MP. Consequentially, there is a price to pay for using simplified models to represent complex systems.

Two factors that contribute to the complexity of the models are: (1) stochastic behavior; and (2) competition among the performance indicators (e.g., multiple criteria/objectives). Solving multi-objective stochastic programming models may require complex solution techniques and considerable execution time. This becomes even more complex when the formulations include non-linear expressions, for instance, quadratic programming (QP). Instances that remain treatable within sensible execution time may be limited to those of highly constrained size. Ignoring stochastic behavior and merging competing objectives into a single function are usual practices for simplifying these models in many applications. These simplifications occur at the expense of detail in the model which may jeopardize its value as a representation of the SC.

One approach to dealing with multiple objectives consists of simplifying the problem by merging all objectives of interest into a single objective (i.e., a single mathematical expression). However, in many real cases in the bioenergy sector, the decision maker(s) (whether it is a farmer or a biorefinery) is interested in individual production, harvest, storage, handling, and transportation costs or some other performance measures such as dry matter loss, percentage of moisture, percentage of ash, probability of failing to meet the specifications at the conversion facility, instead of the single estimation of the overall costs (i.e., delivered cost). In supply chains, it is not unusual to find that the optimization of some performance measures frequently competes with the optimization of others [8]. Merging multiple objectives into a single performance measure prevents the decision maker(s) from conducting individual evaluations on each objective. An even more important, information on the behavior of each measure and interactions among them is lost when competing objectives are merged. In any case, a set of solutions for different scenarios and priorities provide the decision maker(s) with a better understanding on the behavior and interactions among individual objectives. This has motivated the development of models that can be used for the optimization of multiple competing performance measures in biomass SCs.

The applicability of these models, whether simplified or complex, to specific situations depend on the information required by the decision maker(s), the required level of detail in the modeling of the SC, the required level of resolution on the solutions, available input data, and available time for the execution of the model.

Many models in the literature are readily applicable to the minimization of the collection, storage, and transportation costs derived from supplying biomass to biorefineries for the production of biofuels. However, an appropriate design and integral evaluation of a SC must also consider the cost from having imperfect quality of feedstock. Table 1 mentions some papers that discuss the development and applications of models developed for the design and improvement of biomass SCs. The last column is reserved for indicating papers that are relevant to the development and implementation of the IBSAL model.

Table 1. Models for biomass SCs.

Title	Author	Objective/Feedstock	Method	IBSAL
		Biomass Supply Chains		
Development of a multicriteria assessment model for ranking biomass feedstock collection and transportation systems	Kumar et al. [9]	Rank alternatives for the collection and transportation of biomass. Feedstock: Biomass (corn stover).	Multicriteria assessment methodology (PROMETHEE I, II) and IBSAL	√
Switchgrass (*Panicum vigratum*, L.) delivery to a biorefinery using integrated biomass supply analysis and logistics (IBSAL) model	Kumar and Sokhansanj [10]	Evaluate a collection systems of switchgrass and analysis of the related costs, energy input, and carbon emissions. Feedstock: Switchgrass	Discrete event simulation	√
Cotton logistics as a model for a biomass transportation system	Ravula et al. [11]	Determine the operating parameters for reducing the transportation cost of biomass. Feedstock: Cotton gin (analogously to biomass).	Discrete event simulation	
The impact of agricultural residue yield range on the delivered cost to a biorefinery in the Peace River region of Alberta, Canada	Stephen et al. [12]	Evaluate the impact of residue yield on the biomass delivered cost. Feedstock: Straw and chaff from wheat, barley, and oats.	Discrete event simulation	√
Techno-economic analysis of using corn stover to supply heat and power to a corn ethanol plant-Part 1: Cost of feedstock supply logistics	Sokhansanj et al. [13]	Propose a supply chain for corn stover to produce heat and power for a dry mill ethanol plant. Feedstock: Corn stover.	Discrete event simulation	√
A mathematical model to design a lignocellulosic biofuel supply chain system with a case study based on a region in Central Texas	An et al. [14]	Maximize the profit of a lignocellulosic biofuel supply chain. Feedstock: Switchgrass.	Multi-commodity flow mathematical model	
Impact of distributed storage and pre-processing on Miscanthus production and provision systems	Shastri et al. [15]	Analyze the cost reduction from implementing distributed storage and pre-processing at satellite storage locations. Feedstock: Miscanthus.	Mixed integer linear programming (MILP) (BioFeed) optimization model	
Economic and energy evaluation of a logistics system based on biomass modules	An and Searcy [16]	Minimize the feedstock costs in a logistic system by maximizing highway load and minimizing load/unload times. Feedstock: Cotton (analogously to biomass)	Discrete event simulation	√
An analysis of logistic costs to determine optimal size of a biofuel refinery	Larasati et al. [17]	Analyze the impact of logistic costs on the size of cellulosic ethanol biorefineries. Feedstock: Switchgrass.	Variable distance on grids approach	

Table 1. *Cont.*

Title	Author	Objective/Feedstock	Method	IBSAL
Biomass Supply Chains				
Development of an integrated tactical and operational planning model for supply of feedstock to a commercial-scale bioethanol plant	Ebadian et al. [18]	Integrate the tactical and operational levels in the biomass supply chain for a commercial-scale cellulosic ethanol plant. Feedstock: Multi-biomass.	Discrete event simulation/Mixed-integer linear programming	
Biomass round bales infield aggregation logistics scenarios	Igathinathane, et al. [19]	Evaluate logistic scenarios for aggregating biomass bales to a field-edge stack or a storage. Feedstock: Biomass.	Computer simulation	
Evaluation of a modular system for low-cost transport and storage of herbaceous biomass	Searcy and Hartley [20]	Optimize the harvest of energy sorghum in the humid southern region by using a modified cotton module builder for the formation of modules. Feedstock: Energy sorghum.	Discrete event simulation	√
Analyzing and comparing biomass feedstock supply systems in China: corn stover and sweet sorghum case studies	Ren et al. [21]	Analyze the harvest, collection, storage, transportation, preprocessing, handling, and queuing operations in the rural China biomass supply system. Feedstock: Corn stover and sweet sorghum.	System's dynamics. Biomass Logistic Model (BLM) framework (BLM Sino-Feedstock Supply (FS))	
Simulation-based multi-objective model for supply chains with disruptions in transportation	Chavez et al. [22]	Optimize the supply of agricultural products by using a simulation-based optimization model. Feedstock: Agricultural products.	Simulation-based Optimization	
Modelling a biomass supply chain through discrete-event simulation	Pinho et al. [23]	Providing a decisions tool for the biomass supply chain management services. Feedstock: Wood chip supply to a co-generation power plant.	Discrete event simulation (SIMEVENTS)	
Quantifying the impact of feedstock quality on the design of bioenergy supply chain networks	Castillo-Villar, Minor-Popocatl, and Webb [24]	Present a mixed-integer quadratically constrained programming model that minimizes the bioethanol SC, considering quality related costs. Feedstock: Logging residues.	Mixed-integer quadratic programming	

Table 1. *Cont.*

Title	Author	Objective/Feedstock	Method	IBSAL
Development of the IBSAL model				
Development and implementation of integrated biomass supply analysis and logistics model (IBSAL)	Sokhansanj et al. [7]	Describe the development of the IBSAL model. Feedstock: Corn stover.	Discrete event simulation	√
Development of the Integrated Biomass Supply Analysis and Logistics Model (IBSAL)	Sokhansanj, Turhollow, and Wilkerson [25]	Detailed description of the IBSAL model. Feedstock: Corn stover.	Discrete event simulation	√
A new simulation model for multi-agricultural biomass logistics system in bioenergy production	Ebadian et al. [26]	Propose a model for the supply of a mixture of feedstock to a cellulosic ethanol plant. Feedstock: Multi-agricultural.	Discrete event simulation (IBSAL-MC (multi-crop))	√
Methods to optimise the design and management of biomass-for-bioenergy supply chains: A review	Meyer et al. [27]	Present an overview of the optimization methods and model focusing on the design of biomass SCs. Feedstock: Biomass.	Various methods	
The ExtendSim Optimizer	Diamond [28]	Describe the technique used by the ExtendSim Optimizer®™ to find the best set of parameters for a simulation model. Feedstock: N/A.	Simulation optimization	
Availability of feedstock and technologies				
Availability of corn stover as a sustainable feedstock for bioethanol production	Kadam and McMillan [29]	Analyze the potential long-term amount of corn stover available to ethanol plants. Feedstock: Corn stover.	Analysis of a generalized model for stover production and removal	
Large-scale production, harvest and logistics of switchgrass (Panicum virgatum L.)—Current technology and envisioning a mature technology	Sokhansanj et al. [30]	Evaluation of technologies for the production, harvest, storage, and transportation. Feedstock: Switchgrass.	Empirical	
Evaluation of a modular system for low-cost transport and storage of herbaceous biomass	Searcy and Hartley [20]	Optimize the harvest of energy sorghum in the humid southern region by using a modified cotton module builder for the formation of modules. Feedstock: Energy sorghum.	Discrete event simulation	√
Influence of weather on the predicted moisture content of field chopped energysorghum and switchgrass	Popp et al. [31]	Determine the effect of weather on harvested moisture content. Feedstock: Switchgrass and energy sorghum.	Discrete event simulation	√

Some of the papers included Table 1 discuss the implementation of simulation-based methods for problems that consider multiple performance measures. The IBSAL model is used in some of these papers for: (1) searching for acceptable solutions; or (2) finding the value of some variables that serve as input parameters to other models proposed in that paper. Kumar et al. [9], and Kumar and Sokhansanj [10] present methodologies that use the IBSAL model for assessing alternatives for biomass collection and transportation systems while considering multiple performance measures (e.g., cost of delivered biomass, quality of biomass supplied, emissions during collection, energy input to the chain operation, and maturity of supply system technologies). Sokhansanj et al. [13] use the IBSAL model to estimate the economics (e.g., cost of collection, cost of pre-processing, transportation cost, on-site fuel storage and preparation costs, and overall delivered cost) of the supply of corn stover to produce heat and power for a drill mill ethanol plant. An and Searcy [16] use the IBSAL model to evaluate multiple performance measures (e.g., collection and processing cost, transportation cost, energy consumption, and CO_2 emissions) in a biomass logistic system. Searcy and Hartley [20] utilize the IBSAL for evaluating a system of transportation and storage of herbaceous biomass in terms of volumetric percentages of different biomass quality classes.

Other works do not necessarily use the IBSAL model, but present an implementation of Discrete Event Simulation (DES) models for designing biomass SCs. Ravula et al. [11] use a DES model to determine the operation parameters of a cotton gin SC while considering multiple performance measures. They describe in detail some key components that cotton gin transportation systems share with biomass transportation systems. Igathinathane et al. [19] propose a simulation model to determine the bale storage configuration that minimizes the required time, fuel, and incurred cost from handling bales at the storage facility or field-stack. Pinho et al. [23] use SIMEVENTS®™ to build a DES model for a typical biomass production chain. The performance measures that are considered in this model include transportation time, chipping time, and idle time among others.

Some other papers in Table 1 propose simulation-based methods but focus more emphatically on a single performance measure (e.g., overall delivered cost). For instance, Stephen et al. [12] use the IBSAL model to assess the impact of agricultural residue yield on the delivered cost. Variability on physical properties of feedstock has a considerable effect on yield.

A second group includes papers that propose models based on non-simulation-based approaches, such as MP. An et al. [14] propose a MP model to maximize the profit of a lignocellulosic biofuel supply chain. The scope of the formulation covers from the feedstock suppliers to the biofuel customers. Shastri et al. [15] formulate a MILP model (e.g., BioFeed) to maximize the profit of the system (i.e., Miscanthus production and provision system). Other papers such as Larasati et al. [17] explain the dynamics between competing performance measures, such as the trade-off between switchgrass transportation cost and the economic scale of cellulosic ethanol production. Castillo-Villar et al. [24] use a QP model that merges the total cost of transportation, total cost of the harvest processes, cost for opening a collection of facilities, cost for opening a collection of biorefineries, mechanical drying cost, ash disposal cost, screening cost, grinding cost, and a penalty cost for reduced oil yield due to high ash content into a single cost function to be minimized.

Finally, a third group consists of those papers that propose integration between simulation and optimization. Ebadian et al. [18] integrate an optimization model and a simulation model for designing and planning of a bioethanol plant in Canada. They describe the limitations of the integral model and the effect of the termination criterion on its ability to meet a global optimal solution. Ren et al. [21] use a systems dynamics (SD) approach to analyze the corn stover and sweet sorghum supply systems and estimate the logistic costs. Chavez et al. [22] propose a simulation-based optimization model for minimizing two competing objectives in a supply chain of perishable commodities.

The papers mentioned in section "Development of the IBSAL Model" in Table 1 describe in detail the IBSAL model, new versions of the IBSAL (e.g., Ebadian et al. [26]), or a methodology that the authors of this paper found relevant in successfully integrating an optimization methodology with the IBSAL model.

Section "Availability of Feedstock and Technologies" in Table 1 includes papers that provide insight into the effect of weather and time on the physical properties of feedstock as it moves along the supply chain. It also includes a paper that provides solid ground for the selection of the case study. Kadam and McMillan [29] state that about 60–80 million dry tonne/yr of stover can be potentially available for ethanol production. The importance of corn stover as a source of biomass (particularly in some regions) supports the selection of the case study presented in this paper.

3. Methodology

3.1. The Impact of Feedstock Quality

In this study, two physical characteristics of the feedstock are used to quantify the quality of feedstock: (1) percentage of moisture content; and (2) percentage of ash. The SimMOpt model presented in this paper considers six goals during the optimization: (a) minimization of dry matter loss; (b) minimization of cost related to the implementation of activities intended to improve the physical characteristics of the feedstock (i.e., moisture content and ash content); (c) minimization of the percentage of moisture content; (d) minimization of the percentage of ash content; (e) minimization of the penalization dockage cost derived from failing in meeting the maximum acceptable percentage of ash at the gate of the conversion facility; and (f) minimization of the percentage of feedstock not meeting the moisture and ash specifications at the conversion process. All of these objectives are affected by these two quality-related physical characteristics. Hence, it is crucial to improve these physical characteristics.

The percentage of moisture content is a common variable used for quantifying the quality of the feedstock. From the conversion point of view, the moisture content does not significantly contribute to the cost [32]. However, the supply system is more sensitive to the costs derived from having imperfect moisture content [32]. Moisture content can have a significant impact on the transportation, preprocessing, and feedstock handling costs [33]. An appropriate moisture content (i.e., below 20% for corn stover) has a positive effect on the degradation and consumption of structural carbohydrates during prolonged storage [32].

Ash content is another physical characteristic of the feedstock that has a significant effect on the supply system. High content of ash reduces the pretreatment efficiency, increases the deterioration of handling systems, increases operational costs such as the usage of water, treatment of the waste stream, and ultimately disposing of the excess ash [33]. Bonner et al. [34] found that two thirds of the increase on the cost of the supply system for corn stover with ash content ranging from 10% to 25% was due to the feedstock replacement needed to maintain the supply of convertible biomass. The other third comes mainly from the ash disposal operation. Bonner et al. [34] assume the specification for ash content at 5%.

The variability on the physical characteristics of the feedstock generates the need for models that can accurately represent the stochastic behavior followed by interacting elements in complex biomass SCs.

3.2. Interaction Between the Extended IBSAL and the SimMOpt Optimization Method

The most powerful feature of DES is that it allows modeling complex systems with numerous interacting elements that follow stochastic behavior. DES models have been extensively used in the bioenergy industry. For instance, the IBSAL model has been used to represent the interactions among different operations in the biomass SCs and the variability on the behavior of the system with more accuracy than most MP formulations are capable of. However, this model has been consistently used as a manual what-if scenario tool.

The proposed IBSAL-SimMOpt is a method for finding near-optimal solutions for multi-objective stochastic problems. The IBSAL model iteratively interacts with the proposed SimMOpt method which is used for optimizing the set of competing objectives.

The IBSAL-SimMOpt method is a two-phase procedure that searches a near-optimal set of solutions. During the first phase (i.e., optimization using the SimMOpt) the method is initialized and particular values are given to the decision variables (i.e., perturbed solution). The SimMOpt method uses the Gaussian distribution to perturb the decision variables. A discretization of the Gaussian distribution allows exploring with a high probability those integer values close to the value of the selected variable in the best solution at a given iteration. This perturbation technique is intended to improve the exploitation conditions during local searches. The appropriate selection of the standard deviation improves the exploration conditions. This perturbed solution becomes the input to the IBSAL model during the second phase (i.e., simulation using the extended IBSAL). During the second phase, several replications are computed for evaluating the competing objectives. These evaluations become feedback to a new iteration. Based on these evaluations, a new perturbed solution is computed. This process is repeated until the stopping conditions are met. The stochastic behavior of the system is represented by executing multiple times (i.e., replications) the extended IBSAL model using ExtendSim®™.

The proposed IBSAL-SimMOpt method can be described as follows:

- Extended IBSAL: This research is especially concerned with the costs derived from imperfect quality of feedstock. The IBSAL model is extended so that the cost of imperfect quality of the feedstock is estimated for the corn stover-to-bioethanol SC. For this purpose, blocks were added to the baseline IBSAL for estimating these quality-related costs and other performance measures of interest (e.g., dry matter loss at operations designed to improve the physical characteristics of the feedstock, percentage of moisture, percentage of ash, and probability of having non-conforming stover at the gate of the biorefinery). Extended IBSAL accurately represents the stochastic behavior of elements intervening in the SC and estimates the costs derived from collection, storage, and transportation operations and from having imperfect quality of feedstock.

- SimMOpt: The SimMOpt method is inspired on the procedure used by the stochastic Multi-Objective Simulated Annealing (MOSA). It can be expected that the solutions obtained with this method converge to global optima given the appropriate SA schedule. However, just as with any metaheuristic, optimality is not guaranteed. The procedure can be described as follows:

    (a)   At the initial iteration, a large set of random solutions is computed. The best solutions for each objective of interest determine a cell in the grid of the Pareto front. The initial temperatures are computed.

    (b)   Solutions are perturbed and several replications of the evaluation of the objectives are computed. This part is where the extended IBSAL connects with the SimMOpt. The extended IBSAL model is used to perform the replication and return the evaluations of the performance measures to the SimMOpt.

    (c)   The SimMOpt method uses some hypothesis testing techniques and the Pareto Archive Evolution Strategy (PAES) with the evaluations of the performance measures from the extended IBSAL to find good solutions and perturb the solution that will be used in the following iteration.

    (d)   The procedure is repeated until some conditions are met (e.g., minimum temperature, value of the objectives, maximum number of iterations, maximum number of rejects, and maximum number of accepts).

- The extended IBSAL and the SimMOpt method are coupled (extended IBSAL-SimMOpt) and used to minimize the set of competing objectives.
- Non-dominated solutions and Pareto-optimal fronts constructed from these solutions are obtained for multiple SA schedules for the problem in the case study.
- Performances of the Pareto-optimal fronts are evaluated using the hypervolume indicator approach.

Figure 1 describes at a high level the interaction between the SimMOpt optimization method (FIRST PHASE) and the extended IBSAL model (SECOND PHASE). This figure shows how decision variables are input to the extended IBSAL from the SimMopt optimization method, and how objective evaluations considering these decision variables are input back to the SimMOpt from the replications performed on the extended IBSAL. The six objectives of interest and four types of decision variables considered in the case study are shown in Figure 1. The definitions of all the abbreviations, symbols and variables used in this paper (figures, tables, and equations) are enlisted in Appendix A.

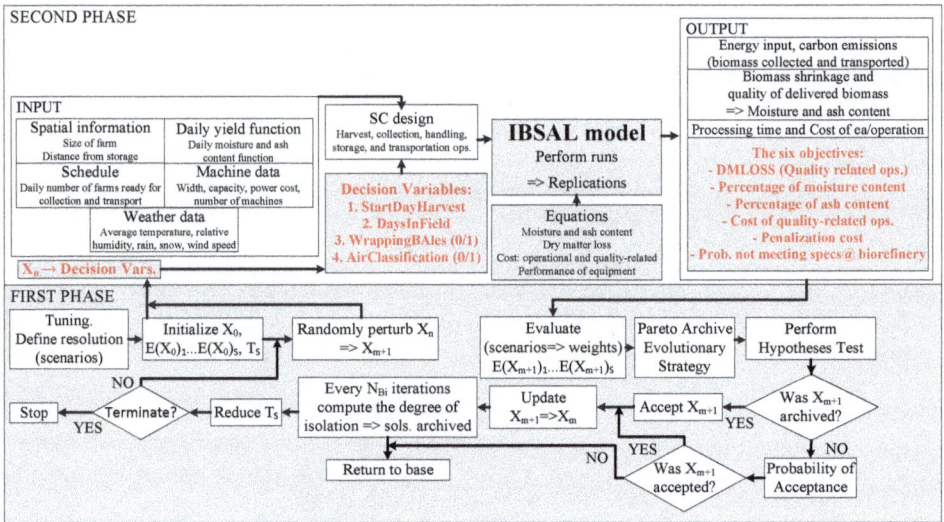

Figure 1. Extended IBSAL-SimMOpt method.

In the case study presented in Section 4, the types of decisions variables are: (1) initial day of harvest; (2) number of day allowed for field drying; (3) wrap the bales (binary variable); and (4) perform air classification operation (binary variable). The first two are manipulated by the SimMOpt method by perturbation with the purpose of reducing the percentage of moisture uniformly. The third is related to an activity designed to reduce dry matter loss after field drying. The fourth is related to an activity intended to reduce the percentage of ash content. One block is added to the extended IBSAL to represent the opportunity cost derived from failing in meeting the ash content specifications at the conversion facility (i.e., ash dockage). Current biorefineries operating at commercial scale have reported losses due to non-flowable poor quality of biomass and penalization fees [24].

The first and second types of decision variables are defined for each "farm land" (i.e., the amount of land that can be harvested based on the windrower capacity) passing through the Shred/Windrow Module in the extended IBSAL. The third type is defined for each bale passing though the Field-Side Storage Module. The fourth is defined for each truckload leaving the last module of the extended IBSAL model.

3.3. Detailed Description of the SimMOpt Optimization Method

The SimMOpt method is based on SA, which is analogous to the metal annealing process [35,36]. Although SA can be considered slow in converging to a global optimum when compared to other metaheuristics [35], it exhibits one extremely desirable feature. Given the appropriate schedule, which implies a slow cooling rate, convergence to a global minimum is ensured in most applications [35,37]. Figure 2 shows the details of the SimMOpt method and points where it connects with the extended IBSAL.

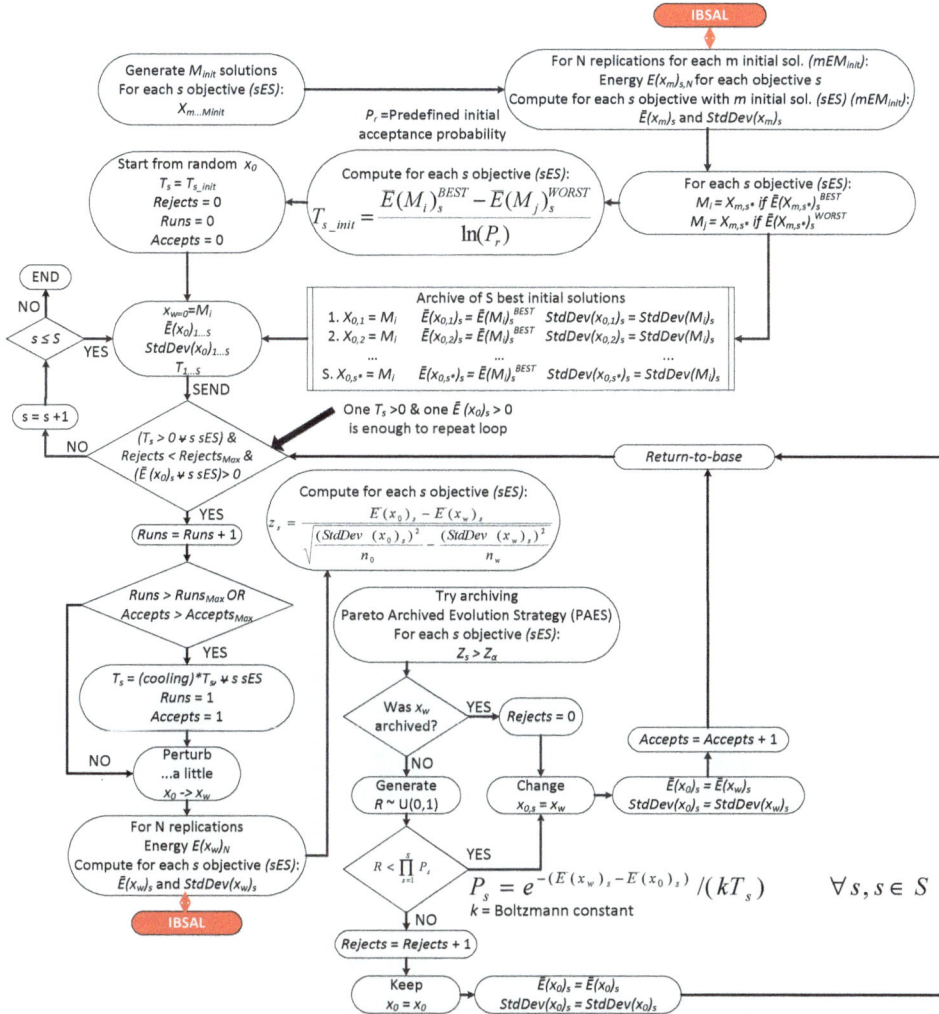

IBSAL

Generate M_{init} solutions
For each *s* objective *(sES)*:
$X_{m...Minit}$

For N replications for each m initial sol. *(mEM_{init})*:
Energy $E(x_m)_{s,N}$ for each objective *s*
Compute for each *s* objective with *m* initial sol. *(sES) (mEM_{init})*;
$\bar{E}(x_m)_s$ and $StdDev(x_m)_s$

P_r =Predefined initial acceptance probability

Start from random x_0
$T_s = T_{s\_init}$
Rejects = 0
Runs = 0
Accepts = 0

Compute for each *s* objective *(sES)*:
$$T_{s\_init} = \frac{E(M_i)_s^{BEST} - E(M_j)_s^{WORST}}{\ln(P_r)}$$

For each *s* objective *(sES)*:
$M_j = X_{m,s^*}$ if $\bar{E}(X_{m,s^*})_s^{BEST}$
$M_j = X_{m,s^*}$ if $\bar{E}(X_{m,s^*})_s^{WORST}$

END

NO

$s \leq S$ — YES

$s = s +1$

$x_{w=0} = M_i$
$\bar{E}(x_0)_{1...s}$
$StdDev(x_0)_{1...s}$
$T_{1...s}$
SEND

Archive of S best initial solutions

1. $X_{0,1} = M_i$ $\bar{E}(x_{0,1})_s = \bar{E}(M_i)_s^{BEST}$ $StdDev(x_{0,1})_s = StdDev(M_i)_s$
2. $X_{0,2} = M_i$ $\bar{E}(x_{0,2})_s = \bar{E}(M_i)_s^{BEST}$ $StdDev(x_{0,2})_s = StdDev(M_i)_s$
...
S. $X_{0,s^*} = M_i$ $\bar{E}(x_{0,s^*})_s = \bar{E}(M_i)_s^{BEST}$ $StdDev(x_{0,s^*})_s = StdDev(M_i)_s$

One $T_s >0$ & one $\bar{E}(x_0)_s > 0$ is enough to repeat loop

Return-to-base

NO

$(T_s > 0 \lor s \, sES)$ &
Rejects < Rejects$_{Max}$ &
$(\bar{E}(x_0)_s \lor s \, sES) > 0$

YES

Runs = Runs + 1

Compute for each *s* objective *(sES)*:
$$z_r = \frac{E(x_0)_s - E(x_w)_s}{\sqrt{\frac{(StdDev\,(x_0)_s)^2}{n_0} - \frac{(StdDev\,(x_w)_s)^2}{n_w}}}$$

Runs > Runs$_{Max}$ OR
Accepts > Accepts$_{Max}$

Try archiving
Pareto Archived Evolution Strategy (PAES)
For each *s* objective *(sES)*:
$Z_s > Z_\alpha$

YES

T_s = (cooling)$*T_s \lor s \, sES$
Runs = 1
Accepts = 1

Was x_w archived? — YES — Rejects = 0

Accepts = Accepts + 1

NO

Perturb
...a little
$x_0 \rightarrow x_w$

NO

Generate
$R \sim U(0,1)$

Change
$x_{0,s} = x_w$

$\bar{E}(x_0)_s = \bar{E}(x_w)_s$
$StdDev(x_0)_s = StdDev(x_w)_s$

For N replications
Energy $E(x_w)_N$
Compute for each *s* objective *(sES)*:
$\bar{E}(x_w)_s$ and $StdDev(x_w)_s$

IBSAL

YES

$R < \prod_{s=1}^{s} P_s$

$$P_s = e^{-(E(x_w)_s - E(x_0)_s)/(kT_s)} \qquad \forall s, s \in S$$
k = Boltzmann constant

NO

Rejects = Rejects + 1

Keep
$x_0 = x_0$

$\bar{E}(x_0)_s = \bar{E}(x_0)_s$
$StdDev(x_0)_s = StdDev(x_0)_s$

Figure 2. SimMOpt optimization method.

3.4. Extended IBSAL-SimMopt: Quality Related Blocks

Some elements necessary for the estimation of quality-related costs have been added to the extended IBSAL model. Preliminary versions of the new quality-related blocks were described in [38]. Appendix B shows a diagram block of a section of the original Shred/Windrow Module. The computations performed in the added blocks are discussed in this section.

Initial moisture content of the grain (variable "MoistGrainInit") (decimal): Moisture content of corn stover after harvest ranges from 35% to 50% when the grain moisture content is at 25%; this level is optimum for harvest. For the case presented by Sokhansanj et al. [7], initial moisture content of stover stalks is assumed at 72% when the grain moisture content is at about 40%. Harvest moisture content of grain at 40% is limited to cold and humid northern regions. In this study, the initial moisture content of grain is assumed to be uniformly distributed (0.35, 0.40) (decimal). Refer to block (83, 24) in the Biomass Field Module—extended IBSAL.

Windrow density (variable "WindrowDensity") (kg/m²): Khanchi and Birrel [39] evaluate the effect of rainfall and swath density on dry matter, and the composition change during the drying of corn stover. They assume three possible densities 0.8, 1.3, and 2.6 kg/m² (low density—LD, medium density—MD, and high density—HD, respectively). For this study, the windrow density is assumed to be uniformly distributed (0.8 and 2.6) (decimal). Refer to block (195, 15) in the Shred/Windrow Module—extended IBSAL.

Stover moisture content after harvest (variable "MoistContWet") (decimal): Moisture content of stover after harvest is determined based on the equations described by Sokhansanj et al. [7]. Refer to block (881, 74) in the Shred/Windrow Module—extended IBSAL.

$$\text{MoistEq}(t) = \left(\frac{1}{100}\right)\left(\frac{\log(1 - \text{RH}(t))}{-k_1(\text{Temp}(t) + k_2)}\right)^{k_3} \tag{1}$$

The coefficients of ear corn are $k_1 = 6.4424 \times 10^{-5}$, $k_2 = 22.15$, and $k_3 = 0.4795$ [7,25].

"RH" and "Temp" depend on "StartDayHarvest", which is one type of the decision variables that are iteratively evaluated and sent to the extended IBSAL by the SimMOpt optimization tool.

$$P(t) = \frac{\text{CumHarvestAreaDay}}{\text{TotalPlantingArea}} \tag{2}$$

"P(t)" is computed in block [211, 44] in the Shred/Windrow Module—extended IBSAL.

$$\text{MoistGrainWet}(t) = (\text{MoistGrainInit})(1 - P(t)) + (\text{MoistEq}(t))(P(t)) \tag{3}$$

$$\text{MoistContWet} = 0.73(1 - 3.9451e^{-18.8397(\text{MoistGrainWet}(t))}) \tag{4}$$

Initial ash content after harvest (variable "AshCont") (decimal): Schon and Darr [40] enlist the total ash content and all elemental components of ash for the windrowing method and single and multi-pass systems. From the multi-pass bale, this study assumed that the ash content is uniformly distributed (0.08, 0.12). Refer to block [881, 74] in the Shred/Windrow Module—extended IBSAL.

Stover moisture content after field drying (variable "MoistContWet") (decimal): Moisture content of stover after field drying is determined based on equations described by Khanchi and Birrel [41]. Refer to block [882, 52] in the Shred/Windrow Module—extended IBSAL.

$$\text{Mo} = \text{MoistContWet} \text{ (computed in block (881, 74)} \tag{5}$$

$$\text{Me} = (10.9173 - 0.0746 \, \text{Temp}(t))\left(\frac{\text{AvgRH}(t)}{1 - \text{AvgRH}(t)}\right)^{0.4147} \tag{6}$$

$$\text{VPD}(t) = \left(1 - \frac{\text{AvgRH}(t)}{100}\right)\left(6.11 \, e^{\frac{17.47 \, \text{Temp}(t)}{239 + \text{Temp}(t)}}\right) \times 100 \tag{7}$$

$$k = e^{(-2.5238 + 0.005564 \times \text{Rad}(t) - 0.1430 \times \text{WS}(t) + 0.0001081 \times \text{VPD}(t)}$$
$$^{-0.2212 \times \text{WindrowDensity} - 0.00074 \times \text{Rad}(t) \times \text{WS}(t))} \tag{8}$$

$$\text{MoistContWet} = \text{Me} + (\text{Mo} - \text{Me})e^{(-k \times t)} \tag{9}$$

Variable "t" depends on "StartDayHarvest" and "DaysInField", which are two types of decision variables that are iteratively evaluated sent to the extended IBSAL by the SimMOpt optimization tool.

Ash content after field drying (variable "AshCont") (decimal): A linear regression was performed on the data provided by Khanchi and Birrel [39]. The resulting equation is used for computing the percentage of ash content after field drying. (Refer to block [882, 52] in the Shred/Windrow Module—extended IBSAL).

$$AshCont = \frac{1.94 - 0.0027 \, \text{Rainfall}(t) + 2.37 \, \text{WindrowDensity}}{100} \qquad (10)$$

"Rainfall" variable depends on "StartDayHarvest" and "DaysInField".

Average moisture and ash content (i.e., average "MoistContWet" and "AshCont") are minimized by two objectives considered during the optimization with the SimMOpt method.

Dry matter loss during field drying (variable "DMLOSS") (decimal): Linear regression was performed on the data provided by Khanchi and Birrel [39]. The resulting equation is used to compute the percentage of dry matter lost during the field drying period. "0.35" in the numerator of Equation (11) is a multiplier factor used for predicting dry matter loss while biomass is in a "queue" [7,25]. Refer to block [882, 52] in the Shred/Windrow—extended IBSAL.

$$DMLOSS = \frac{0.35 \times (9.91 - 0.0191 \times \text{Rainfall}(t) - 0.444 \times \text{WindrowDensity})}{100} \qquad (11)$$

Items (i.e., farm lands) are delayed in a block a time equal to "DaysInField".

"WrappingBales" (binary: wrapping/no wrapping): This is the third type of decision variable. It is defined for each bale coming from a particular field. The cost of wrapping each bale is $3/bale and the processing time is 2 min/bale. Bales are delayed in a block during a time equal to the processing time. The dry matter loss percentage for wrapped ("DMLOSSWrap") bales is 3% [42]. Therefore, if the bales coming from the same farm are wrapped, then DMLOSS = DMLOSSWrap. The operation of wrapping bales reduces the percentage of dry matter loss. The cost of quality-related operations ("CostQual") is increased by the cost of wrapping (if applicable). Refer to block [889, 22] in the Field-Side Storage Module—extended IBSAL.

Total cost derived from quality-related operations and amount of dry matter loss (i.e., "CostQual" and "DMLOSS") are minimized by two objectives considered during the optimization with the SimMOpt method.

"AirClassification" (binary: classification/no classification): This is the fourth type of decision variables. It is defined for each truckload arriving to the conversion facility. The air classification operation is designed to reduce the content of ash. Processing time was assumed to be 3.4 ton/h [43]. The whole ash concentrations in air classified fractions was assumed that corresponding to the 15 Hz light sieve, which results in a uniformly distributed ash content of 6.44 ± 2.54% [43] (i.e., "AshCont" = U(0.0391, 0.0897). Total air classification cost is assumed at $1.05/ton of biomass [43]. The "CostQual" variable is increased accordingly. Refer to block [1267, 120] in the Unload (Facility) Module—extended IBSAL.

"Dockage cost" (CostNoQual): This variable is related to the fifth objective considered during the optimization with the SimMOpt method. This variable is used to compute the cost from having truckloads with bales that have ash content over 5%. These corresponding computations assume that dockage cost is calculated as $1.90/ton for every percentage point of ash above 5% [43]. Refer to block [1262, 114] in the Unload (Facility) Module—extended IBSAL.

In the last block in the last module (i.e., Unload (Facility) Module) in the extended IBSAL, it is possible to identify the items that do not meet the specifications of a maximum of 20% and 5% for the percentage of moisture and ash content, respectively. The minimization of the percentage of items failing to meet these requirements represents the sixth and last objective considered during the optimization. Refer to block [1262, 114] in the Unload (Facility) Module—extended IBSAL.

4. Case Study

The scope of the case study includes field operations, just in time delivery, and storage delivery of corn stover supplied for the production of bioethanol. The starting point of SC under analysis is the "creation of the corn field." The exit module is the "unload at the biorefinery." Figure 3 shows the modules of the corn stover SC for the production of bioethanol described in IBSAL for this case study.

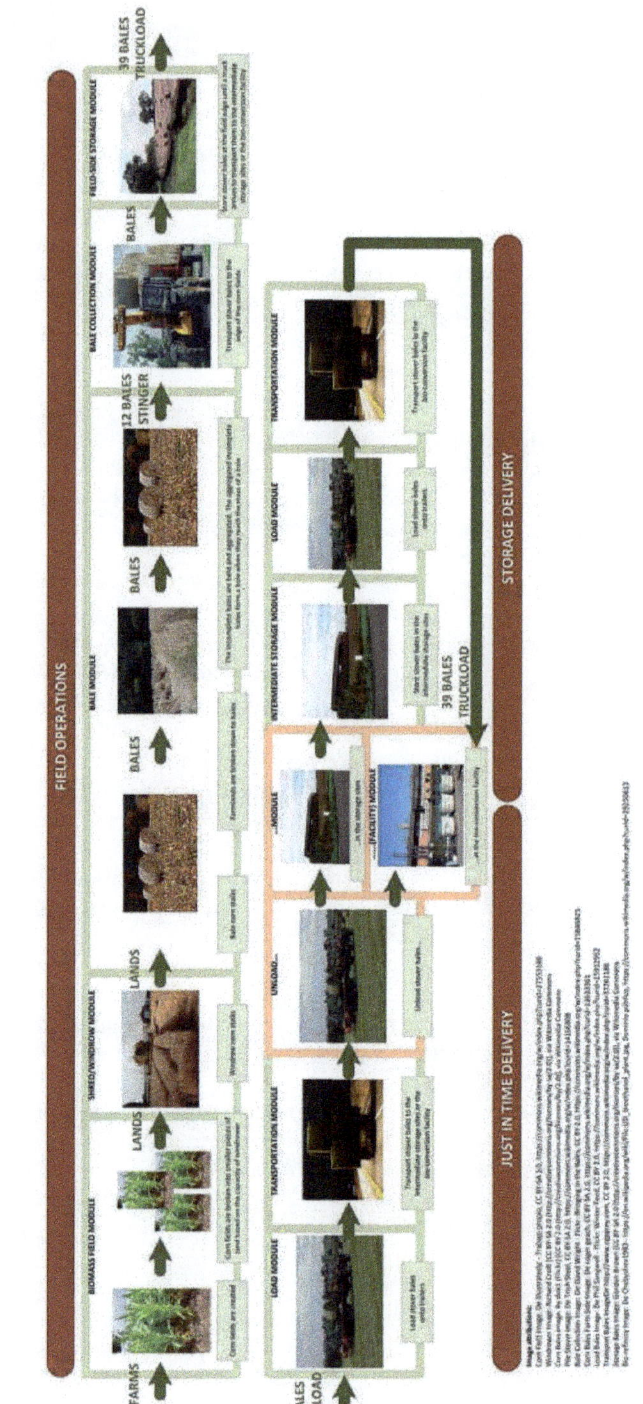

Figure 3. Corn stover SC for the production of bioethanol (IBSAL).

Some of the characteristics of the model used in this case study for representing a corn stover SC include: specific dimensions of the geographical implementation, geographical location of the biorefineries, distances from stacking (storage) locations, crop availability and yield, local weather data, fixed biorefinery and storage facility capacities, equipment available for collection and transportation, equipment-related costs (e.g., purchase price, maintenance cost, among others), analytical expressions for the estimation of the residue feedstock delivery price, initial values of the physical characteristics for the corn grain and formulas to compute estimations for the corresponding properties for the stover, and expressions for computing the cost derived from having imperfect feedstock quality in the SC.

4.1. Geographical Regions and Farm Input Data

The case study considers 1961 farms distributed in the counties of Lambton, Chatham-kent, Middlesex, and Huron, in the regions of Southern and Western Ontario, Canada. Input information for each farm was entered into IBSAL: size (ac), corn yield (bu/ac), produced corn stover (dry tonne/ac), sustainable corn stover removal (dry tonne/ac), total harvestable corn stover (dry tonne), satellite storage index (zone; only one zone was considered), distance from the satellite storage (km), and distance from the cellulosic sugar plant (km). Figure 4 shows the geographical location of the region considered in the case study.

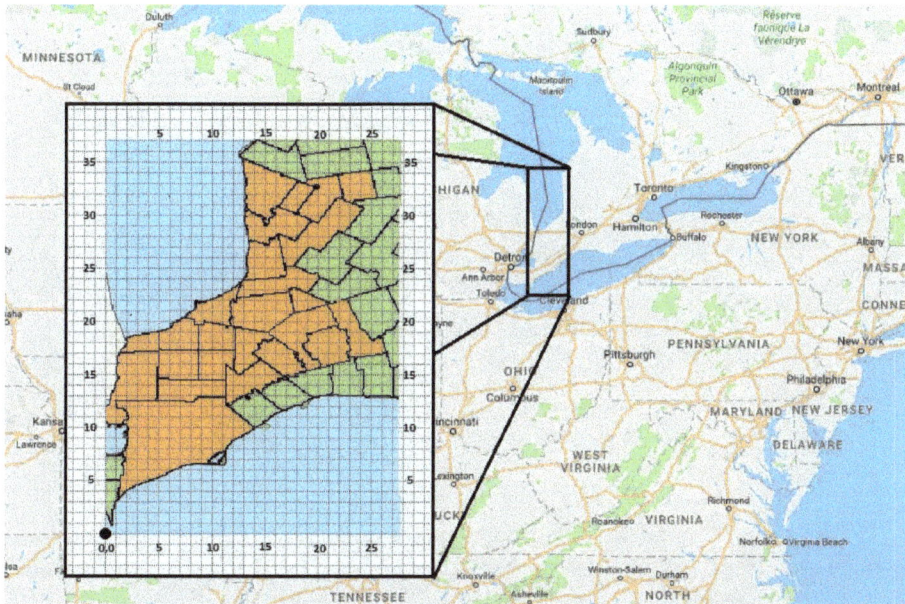

Figure 4. Geographical location (based on [44]).

4.2. Local Weather Input Data

Local weather input data include: daily average temperature (°C), daily snow on the ground (mm), monthly maximum humidity (decimal), monthly minimum humidity (decimal), daily evaporation (mm), daily rainfall (mm), monthly wind speed (m/s) (most common daily wind speed for each month), and monthly average radiation intensity for the hour (watt/m^2).

4.3. Machinery Input Data

Table 2 shows the equipment input data provided to the IBSAL model.

Table 2. Equipment input data.

Equipment	Input data
Mower-Conditioner	Width (ft), field speed average (mph), field efficiency (dec. fraction), machine efficiency, horsepower (hp), custom cost ($/h), annual fixed cost ($), variable cost ($/h), labor cost ($/h), purchase price ($), number of equipment, number of operators, machine biomass loss (decimal), and moisture content with zero machine loss (decimal w.b.)
Windrow-Tractor	Power, (hp), total cost ($/h), annual fixed cost ($), variable cost ($/h), labor cost ($/h), and purchase price.
Rectangular Baler	Baling pick up width (ft), field average speed (mph), field efficiency (dec. fraction), machine efficiency, power requirement (hp), custom cost ($/h), annual fixed cost ($), variable cost ($/h), labor cost ($/h), purchase price ($), number of equipment, number of operator, machine biomass loss (decimal), moisture content with zero machine loss (decimal w.b.), bale dimension width (ft), bale dimension height (ft), bale dimension length (ft), bale density (lb/ft^3), bales mass (ton), and cost of twine per bale ($/bale).
Tractor-Baler	Power (hp), total cost ($/h), annual fixed cost ($), variable cost ($/h), labor cost ($/h), and purchase price (S)
Infield-Transporter	Bale size 3x4x8 (volume ft^3), bale density (lb/ft^3), number of bale per load, load tie time—securing load (min/bale), load efficiency, travel speed full (mph), travel speed empty (mph), efficiency travel, unloading weight and inspection time (min/bale), efficiency unload, power (hp), custom cost ($/h), annual fixed cost ($), variable cost ($/h), labor cost ($/h), purchase price ($), number of equipment, number of operators, machine biomass loss (decimal), moisture content with zero machine loss (decimal w.b.), and winding factor.
Loader Storage	Number of bales per load, weight of each bale (ton), loading time per load (min), unloading time per load (min), efficiency, horsepower (hp), custom cost ($/h), annual fixed cost ($), operating cost ($/h), labor cost ($/h), purchase price ($), number of equipment, number of operators, machine biomass loss (decimal), and moisture content with zero machine loss (decimal w.b.)
Flatbed Trailer	Bale size 3x4x8 (volume ft^3), bale density (lb/ft^3), number of bale per load, load tie time—securing load (min/bale), load efficiency, travel speed full (mph), travel speed empty (mph), efficiency travel, unloading weight and inspection time (min/bale), efficiency unload, power (hp), custom cost ($/h), annual fixed cost ($), variable cost ($/h), labor cost ($/h), purchase price ($), number of equipment, number of operators, machine biomass loss (decimal), moisture content with zero machine loss (decimal w.b.), and winding factor.
Truck-Tractor	Power (hp), total cost ($/h), annual fixed cost ($), variable cost ($/h), labor cost ($/h), and purchase price
Bale Wrapper	Purchase price, stone pad cost ($/m^2), cost of plastic wrap per bale, wrapping time (min/bale)
Tarp	Tarp cost ($/ft^2), labor cost to remove and place tarps ($/ft^2), and tarp useful life (years)

4.4. Storage and Other Operational/Quality Input Data

The storage-related input data include the size of bales, stack width (bale), stack height (bale), stack length (bale), clearance between bale stack in a row (ft), clearance between bale stacks in a column (ft), and land charge for storage ($/ac).

Other input data are the daily biomass demand (dry tonne), nutrient replacement cost ($/tonne), daily working hours in the field, and daily working hours performing transportation activities. Input quality-related data consider a user-defined initial moisture content of grain (decimal), and initial ash content (decimal) for the initialization of the simulation model.

4.5. IBSAL Set Up and Tuning

The simulation setup of the extended IBSAL model was determined from literature [7,25]. The "end Time" was set at 8640 h (i.e., one year). The number of replications was set to 30 so it was possible to assume normality (based on the Central Limit Theorem) when performing the hypotheses tests during the optimization phase.

The tuning of the parameters for each SA schedule was determined by using a systematic screening process based on previous experiments with a simplified version of the IBSAL model [38]. Table 3 provides the levels of the parameters corresponding to the SA schedules considered in the screening process. Haddock et al. [35] further discuss the importance of the appropriate setting of these parameters.

Table 3. SA schedules.

Sched.	Pr	MinTemp (all)	MinOFVal (all)	$Rejects_{max}$	$Runs_{max}$	$Accepts_{max}$	k	A_{li}	N_{Bi}
*	0.95	0.001	0.01	10	10	10	1	10	100
**	0.6	0.001	0.01	5	5	5	1	10	10
***	0.6	0.001	0.01	10	10	10	1	10	100

A_{li} = Number of solutions in the candidate list from which a solution is selected when the SimMOpt returns-to-base [45]; N_{Bi} = The i-th return-to-base occurs after these many iterations [45].

5. Results and Discussion

The results from solving the case study by using the IBSAL-SimMOpt method are presented in this section. The solutions are evaluated by using a hypervolume indicator [46] to determine the ability of the model for producing Pareto-optimal fronts.

For schedule (*) it was noted that the percentage of feedstock not meeting the moisture and ash specifications at the conversion process remained constant throughout most of the simulation runs. Therefore, the objective of minimizing this probability was not considered for further analysis.

It can be shown that some of the objectives compete with each other, but this is not the case for all objectives under consideration. Some show almost no competition at all. For example, two objectives that change elastically in the same direction are the ash content and the penalization dockage cost. The corresponding Pearson correlation coefficient for the evaluations of the corresponding two objectives using schedule (*) is 0.861. Figure 5 shows the scatter plot of these evaluations. The best solution for each objective is marked with a cross. Other figures in this paper show that not all of objectives behave like this.

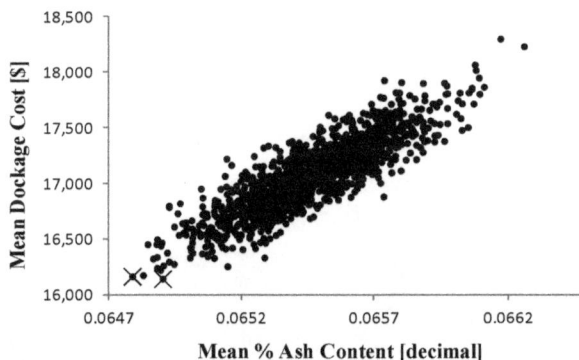

Figure 5. Scatter plot percentage of ash content vs. dockage cost (schedule (*)).

The purpose of the in-field drying operation is to uniformly reduce the percentage of moisture content. In the IBSAL model, the attribute that represents the percentage of moisture content is passed on to the farmlands, and to the bales. These attributes are averaged before forming the batches to be loaded to Stinger®™ equipment, and finally to the truck. This process results in values of this attribute uniformly distributed within a small range about values that depends on the harvest day and the number of days that the stover is left on the field to dry. In this case study, the initial moisture content after harvest is uniformly distributed with values close to 73% (Equation (4)). In Figures 6–8, it can be observed that this value is brought down by performing operations designed to improve the physical properties of feedstock to values in the range from 45% to 53%.

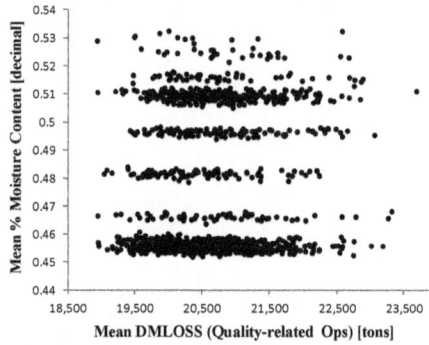

Figure 6. Average MC vs. DMLOSS (schedule (*)).

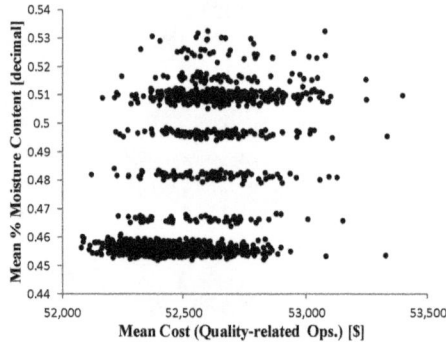

Figure 7. Average ash content vs. cost (schedule (*)).

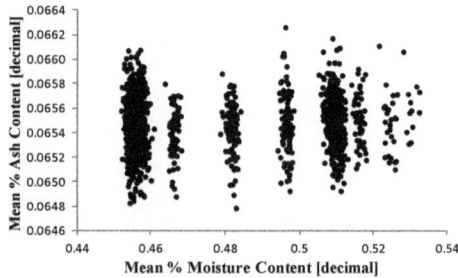

Figure 8. Average ash content vs. MC (schedule (*)).

In Figure 6, the correlation between the two objectives is close to zero. Moisture content depends on the relative humidity and temperature corresponding to the harvesting day and number of days the stover is left in the field (Equations (1) and (6)–(8)). In addition, dry matter loss depends on the amount of rainfall (Equation (11)). Due to lack of weather data for the region, the relative humidity was averaged for the entire month, while the rainfall was allowed to fluctuate during the month. This caused different moisture contents for fixed values of dry matter loss. Similarly, the percentage of ash depends on the amount of rainfall (Equation (10)). In Figure 8, the correlation between the average moisture content and average ash content is also close to zero.

The IBSAL computes the total tons (fraction of ton) per bale as follows (Refer to block [318, 11] in the Bale Module—extended IBSAL:

$$Lm = Lmmax \left(1 - \frac{MoistContWet}{Mwmax}\right) \tag{12}$$

$$DMLossMach = (Item\_Ton)(Lm) \tag{13}$$

$$Item\_Ton = Item\_Ton - DMLossMach \tag{14}$$

Similar formulas are used to compute the tonnage per farm land (Refer to block [195, 14] in the Shred/Windrow Module). Notice that larger values for moisture content result in more tons of dry matter per item (e.g., bales). This relation explains the behavior shown in Figure 7.

In Figure 7, the corresponding Pearson correlation coefficient is 0.3921. The cost derived from quality-related operations is computed from two sources: (1) the cost from wrapping bales; and (2) the cost from air classifying the truckloads. These costs increase with the number of bales, and tonnage per truckload, respectively. This is supported by the results from IBSAL. These results show that the cost derived from performing quality-related operations has a slight overall positive correlation with moisture content (Figure 7). However, selecting the non-dominated solutions for these objectives implies some trade-offs. Incurring low cost performing operations intended to improve quality results in high moisture content (Figure 9f).

For any pair of objectives in Figures 6–8, no single solution optimizes both objectives. Figure 9b,f,h shows the Pareto-optimal fronts for pairs of objectives corresponding to Figures 6–8, respectively. The remaining Pareto-optimal fronts for objective pairs are also show in Figure 9.

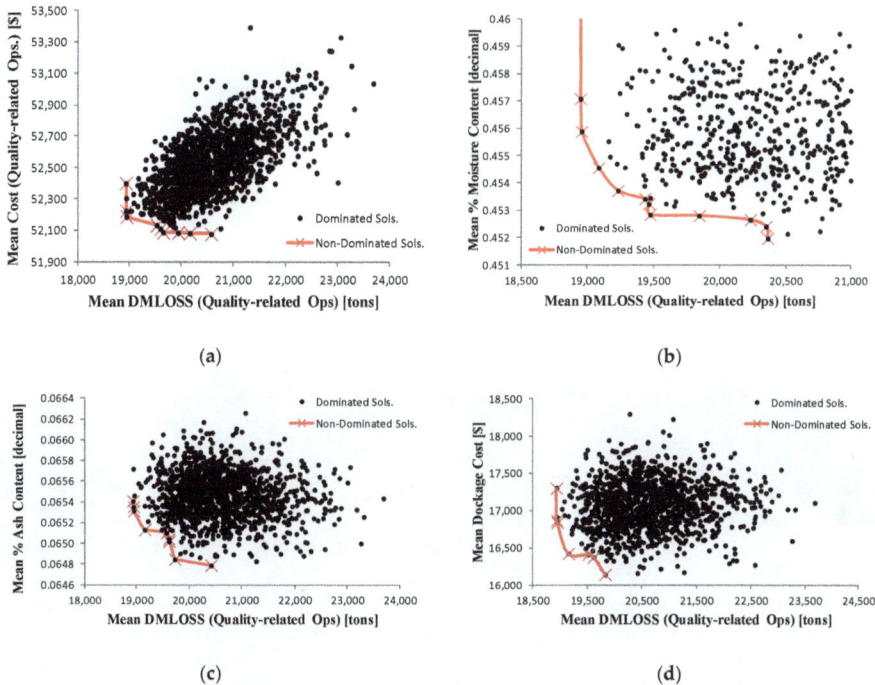

Figure 9. *Cont.*

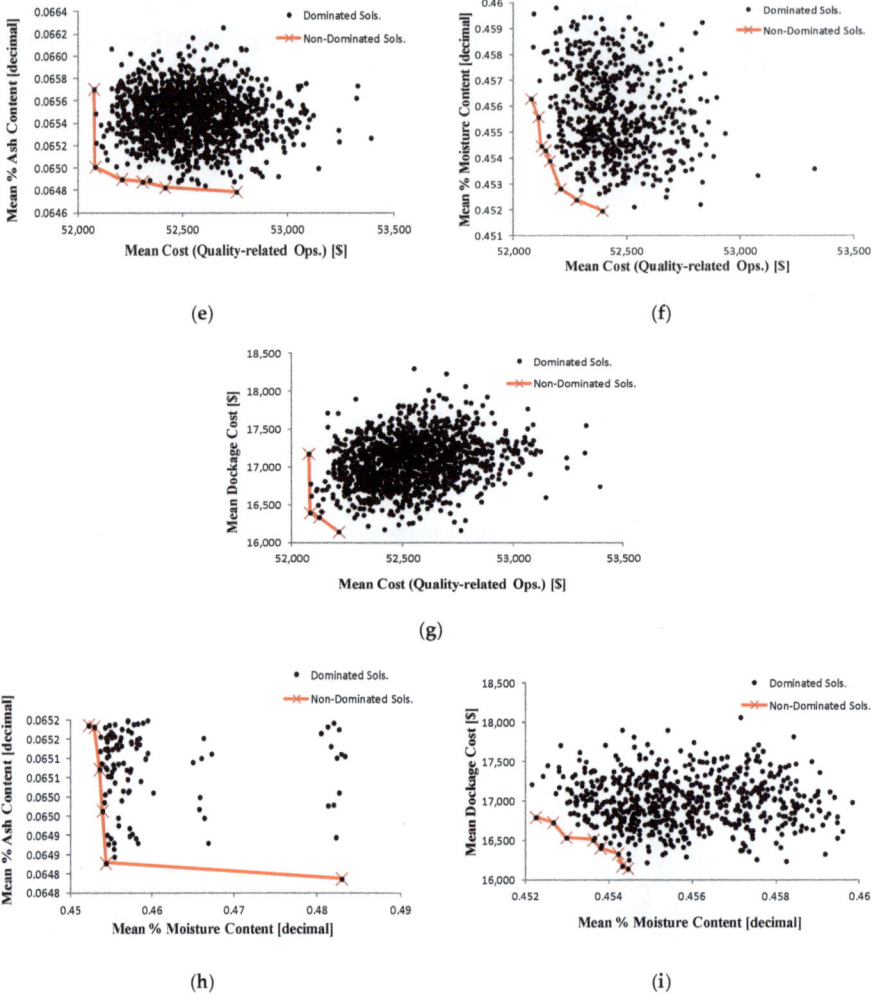

Figure 9. Pareto fronts (schedule (*)): (**a**) mean DMLOSS vs. mean cost; (**b**) mean DMLOSS vs. mean percent moisture; (**c**) mean DMLOSS vs. mean percent ash; (**d**) mean DMLOSS vs. mean dockage cost; (**e**) mean cost vs. mean percent ash; (**f**) mean cost vs. mean percent moisture; (**g**) mean cost vs. mean dockage cost; (**h**) mean percent moisture vs. mean percent Ash; and (**i**) mean dockage cost vs. mean percent moisture.

From the total of computed non-dominated solutions, 31 appeared in one out of the nine fronts (Figure 9), 10 solutions appeared in two, six solutions were present in three, and one solution is included in four. The IBSA-SimMOpt method computed 1366, 94, and 382 solutions using schedules (*), (**), and (***), respectively.

Not all of the non-dominated solutions shown in Figure 9 remain non-dominated solutions when considering the multi-objective problem (i.e., Many-objective Pareto-optimal front). Figure 10 shows the set of non-dominated solutions for the many-objective problem. It can be observed that no single solution (i.e., point) yields minimum values for all objectives. Implementation of any one particular solution from the set depends on the preferences of the decision-maker.

The hypervolume indicator approach [46] is used to compare the performance of the three schedules. Figure 11 shows the hypervolume indicators for the two schedules with the higher values. The random points are computed using a uniform distribution between the minimum values of the objectives from all the computed solutions (mapped onto a space bounded by axes (0, 1)) and 1. Of course, not all the random points represent feasible solutions. For example, the point where all objectives reach their minimum is not feasible as the objectives compete with each other during the optimization. Even when not all the random points represent feasible solutions, this method provides an estimator of the space where inhabiting points are dominated by a given Pareto front. This method allows the quality of the solutions, obtained from using multiple schedules, to be compared.

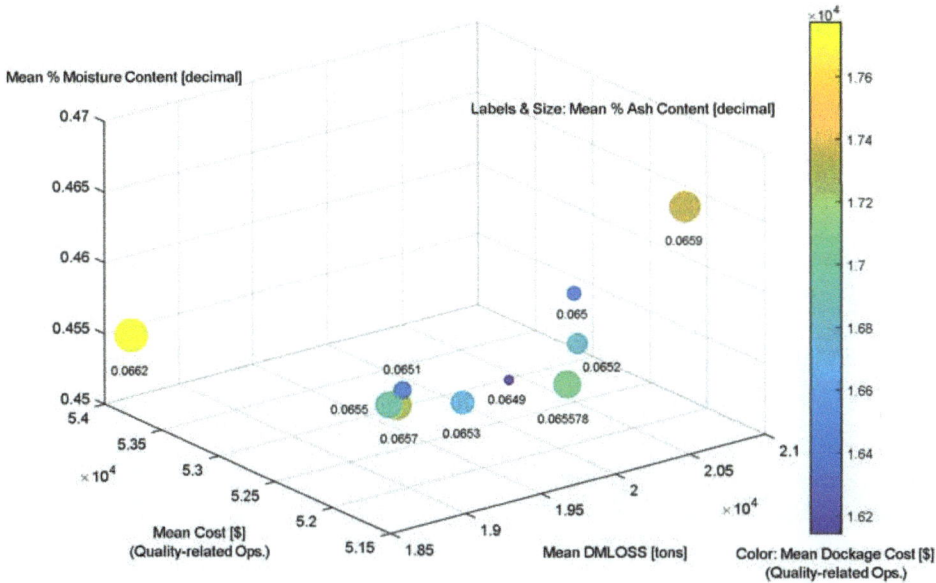

Figure 10. Many-objective set of non-dominated solutions (schedule (*)).

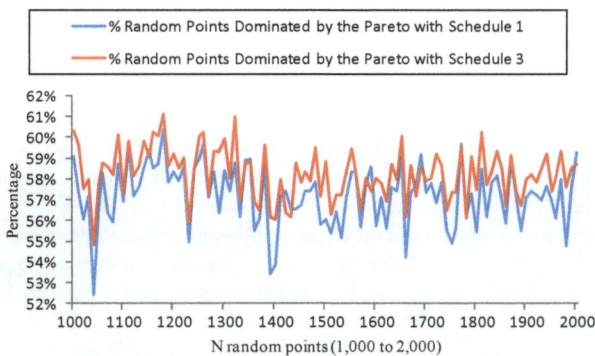

Figure 11. Hypervolume indicator approach (note: schedule 1 = schedule (*)).

According to the hypervolume indicators, schedule 3 (i.e., schedule (***)) outperformed the other two. By using this schedule, it is possible to construct a Pareto front with a greater probability of containing solutions that dominate random solutions. For example, for the experiment where

1390 random solutions (not necessarily all feasible) were generated, the non-dominated solutions obtained with schedule (***) outperformed over 56% (779 out of the 1390 random solutions) of them. Solutions from using schedule (**) only outperformed 53.5%. This difference may not seem significant, but the same conclusion can be reached from over a hundred different experiments. Moreover, the difference becomes greater in several experiments.

The practitioner is provided with a set of non-dominated solutions. These solutions become more or less attractive depending on the priorities determined by the decision-maker. For example, let us assume that the decision-maker is interested in the following objectives: (1) minimization of the cost derived from implementing activities intended to improve physical characteristics of the feedstock (i.e., moisture content and ash content); (2) minimization of dry matter loss; and (3) minimization of the percentage of moisture content. For illustration purposes, two solutions from the set of non-dominated solutions for the many-objective optimization are compared (Figure 12).

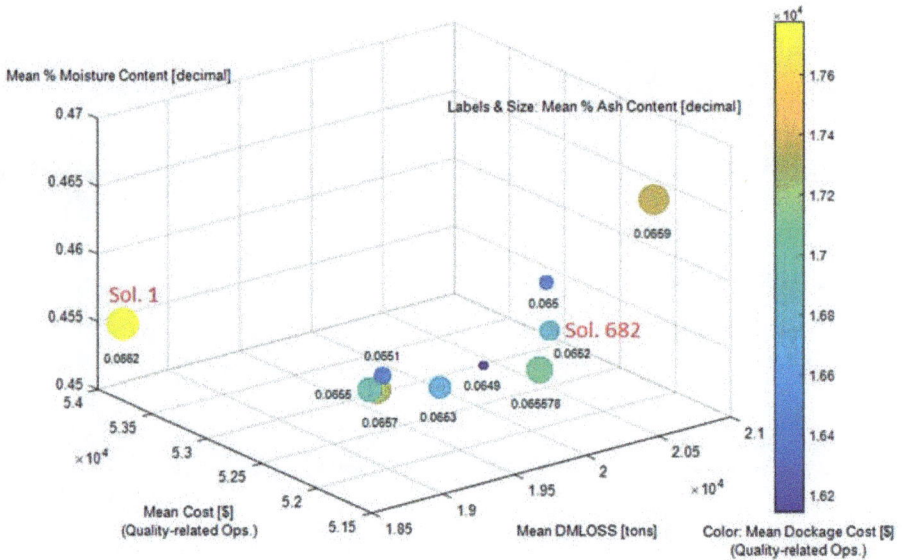

Figure 12. Solutions 1 and Solutions 682 (schedule 1).

Solution 1 implies that only 8.8% of the farm lands were left on the field for drying for five or more days. This solution resulted in an average moisture content of 45.44%. Solution 682 implies that 10.48% of the farm lands were left on the field for five or more days. The corresponding average moisture content remained close to 45%. However, the trade-off between solutions becomes more evident when the mean cost and the mean DMLOSS objectives are analyzed.

Solution 1 implies an average cumulative cost from quality-related operations of approximately $54,000. Solution 682 implies an average cumulative cost of approximately $52,000. Although the difference is modest, the reduction in dry matter loss from investing more resources on wrapping bales and air separating matter is equal to 2099 tons (solution 1).

Dockage cost is elastic with respect to the percentage of ash content. Even when the average percentage of ash content is similar for solutions 1 and solutions 682, the difference between the dockage costs is proportionally greater. Ash content is computed as the average from all the truckloads at the gate of the conversion facility while the dockage cost is computed as the cumulative cost (see Section 3). It is expected that the dockage cost increases as the ash content increases. However, for small increments on the average ash content, it is possible to observe dockage costs that differ considerably. This is also explained by the fact that dockage cost is affected by the mass at the gate of the biorefinery, and consequently by the cumulative dry matter loss.

The characteristics of the workstations used to run the experiments can set limitations on the size of the instances that can be solved by the proposed method. The equipment used for executing experiments is a computer with an Intel(R)®™ CORE®™ i7-3632QM central processing unit (CPU) processor that operates at 2.20 GHz with 8 GB of random access memory (RAM). The extended IBSAL model was modified and executed using Imagine That! Inc.®™ ExtendSim®™ 9 advanced technology (AT). The SimMOpt method was coded using Microsoft®™ (MS) Excel VBA®™.

6. Conclusions

The IBSAL-SimMOpt method provides the bioenergy industry with a tool capable of finding near-optimal solutions to a set of competing objectives, such as: minimization of dry matter loss, moisture content, ash content, costs related to different operations along the SC (e.g., collection, storage, and transportation), and cost derived from having imperfect quality of feedstock while accurately representing the stochastic behavior of the corn stover supply chain for the production of bioethanol. This provided without the need for additional assumptions intended to simplify the modeling of such complex system.

However, the validity of the model as a representation of the real SC is highly dependent on the assumptions related to the input parameters. These parameters attempt to represent, with numeric values and probability distributions, the behavior of elements interacting in the supply chain. Therefore, it is important to perform the corresponding statistical tests to evaluate the accuracy of these parameters. In the future, the equations included in the new blocks added to IBSAL will be validated using data obtained from field experiments and performing the appropriate statistical tests.

The quality of the solutions can be assessed by computing the hypervolume indicator for the Pareto-optimal solutions obtained from using different methods or settings of a particular method. Results show that the quality of the set of non-dominated solutions depends on the SA schedule. Therefore, it is important in real-life applications to allocate time for the execution of the trial runs required for the tuning of the parameters that define the schedule.

The applicability of the IBSAL-SimMOpt method to real biomass SCs largely depends on the computational effort (i.e., time and data resources) required for the execution of the model. The IBSAL-SimMOpt method required considerably large execution time for solving the problem in the case study. The computational time required by using the three schedules mentioned in Section 5 ranged from four hours to over 48 h. Future work involves quantifying the range of problem sizes that can be solved within reasonable time.

The main contributions of the extended IBSAL-SimMOpt method to biomass SC modeling can be summarized by the following:

- Provides a novel method for finding near-optimal solutions to the problem of designing biomass supply chains that improve the physical characteristics of the feedstock at an acceptable cost.
- Provides the bioenergy sector and the research community with a simulation-based optimization model with activity-based costing that is capable of representing the stochastic behavior of some elements intervening in the corn stover SC for the production of bioethanol. This analytical model can be used for decision making as well as educational and training purposes since it allows the user to conduct "what if" scenarios.
- Optimizes a set of competing individual objectives that include the cost of having imperfect corn stover quality for the production of bioethanol. The bioenergy industry will potentially benefit from this methodology, which can be adapted to minimize multiple costs in the biomass supply chain (e.g., collection, storage, transportation, and quality-related) as well as other key output measurements (e.g., dry matter loss, moisture content, and ash content) within a computational effort that is sensible for the solution of real problems. Here, the decision maker assigns the priority of objectives when finding a solution.

- Allows modeling complex systems with multiple intertwined stochastic elements; this is one of the most powerful features. The IBSAL-SimMOpt method captures the variability in the behavior of the elements of the biomass SC. Replications of the IBSAL model are performed to represent the stochastic behavior of the system.
- The SimMOpt method interacts with the output from the extended IBSAL model. There are no limitations in terms of the characteristics of the equalities/inequalities representing the objectives and constraints, which is an important concern when using mathematical programming instead of simulation.
- The IBSAL-SimMOpt method can be modified with relative ease using a top-down or bottom-up approach, as required. Therefore, the method can be easily transferred to applications involving biomass SCs with a wider range of feedstock to larger instances.

Acknowledgments: Funding for this research was provided by the U.S. Department of Energy, Office of Energy Efficiency and Renewable Energy, Bioenergy Technologies Office (4000142556). This material is based upon work that is supported by the National Institute of Food and Agriculture, U.S. Department of Agriculture, under award number 2014-38502-22598 (OK State Sub # 3TC460, 2568930.UTSA1) through South Central Sun Grant Program and the Hispanic-Serving Institutions (HSI) program (2015-38422-24064). The fellowship from the Mexican Council for Science and Technology (CONACYT) is gratefully acknowledged. The support provided by Imagine That!®™ by donating of a full version of ExtendSim®™ through the ExtendSim Research Grant is gratefully acknowledged. The research work on the IBSAL model by Sokhansanj, Turhollow, and Ebadian was relevant for the development of the proposed approach.

Author Contributions: Hernan Chavez developed the IBSAL-SimMOpt solution procedure and solved the instance of the extended IBSAL model. Krystel K. Castillo-Villar supervised and provided theoretical and conceptual advice. Erin Webb provided expert advice on biomass SC and the baseline IBSAL model.

Conflicts of Interest: The authors declare no conflict of interest. This manuscript has been authored by University of Tennessee (UT)-Battelle, Limited Liability Company (LLC) under Contract No. DE-AC05-00OR22725 with the U.S. Department of Energy. The United States Government retains and the publisher, by accepting the article for publication, acknowledges that the United States Government retains a non-exclusive, paid-up, irrevocable, worldwide license to publish or reproduce the published form of this manuscript, or allow others to do so, for United States Government purposes. The Department of Energy (DoE) will provide public access to these results of federally sponsored research in accordance with the DoE Public Access Plan (http://energy.gov/downloads/doe-public-access-plan). The founding sponsors had no role in the design of the study; in the collection, analyses, or interpretation of data; in the writing of the manuscript, and in the decision to publish the results.

Appendix A

Symbols

S	Set of objective
M	Set of solutions
X_m	Solution m (m $\in$ M)
$E(X_m)_s$	Evaluation of objective *s* by considering solution m (s $\in$ S, m $\in$ M)
T_s	Temperature related to objective s (s $\in$ S)
N_{Bi}	The *i*-th return-to-base occurs after these many iterations
N	Set of replications
M_{init}	Set of initial solutions
W	Set of computed solutions
X_m	Solution m (m $\in$ M_{ini})
$M_i = X_{m,s}*$	Best initial solution for objective s (m $\in$ M_{init}, i $\in$ M_{init}, s $\in$ S)
$M_j = X_{m,s}*$	Worst initial solution for objective s. (m $\in$ M_{init}, j $\in$ M_{init}, s $\in$ S)
x_w	Solution w computed by a series of perturbations starting from M_i (w $\in$ W)
$E(X_w)_s$	Evaluations of objective s by considering solution X_w (s $\in$ S, w $\in$ W)
T_{s_init}	Initial temperature
T_s	Temperature
$\bar{E}(X_w)_s$	Mean value from evaluations of objective s by considering solution X_w (s $\in$ S, w $\in$ W)
$StdDev(X_w)_s$	Standard deviation from evaluations of objective s by considering solution X_w (s $\in$ S, w $\in$ W)
$Runs_{max}$	Maximum number of runs (SA schedule)
$Rejects_{max}$	Maximum number of rejects (SA schedule)

Accepts$_{max}$	Maximum number of accepts (SA schedule)
z_s	Critical value (Hypotheses testing on the mean of evaluations of objective s for two large samples) ($s \in S$)
n_w	Size of sample consisting of the evaluations of objectives while considering solution X_w ($w \in W$)
z_α	Value from standard normal table corresponding to significance level α.
R	Pseudo-random number ~Uniform(0, 1)
P_s	Probability of acceptance of solution X_w based on the mean of the evaluations of objective s ($s \in S$)
K	Boltzmann constant
Sched.	SA Schedule
Pr	Probability of acceptance of neighbor solution
MinTemp	Minimum Temperature—stopping criterion
MinOFVal	Minimum value of the objective evaluation—stopping criterion
A_{li}	Number of solutions in the candidate list from which a solution is selected when the SimMOpt returns-to-base
Lm	Zero dry matter loss (due to machine) at 80% moisture content
Lmmax	Machine biomass loss (decimal)
MoistContWet	Moisture content in stover (decimal)
Mwmax	Moisture content with zero machine loss (decimal wb)
DMLoss	Loss tonnage due to machine (tonne)
ItemTom	Tonnage of item (e.g., land)
MoistGrainInit	Initial moisture content of the grain (decimal)
WindrowDensity	Windrow density (kg/m^2)
MoistEq(t)	Equilibrium moisture content on the grain (cob) at time t (decimal)
RH(t)	relative humidity at time t (decimal)
Temp(t)	Ambient air temperature at time t (°C)
P(t)	Cumulative harvested fraction at time t (decimal)
CumHarvestAreaDay	Cumulative harvested area (considered for all farm lands harvested on the same day) (ac)
MoistGrainWet(t)	Grain kernel moisture content at time t (decimal). Assumed ≥ 0.08
MoistContWet	Moisture content of the stover stalk (w.b.) (decimal)
Mo	Initial moisture content (decimal)
Me	Equilibrium moisture calculated by modified Oswin equation (decimal)
K	Drying rate constant (applies to Equations (8) and (9))
AvgRH(t)	Average relative humidity at time t (decimal)
VPD(t)	Vapor pressure density at time t
Rad(t)	Radiation intensity for the hour at time t (watt/m^2)
WS(t)	Wind speed at time t (m/s)
T	Each day the stover is left on the field for dying (days)
StartDayHarvest	Start day of harvest (type of decision variable)
DaysInField	Number of days left on the field for drying (type of decision variable)
AshCont	Ash content after field drying (decimal)
Rainfall(t)	Amount of rain at time t (mm)
DMLOSS	Dry matter loss during field drying (decimal)
WrappingBales	Binary: wrapping/no wrapping (type of decision variable)
DMLOSSWrap	Dry matter loss percentage for wrapped bales (decimal)
CostQual	Cost of quality-related operations ($)
AirClassification	Binary: classification/no classification
CostNoQual	Dockage cost ($)

Abbreviations

IBSAL	Integrated Biomass Supply Analysis and Logistics
SimMOpt	Simulation-based Multi-Objective Optimization
SA	Simulated Annealing
NREL	National Renewable Energy Laboratory
DES	Discrete Events Simulation
SC	Supply Chain

ORNL	Oak Ridge National Laboratory
DoE	Department of Energy
MP	Mathematical Programming
QP	Quadratic Programming
PROMETHEE	Preference Ranking Organization Method for Enrichment of Evaluations
MILP	Mixed-Integer Linear Programming
BLM	Biomass Logistic Model
FS	Feedstock Supply
IBSAL-MC	IBSAL Multi-Crop
SD	Systems Dynamics
MOSA	Multi-Objective Simulated Annealing
PAES	Pareto Archive Evolution Strategy
CTL	Central Limit Theorem
w.b.	Wet Basis

Appendix B

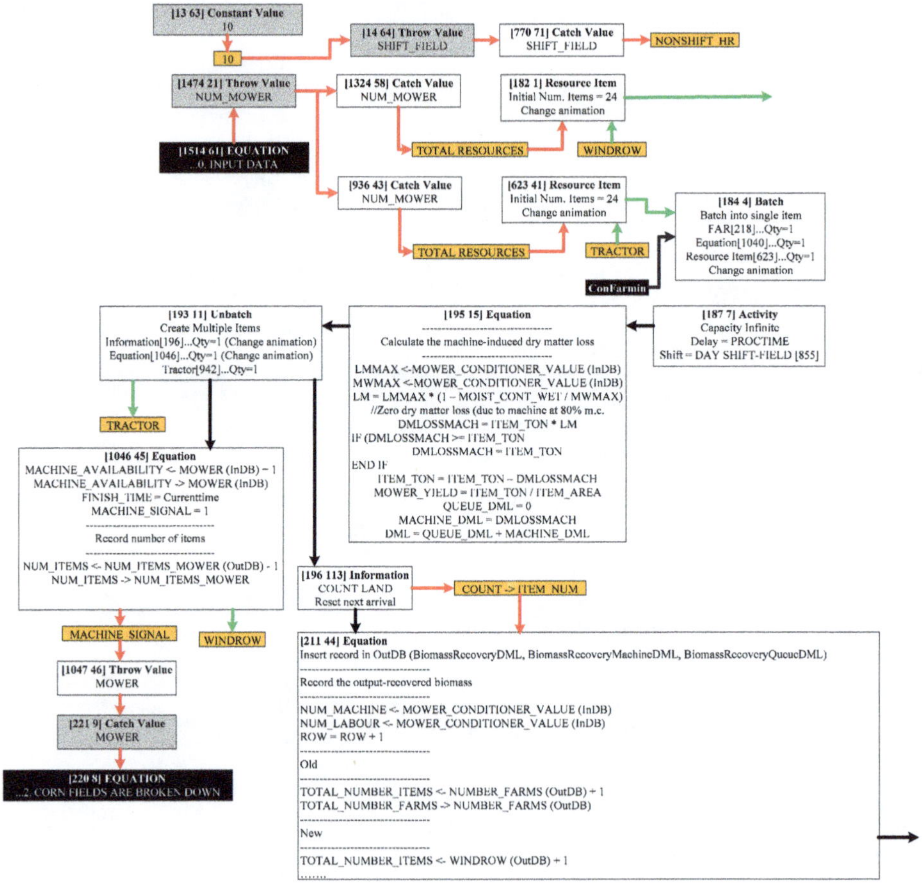

Figure A1. *Cont.*

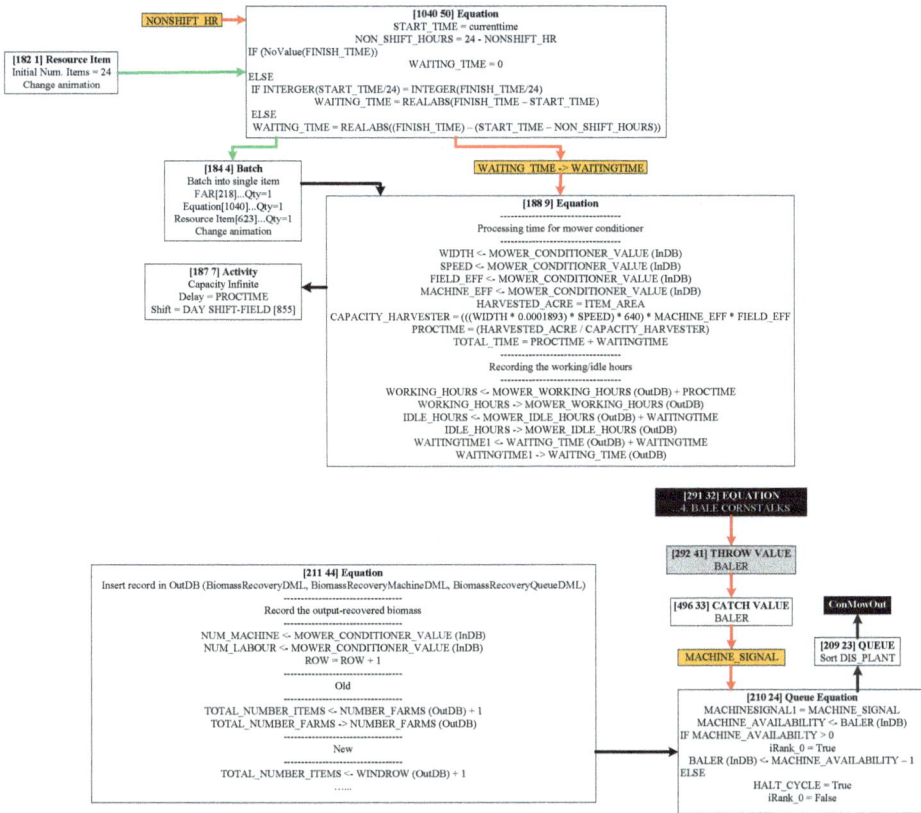

Figure A1. Section of the original Shred/Windrow module.

Bibliography

1. Office of Energy Efficiency & Renewable Energy. Available online: http://www.energy.gov/eere/bioenergy/biomass-feedstocks (accessed on 2 August 2017).
2. Langholtz, M.H.; Stokes, B.J.; Eaton, L.M. *2016 Billion-Ton Report: Advancing Domestic Resources for a Thriving Bioeconomy, Volume 1: Economic Availability of Feedstocks*; Oak Ridge National Laboratory: Oak Ridge, TN, USA, 2016.
3. Caputo, A.C.; Palumbo, M.; Pelagagge, P.M.; Scacchia, F. Economics of biomass energy utilization in combustion and gasification plants: Effects of logistic variables. *Biomass Bioenergy* **2015**, *28*, 35–51. [CrossRef]
4. Leistritz, F.L.; Hodur, N.M.; Senechal, D.M.; Stowers, M.D.; McCalla, D.; Saffron, C.M. Biorefineries Using Agricultural Residue Feedstock in the Great Plains. Available online: http://ageconsearch.umn.edu/bitstream/7323/2/ae070001.pdf (accessed on 2 August 2017).
5. Sadhukhan, J.; Ng, K.S.; Hernandez, E.M. *Biorefineries and Chemical Processes: Design, Integration and Sustainability Analysis*; John Wiley & Sons: Hoboken, NJ, USA, 2014.
6. Stephen, J.D. Biorefinery Feedstock Availability and Price Variability: Case Study of the Peace River Region, Alberta. Ph.D. Thesis, University of British Columbia, Vancouver, BC, Canada, 2008.
7. Sokhansanj, S.; Kumar, A.; Turhollow, A.F. Development and implementation of integrated biomass supply analysis and logistics model (IBSAL). *Biomass Bioenergy* **2006**, *30*, 838–847. [CrossRef]

8. Aslam, T.; Amos, H.N. Multi-objective optimization for supply chain management: A literature review and new development. In Proceedings of the 2010 8th International Conference on Supply Chain Management and Information Systems (SCMIS), Hong Kong, China, 6–8 October 2010; pp. 1–8.
9. Kumar, A.; Sokhansanj, S.; Flynn, P.C. Development of a multicriteria assessment model for ranking biomass feedstock collection and transportation systems. *Appl. Biochem. Biotechnol.* **2006**, *129*, 71–87. [CrossRef]
10. Kumar, A.; Sokhansanj, S. Switchgrass (*Panicum vigratum*, L.) delivery to a biorefinery using integrated biomass supply analysis and logistics (IBSAL) model. *Bioresour. Technol.* **2007**, *98*, 1033–1044. [CrossRef] [PubMed]
11. Ravula, P.P.; Grisso, R.D.; Cundiff, J.S. Cotton logistics as a model for a biomass transportation system. *Biomass Bioenergy* **2008**, *32*, 314–325. [CrossRef]
12. Stephen, J.; Sokhansanj, S.; Bi, X.; Sowlati, T.; Kloeck, T.; Townley-Smith, L.; Stumborg, M. The impact of agricultural residue yield range on the delivered cost to a biorefinery in the Peace River region of Alberta, Canada. *Biosyst. Eng.* **2010**, *105*, 298–305. [CrossRef]
13. Sokhansanj, S.; Mani, S.; Tagore, S.; Turhollow, A. Techno-economic analysis of using corn stover to supply heat and power to a corn ethanol plant–Part 1: Cost of feedstock supply logistics. *Biomass Bioenergy* **2010**, *34*, 75–81. [CrossRef]
14. An, H.; Wilhelm, W.E.; Searcy, S.W. A mathematical model to design a lignocellulosic biofuel supply chain system with a case study based on a region in Central Texas. *Bioresour. Technol.* **2011**, *102*, 7860–7870. [CrossRef] [PubMed]
15. Shastri, Y.N.; Rodriguez, L.F.; Hansen, A.C.; Ting, K. Impact of distributed storage and pre-processing on Miscanthus production and provision systems. *Biofuels Bioprod. Biorefining* **2012**, *6*, 21–31. [CrossRef]
16. An, H.; Searcy, S.W. Economic and energy evaluation of a logistics system based on biomass modules. *Biomass Bioenergy* **2012**, *46*, 190–202. [CrossRef]
17. Larasati, A.; Liu, T.; Epplin, F.M. An analysis of logistic costs to determine optimal size of a biofuel refinery. *Eng. Manag. J.* **2012**, *24*, 63–72. [CrossRef]
18. Ebadian, M.; Sowlati, T.; Sokhansanj, S.; Smith, L.T.; Stumborg, M. Development of an integrated tactical and operational planning model for supply of feedstock to a commercial-scale bioethanol plant. *Biofuels Bioprod. Biorefining* **2014**, *8*, 171–188. [CrossRef]
19. Igathinathane, C.; Archer, D.; Gustafson, C.; Schmer, M.; Hendrickson, J.; Kronberg, S.; Faller, T. Biomass round bales infield aggregation logistics scenarios. *Biomass Bioenergy* **2014**, *66*, 12–26. [CrossRef]
20. Searcy, S.W.; Hartley, B.E.; Thomasson, J.A. Evaluation of a modular system for low-cost transport and storage of herbaceous biomass. *BioEnergy Res.* **2014**, *7*, 824–832. [CrossRef]
21. Ren, L.; Cafferty, K.; Roni, M.; Jacobson, J.; Xie, G.; Ovard, L.; Wright, C. Analyzing and comparing biomass feedstock supply systems in China: Corn stover and sweet sorghum case studies. *Energies* **2015**, *8*, 5577–5597. [CrossRef]
22. Chávez, H.; Castillo-Villar, K.K.; Herrera, L.; Bustos, A. Simulation-based multi-objective model for supply chains with disruptions in transportation. *Robot. Comput. Integr. Manuf.* **2017**, *43*, 39–49. [CrossRef]
23. Pinho, T.M.; Coelho, J.P.; Moreira, A.P.; Boaventura-Cunha, J. Modelling a biomass supply chain through discrete-event simulation. *IFAC PapersOnLine* **2016**, *49*, 84–89. [CrossRef]
24. Castillo-Villar, K.K.; Minor-Popocatl, H.; Webb, E. Quantifying the impact of feedstock quality on the design of bioenergy supply chain networks. *Energies* **2016**, *9*, 203. [CrossRef]
25. Sokhansanj, S.; Turhollow, A.; Wilkerson, E. *Development of the Integrated Biomass Supply Analysis and Logistics Model (IBSAL)*; Oak Ridge National Laboratory: Oak Ridge, TN, USA, 2008.
26. Ebadian, M.; Sowlati, T.; Sokhansanj, S.; Stumborg, M.; Townley-Smith, L. A new simulation model for multi-agricultural biomass logistics system in bioenergy production. *Biosyst. Eng.* **2011**, *110*, 280–290. [CrossRef]
27. De Meyer, A.; Cattrysse, D.; Rasinmäki, J.; Van Orshoven, J. Methods to optimise the design and management of biomass-for-bioenergy supply chains: A review. *Renew. Sustain. Energy Rev.* **2014**, *31*, 657–670. [CrossRef]
28. Diamond, B. *The ExtendSim Optimizer*; Imagine That Inc.: San Jose, CA, USA, 2003.
29. Kadam, K.L.; McMillan, J.D. Availability of corn stover as a sustainable feedstock for bioethanol production. *Bioresour. Technol.* **2003**, *88*, 17–25. [CrossRef]
30. Sokhansanj, S.; Mani, S.; Turhollow, A.; Kumar, A.; Bransby, D.; Lynd, L.; Laser, M. Large-scale production, harvest and logistics of switchgrass (*Panicum virgatum* L.)—Current technology and envisioning a mature technology. *Biofuels Bioprod. Biorefining* **2009**, *3*, 124–141. [CrossRef]

31. Popp, M.P.; Searcy, S.W.; Sokhansanj, S.; Smartt, J.B.; Cahill, N.E. Influence of weather on the predicted moisture content of field chopped energy sorghum and switchgrass. *Appl. Eng. Agric.* **2015**, *31*, 179–190.
32. Kenney, K.L.; Cafferty, K.G.; Jacobson, J.J.; Bonner, I.J.; Gresham, G.L.; Hess, R.J.; Ovard, L.P.; Smith, W.A.; Thompson, D.N.; Thompson, V.S.; et al. *Feedstock Supply System Design and Economics for Conversion of Lignocellulosic Biomass to Hydrocarbon Fuels: Conversion Pathway: Biological Conversion of Sugars to Hydrocarbons*; Idaho National Laboratory: Idaho Falls, ID, USA, 2013.
33. Kenney, K.L.; Smith, W.A.; Gresham, G.L.; Westover, T.L. Understanding biomass feedstock variability. *Biofuels* **2013**, *4*, 111–127. [CrossRef]
34. Bonner, I.J.; Smith, W.A.; Einerson, J.J.; Kenney, K.L. Impact of harvest equipment on ash variability of baled corn stover biomass for bioenergy. *BioEnergy Res.* **2014**, *7*, 845–855. [CrossRef]
35. Haddock, J.; Mittenthal, J. Simulation optimization using simulated annealing. *Comput. Ind. Eng.* **1992**, *22*, 387–395. [CrossRef]
36. Van Laarhoven, P.J.; Aarts, E.H. Simulated annealing. In *Simulated Annealing: Theory and Applications*; Springer: Dordrecht, The Netherlands, 1987; pp. 7–15.
37. Geman, S.; Geman, D. Stochastic relaxation, Gibbs distributions, and the Bayesian restoration of images. *IEEE Trans. Pattern Anal. Mach. Intell.* **1984**, *6*, 721–741. [CrossRef] [PubMed]
38. Chavez, H.; Castillo-Villar, K.K.; Webb, E. Simulation-based approach for the optimization of a biofuel supply chain. In Proceedings of the Industrial and Systems Engineering Research Conference (ISERC), Pittsburgh, PA, USA, 20–23 May 2017.
39. Khanchi, A.; Birrell, S.J. Effect of rainfall and swath density on dry matter and composition change during drying of switchgrass and corn stover. *Biosyst. Eng.* **2017**, *153*, 42–51. [CrossRef]
40. Schon, B.; Matt, D. *Corn Stover Ash*; Iowa State University: Ames, IA, USA, 2014.
41. Khanchi, A.; Birrell, S. Drying models to estimate moisture change in switchgrass and corn stover based on weather conditions and swath density. *Agric. Forest Meteorol.* **2017**, *237*, 1–8. [CrossRef]
42. Vadas, P.A.; Digman, M.F. Production costs of potential corn stover harvest and storage systems. *Biomass Bioenergy* **2013**, *54*, 133–139. [CrossRef]
43. Thompson, V.S.; Lacey, J.A.; Hartley, D.; Jindra, M.A.; Aston, J.E.; Thompson, D.N. Application of air classification and formulation to manage feedstock cost, quality and availability for bioenergy. *Fuel* **2016**, *180*, 497–505. [CrossRef]
44. Map Data @ 2017 Google United States. Available online: https://www.google.com/maps/@42.2748687,-82.2910045,7.5z (accessed on 4 April 2017).
45. Suppapitnarm, A.; Seffen, K.A.; Parks, G.T.; Connor, A.M.; Clarkson, P.J. Multiobjective optimisation of bicycle frames using simulated annealing. In Proceedings of the 1st ASMO/ISSMO Conference on Engineering Design Optimization, Ilkley, UK, 8–9 July 1999.
46. Cao, Y. Hypervolume Indicator. Available online: https://www.mathworks.com/matlabcentral/fileexchange/19651-hypervolume-indicator (accessed on 2 August 2017).

MDPI
St. Alban-Anlage 66
4052 Basel
Switzerland
Tel. +41 61 683 77 34
Fax +41 61 302 89 18
www.mdpi.com

Energies Editorial Office
E-mail: energies@mdpi.com
www.mdpi.com/journal/energies

www.ingramcontent.com/pod-product-compliance
Lightning Source LLC
Chambersburg PA
CBHW051724210326
41597CB00032B/5593